Exploring the Sun

NEW SERIES IN NASA HISTORY

Exploring the Sun

✳

Solar Science since Galileo

Karl Hufbauer

THE JOHNS HOPKINS UNIVERSITY PRESS

Baltimore and London

© 1991 The Johns Hopkins University Press
All rights reserved. Published 1991
Printed in the United States of America on acid-free paper

Second printing, 1993
Softshell Books edition, 1993

The Johns Hopkins University Press
2715 North Charles Street
Baltimore, Maryland 21218-4319
The Johns Hopkins Press Ltd., London

Library of Congress Cataloging-in-Publication Data

Hufbauer, Karl.
Exploring the sun : solar science since Galileo / Karl Hufbauer.
p. cm. — (New series in NASA history)
Includes bibliographical references and index.
ISBN 0-8018-4098-8 (alk. paper) ISBN 0-8018-4599-8 (pbk.)
1. Sun. 2. Astronomy—History—19th century. 3. Astronomy—
History—20th century. I. Title. II. Series.
QB521.H93 1991   90-44033
523.7'09—dc20

A catalog record for this book is available from the British Library.

CONTENTS

PREFACE

Few things in the universe are more widely known to our species than the sun. None has played a greater role in our evolution, history, and ongoing activities. Since prehistoric times, indeed, many cultures have regarded the sun as a deity. And since the ancient Greeks, thinkers interested in nature have considered the sun a fit object for inquiry. Among them was Galileo Galilei, who wrote an epistolary treatise in 1612 about his telescopic studies of sunspots. He argued that the motions of sunspots proved the sun rotated and that their waxing and waning demonstrated that, contrary to Aristotelian doctrine, the sun was changeable. But what, exactly, were sunspots? And why did they appear only in the sun's equatorial zone, avoiding the higher latitudes and the poles? Galileo's treatise, which was exceptionally successful both in drawing important conclusions and in posing recalcitrant questions, inaugurated modern solar science.

Starting with Galileo and his contemporaries, three trends have dominated the history of solar science. Scientists have acquired an ever more diverse and precise set of observational and theoretical tools for investigating the sun. The scientists who have used these tools have constituted an increasingly interactive and differentiated collectivity. All the while, as their tool kit has become more powerful and their social organization more complex, scientists studying the sun have developed an increasingly detailed and robust picture of its structure, workings, and evolution. The mutual interactions of these three trends imbue the history of solar science since the early seventeenth century with a special interest.

This book seeks to characterize the development of modern solar science. Since Galileo, the field has passed through three epochs and may now be nearing the end of a fourth. After an introductory Prologue, part 1 focuses on the completed epochs. During the first epoch, 1610–1810, natural philosophers and astronomers came to think of our sun as a star with an attendant planetary system, and they determined its rotation, distance, size, and mass to within 10 percent of today's values. They also investigated the sun's motion through space and speculated about its physical constitution. In the second epoch, 1810–1910, a medley of scientists first developed instruments, techniques, and theories capable of providing new information about the sun and insights into its nature. Then a few pioneering astronomers used these new tools to found "solar physics" and began acquiring support for their distinctive observing programs. Finally, galvanized by George Ellery Hale's advocacy of the idea that the sun was a Rosetta stone for interpreting the stars, solar physicists aspired to leadership of astronomy as a whole. During the third epoch, 1910–40, solar physicists found they could not sustain their bid for leadership of the discipline. While they were refining and articulating the work of their predecessors, outsiders—scientists who had done little or no prior research on the sun—were revolutionizing views on the sun's physical constitution, chemical composition, energy generation, and coronal temperature.

Part 2 describes the progress of solar observational capabilities since the beginning of World War II. Between 1939 and the start of the International Geophysical Year in 1957, solar and radio scientists established the military value of solar monitoring, thereby strengthening the case for support of solar research. Meanwhile, physicists and engineers inaugurated programs for observing the sun's invisible radiations by using antennas as telescopes for studying its radio emissions and rockets as platforms for studying its ultraviolet and X-ray emissions. The Soviet Union's launch of *Sputnik* early in the International Geophysical Year hastened the growth of solar observational capabilities. Until the mid-1970s, patronage was abundant not only for the development of spacecraft as solar observing platforms but also for the construction of new ground-based solar telescopes, the acquisition and refinement of auxiliary instrumentation, and even the establishment of subterranean facilities for detecting solar neutrinos. As their capabilities and numbers grew, solar physicists transformed their specialty into a subdiscipline with its own journal and associations. Since the mid-1970s, solar physics, like most other scientific fields, has had to cope with tight budgets. Still, solar physicists have continued to increase the resolving power of both space-based and ground-based instruments. The upshot has been that, despite some decline in monitoring programs during the past decade or so, their observational capabilities are more puissant than ever.

Part 3 uses two case studies to illuminate the conduct of solar research since *Sputnik*. The focus is on phenomena that, on account of the terrestrial atmosphere, were regarded as hypothetical until instruments aboard spacecraft brought them under direct observation. In particular, the case studies examine the conjectural and technical backgrounds of the space-based observations that confirmed the existence of the solar wind and the variability of the sun's radiative output. Then they trace the early stages of research on these phenomena. These narratives show that ground-based observations have played, and promise to continue playing, a major role in the interpretation of phenomena that can be directly observed only from space. They also illustrate the continuing role of outsiders in enlarging the domain of solar physics.

As this preview indicates, my treatment of the history of solar science is selective. Any attempt at exhaustive coverage is foreclosed by the immensity of the field. Two main considerations have guided my choice of topics. First, it seemed essential to describe the initiation of research on all those solar properties that are at present regarded as fundamental solar character-istics—the sun's size, mass, rotation, radiative output, chemical composi-tion, magnetic field, and activity cycle. To sustain the narrative pace, how-ever, I have only occasionally devoted space to recounting how pioneering observations and theories were followed up. Second, it seemed desirable to focus on subjects that lent themselves to visual as well as written presenta-tion and hence would be accessible to readers unfamiliar with solar science. Therefore, I have generally avoided issues that solar scientists find it difficult to discuss without invoking a good deal of mathematics.

I believe that scientific knowledge of the sun has progressed since the early seventeenth century. Some readers will question this stance—es-pecially those who, shaken by our century's tyrannies, wars, and holo-causts, share the view of historians and philosophers of science such as Thomas S. Kuhn (1962) and Larry Laudan (1977) who question naive notions of scientific progress. When I write that scientific knowledge of the sun has progressed, I have in mind two interdependent trends. First, I mean that the successive generations of scientists who have studied the sun have included a growing number of distinct phenomena and increasingly precise parameters in their descriptions of its structure and behavior. Second, I mean that these successive generations have developed theories of solar phenomena that have both connected ever more phenomena and param-eters and also withstood increasingly severe criticism. I am claiming, in other words, that scientific knowledge of the sun has become ever richer and more robust since Galileo's day.

Is this stance on the progress of solar science tantamount to the view that it has been getting continually closer to a perfect representation of the sun's true nature? Most scientists who have studied the sun have certainly thought it has. Their viewpoint is buttressed by the way the values of vir-tually all solar parameters have closed in on today's values. For instance, since 1687 when Isaac Newton first estimated the sun's mass as 28,700 times the earth's mass, this parameter's value has approached the present value along a damped oscillatory path. Trusted estimates of the sun's mass have fallen within 10 percent of today's value since the 1770s, within 1 percent since the mid-1890s, within 0.1 percent since the mid-1950s, with-in 0.01 percent since 1964, and within 0.001 percent since 1967 (see table 1.2 below). This typical pattern of closure differs so markedly from the way most time series—such as for population, prices, or public opinion—have closed in on current figures that scientists have seen it as powerful testimony that they are zeroing in on the truth. Indeed, in accepting a solar mass of 332,946.0 earth masses, the International Astronomical Union endorsed the proposing committee's belief that this value was within 0.0001 percent of the "true" value (Duncombe et al. 1976, 63).

Although, like other scientists, most post-Galilean investigators of the sun have viewed themselves as journeying toward the truth, I agree with Kuhn (1962) and others in preferring a somewhat more modest image of their enterprise. For one thing, the frequency with which generally accepted scientific descriptions and theories have been discarded or drastically modified suggests that scientists can never be sure they have represented nature perfectly. Moreover, there are no compelling grounds for thinking that all scientific knowledge will end up as a set of interlocking parameters each of which, like the sun's mass, is closing in on a certain value. The metaphor of science as a *journey toward a specific destination* is misleading. Not all journeys have destinations. Parties setting out on journeys of exploration, for instance, have often had the goal of getting as far as possible *beyond* all earlier parties. Science is more like such a journey of exploration. Rather than viewing successive generations of solar scientists as moving ever closer to a knowledge of the sun's true nature, I see them as having moved progressively beyond Galileo by producing ever fuller descriptions and sturdier theories of solar phenomena.

But what are the sources of solar science's progress since the early seventeenth century? What has enabled successive generations of observers to discern more and more solar phenomena and to estimate solar parameters with increasing precision? What has allowed theorists to develop interpretations that have connected a growing number of phenomena and parameters and have withstood increasingly rigorous scrutiny? How, in short, have post-Galilean solar scientists managed to enrich humanity's knowledge of the sun with ever more detailed and precise descriptions and more sweeping and robust theories?

I believe the scientists' success in improving descriptions and interpretations of the sun has depended most immediately upon what they have brought to these endeavors—their general outlooks; their access to instruments and assistants; their background in the field; and contacts with others seeking to contribute to its advancement. The specific character of the assets that solar scientists have brought to their research has depended, in turn, on a host of "external" factors. The kind of people who have chosen to contribute to the field has depended on perceptions of its relative importance and prospects. Their ability to bring new or better tools to bear on the field's observational and theoretical problems has depended on antecedent developments in neighboring scientific and technical fields. Their access to funds, assistants, and colleagues has depended on the organization of patronage, training, and employment in science as a whole. At yet another remove, all these factors have depended upon each epoch's dominant values and beliefs, its distribution of wealth, and its struggles for power. Thus the progress of solar science has depended not only upon its contributors' research assets but also upon conditions in other scientific-technical arenas and upon socioeconomic and political relations within and among the main scientific nations.

ACKNOWLEDGMENTS

My first book—*The Formation of the German Chemical Community* (1982)—examined the eighteenth-century origins of one of the first national, discipline-oriented scientific communities. While writing that book, I often wondered how modern scientists, given the strength of disciplinary training and allegiances, sometimes managed to do first-rate interdisciplinary research. In 1978 I began investigating this issue during a sabbatical visit to the University of Pennsylvania. My earlier teaching had acquainted me with the history of nuclear physics, so I first looked into the nuclear physicists' contributions to astrophysics and physiology during the 1930s. I soon realized it would be impossible to go in both directions simultaneously. Since I had a better background in mathematics and physics than in biology, since I knew that Rob Kohler had already done a nice case study of the introduction of isotopic tracers into physiological research, and since I enjoyed ready access to the extensive papers of the astrophysicist Henry Norris Russell at Princeton University, I decided to focus on the work leading up to Hans Bethe's solution of the stellar-energy problem.

In the ensuing years, I finished my book on the German chemical community and continued my investigation of the stellar-energy problem's history. By the time I had completed my second article on the subject, I could see I had hit on a fascinating example of theoretical interdisciplinary science that, contrary to my initial expectations, would require me to look every bit as closely at the astrophysicists as at the nuclear physicists. This would take a book. In late 1983, still in the early stages of writing, I learned from Sylvia Fries that the NASA History Office was seeking a contract historian to do a history of solar science. Eager to enlarge my knowledge of the background and context of the stellar-energy problem, hopeful of reaching a larger audience than I had with my first book, attracted by the prospect of two years' leave from teaching and committees, and assured that I would have complete academic freedom, I submitted a proposal. It was accepted. The present book is the result.

In the course of my research and writing, I have received help from many colleagues in the history of science and technology. David DeVorkin not only has given much of the draft detailed criticisms but also has shared his concurrent research on the history of solar physics with me. Joe Tatarewicz has commented on the entire manuscript, drawing on his knowledge of the history of planetary science to make many constructive suggestions. Lee Saegesser has guided me around the copious collections maintained in the NASA History Office. Steve Brush, Virginia Dawson, Svante Lindqvist, Wilfried Schröder, Robert Smith, Al van Helden, and Craig Waff have all read drafts of one or more chapters. Jim Capshew, Ron Doel, Stew Gillmor, Owen Gingerich, Dieter Herrmann, Norriss Hetherington, Peggy Kidwell, Allan Needell, Trevor Pinch, Bob Seidel, and Woody Sullivan have assisted me with specific subjects and problems. Spencer Weart, once a solar physicist, has served as a reflective informant as well as a commentator. Ted Porter has given generously of his time and insight in the final stages of the project. And, through it all, Sylvia Fries has been a discriminating critic and superb contract manager.

I have also obtained assistance from any number of people who figure either as participants or as close observers of the developments described here. As the work progressed, Kees de Jager, Jack Eddy, Bengt Edlén, Peter Foukal, Harold Glaser, Leo Goldberg, Bob MacQueen, Gene Parker, John Simpson, Conway Snyder, Jan Stenflo, Richard Willson, and Harold Zirin were particularly generous in responding to my requests for information and advice. Grant Athay, Herbert Bridge, Jack Evans, Konstantin Gringauz, John Hickey, Robert Howard, Hugh Hudson, Marcia Neugebauer, Jay Pasachoff, Martin Pomerantz, Ken Schatten, and Zdenek Svestka have also answered questions and commented on drafts. Those who helped me trace the emergence of three-dimensional modeling of the solar wind and the background of and preparations for the long-awaited solar-polar mission—especially Richard Bogart, William Coles, Lennard Fisk, Harold Glaser, Todd Hoeksema, Arthur Hundhausen, Allen Krieger, Marcia Neugebauer, Edgar Page, Barney Rickett, Ken Schatten, Phil Scherrer, Mike Shultz, David Sime, and Leif Svalgaard—will, I fear, be disappointed that I have had to compress these intriguing stories into a brief epilogue to chapter 6. In addition to those listed above, I also want to acknowledge the cooperation of Loren Acton, Harjit Ahluwalia, Horace Babcock, Samuel Bame, Aaron Barnes, Jacques Beckers, David Bohlin, Ronald Bracewell, John Brown, Paul Coleman, Jr., Lawrence Cram, Raymond Davis, F.-L. Deubner, Walter Dieminger, Murray Dryer, Andrea Dupree, A. O. Fokker, Peter Gilman, John Harvey, Carole Jordan, R. W. Kreplin, Rudolf Kröber, Mukul Kundu, Max Kuperus, Alan Lazarus, Eugene Levy, Jeffrey Linsky, William Livingston, Reimar Lüst, Dimitri Mihalas, Norman Ness, Werner Neupert, Gordon Newkirk, Jr., Robert Noyes, Joseph Plamondon, Hans Plendl, Eric Priest, Walter Roberts, Robert Rosner, Ian Roxburgh, David Rust, Rob Rutten, Takao Saito, G. Schmidtke, Andrei Severny, Edward Smith, Peter Sturrock, Einar Tandberg-Hanssen, Dick Thomas, Yutaka Uchida, Max Waldmeier, N. O. Weiss, Olin Wilson,

George Withbroe, Martin Woodard, Jack Zirker, and Kees Zwaan. During the project, I was poignantly reminded of the passage of time by news of the deaths of four of my scientific informants—Leo Goldberg, Gordon Newkirk, Walter Roberts, and Andrei Severny.

I have been fortunate to have the support of several friends. While on leave in Berkeley, I had many provocative talks about postwar space science with Svante Lindqvist, who was visiting from Stockholm. On excursions to Stanford University, I found that my erstwhile mountaineering companion Peter Banks as well as my former Irvine friends Michael Weiss and Edie Gelles were thoughtful listeners. When sojourning in Washington, I could count on a warm welcome from Paul Forman, David DeVorkin, and their colleagues. On many occasions I took my project on vacation to the Truckee River, where Ida and Edgar Braun presided over much fine food, talk, and music. Here at Irvine I received moral support from Jon Dewald, Mike Johnson, Keith Nelson, and Spence Olin.

Finally, my mother, Arabelle Hufbauer, besides being a stimulating companion on a research trip to Utrecht and Zurich, has commented on several chapters. She and my father, Clyde Hufbauer, provided me with a new word processor and printer midway through the project. Joyce and Robert Caproni led the way on many a good walk in the Berkeley hills. Gary, Carolyn, Randall, and Ellen Hufbauer were lively hosts in Washington. My son Benjamin was a resourceful research assistant as I was getting started. My daughter Ruth helped me round up pictures. Like Benjamin and Ruth, my daughter Sarah Beth often interviewed me about "the book" in ways that turned out to be quite constructive. All the while my wife, Sally, has furnished me with sound counsel and spirited encouragement.

Exploring the Sun

✳

PROLOGUE

The Sun in Western Astronomy to 1610

Greek natural philosophers and astronomers developed the first coherent geometric interpretations of celestial phenomena (Dicks 1970, Vlastos 1975). Their success in this endeavor was reflected in the advance of their thinking about the sun's form, size, role in eclipses, and motions between about 450 and 350 B.C. During these years they recognized the sun's sphericity and immensity, explained solar and lunar eclipses, and proposed models for its daily and annual movements. So many of the era's physical and astronomical treatises were lost in antiquity that historians cannot provide full accounts of any of these developments. Yet the writings of Aristotle (384–322 B.C.), who in astronomical matters was primarily a systematizer, make it clear that the new ideas about the sun had a certain currency by the second half of the fourth century.

Rejecting the common view of the sun as a small disk in the heavens, Aristotle followed the astronomers' lead in portraying it as a distant sphere somewhat larger than the earth. Likewise, he followed them in interpreting eclipses not as divine portents, but rather as natural events (fig. 1). A solar eclipse occurred whenever the moon, by passing directly between the sun and the earth, shadowed part of the earth. Similarly, a lunar eclipse occurred whenever the moon, which received its light from the sun, passed directly through the earth's shadow. He also followed the astronomers in thinking that the sun's daily journey across the sky and its annual tour along the ecliptic (fig. 2) were the end result of the rotations of several interconnected spheres that carried the sun around a stationary earth at the center of the universe.

Although Aristotle followed prevailing astronomical opinion on specific issues such as the sun's properties, he took the opportunity presented by the wide divergence among competing philosophies to develop his own cosmology. His manner of embedding astronomical particulars into his own cosmological frame may be illustrated by his case for the sun's sphericity in his treatise on the heavens (Aristotle ca. 340 B.C./1960, 210). There were, in his view, two main reasons for thinking that the sun, moon, planets, and fixed stars were all spheres. First, the phases of the moon provided clear evidence that the nearest heavenly body was a sphere rather than a disk. But if one heavenly body was a sphere, the sun and all other heavenly bodies must, by analogy, possess this form. Second, the heavenly bodies did not need to move *themselves* from one place to another. Consequently nature, which made nothing that was purposeless, would have formed them as spheres—bodies without the slightest vestige of appendages.

Aristotle's case for the sun's sphericity, though it had some observational underpinning, depended for its cogency on his general cosmology. It rested not only on the presumption that nature was purposeful but also on the supposition that all the heavenly bodies were fundamentally similar. He believed, in fact, that they all consisted of the same substance—the quintessence, or *ether*. The fifth element was "more divine, and prior to, . . . the four in our sublunary world" (Aristotle ca. 340 B.C./1960, 15, 23). It had an unending natural circular motion, unlike the sublunary elements, which moved up or down until they reached their natural places. It was "ageless, unalterable and impassive." In short, Aristotle thought the sun and other heavenly bodies all consisted of an immutable substance whose natural motion was circular.

While Aristotle was teaching in Athens, his former pupil Alexander the Great was conquering the Greek city-states, Egypt, Persia, and lands farther to the east. Alexander's conquests ushered in a cosmopolitan era that lasted until the triumph of Christianity in the fourth century A.D. Early Hellenistic astronomers—notably Aristarchus of Samos (fl. 270s B.C.), Apollonius of Perga (fl. 220s B.C.), and Hipparchus of Nicea (fl. 140s B.C.)—went well beyond their predecessors. Thereafter the only significant additions to ancient astronomy were made by Ptolemy of Alexandria (fl. A.D. 140s), who masterfully synthesized prior work.

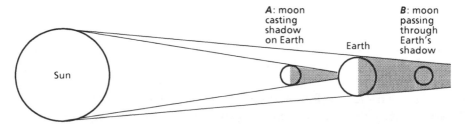

Figure 1. Eclipses: A indicates the moon's position during a solar eclipse and B, its position during a lunar eclipse. Virtually all astronomers have subscribed to this interpretation of eclipses since it was advanced by the Greeks in the fifth century B.C. (Courtesy of NASA History Office.)

Figure 2. The ecliptic line, or the annual path of the sun as projected on the stellar background: S, A, W, V indicate the sun's apparent positions at the summer solstice, autumnal equinox, winter solstice, and vernal equinox; the ecliptic is so named because eclipses can occur only when the moon passes through the ecliptic plane; the stellar constellations through which the ecliptic passes constitute the zodiac. During the fifth and fourth centuries B.C., Greek astronomers ascertained that the obliquity of the ecliptic—the angle between the ecliptic and equatorial planes—was about 24° and that the sun's passage along the ecliptic was slower in the spring and summer than in the fall and winter. (Courtesy of NASA History Office.)

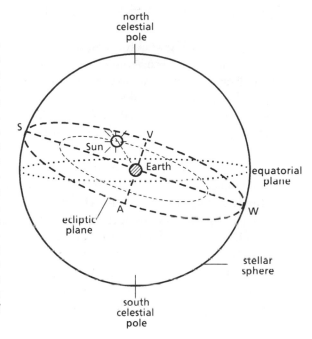

Aristarchus wrote about the sun on two occasions. In one treatise he followed up on earlier work by examining the relative distances and sizes of the sun and moon (Van Helden 1985). His procedure, which exploited knowledge of the geometry of the sun's illumination of the moon and of lunar eclipses, was ingenious; but his measurements of the data needed to carry through the calculations were crude. His resultant values for the sun's distance and diameter—approximately 19 times the moon's distance and about $6\frac{3}{4}$ times the earth's size—probably exceeded earlier estimates. They were, however, still only a small fraction of the modern values. Aristarchus's other treatise dealing with the sun proposed a sun-centered cosmology (Wall 1975). Our chief source on this lost work is a summary by the mathematician Archimedes. Despite the summary's brevity, it is clear that in broadest outline Aristarchus anticipated Copernicus by some eighteen centuries. He accounted for the apparent diurnal revolution of the heavens by hypothesizing that the earth, not the heavens, rotated daily. He accounted for the sun's apparent annual motion around the ecliptic by positing that the earth circled the sun once a year. And he explained away the failure of astronomers to detect changes in the sizes and angular separation of the fixed stars as the earth moved in its orbit by hypothesizing that the celestial sphere was so immense that the changes were imperceptible. Aristarchus's heliocentric system was bold and elegant. But at odds with contemporary cosmologies as well as everyday experience, it had little influence on his principal successors.

Hipparchus, Aristarchus's most creative successor, stayed entirely within the earth-centered cosmological framework. He used geometric devices that had been proposed late in the third century B.C. by Apollonius—eccentric circles and deferent circles with epicycles (Pedersen and Pihl 1974)—to increase astronomy's predictive capabilities. His goal was to develop a quantitative system that could both account for the observational records acquired from Babylonia (Aaboe 1974) and satisfy the elite's mania for horoscopic astrology (Dicks 1960; Charlesworth 1977), also acquired from Babylonia (Sachs 1952; Rochberg-Halton 1984). He devised, among other things, models for the motions of the sun and moon that did a good job of forecasting lunar eclipses. His models also yielded promising results on the much more difficult problem of predicting whether a solar eclipse would be total, partial, or unobservable at a given location (Van Helden 1985).

Some three centuries after Hipparchus, Ptolemy did such an able job of unifying earlier work in astronomy, geography, and optics that his books supplanted most of his forerunners' works. His main astronomical treatises were the highly technical *Syntaxis mathematica*, which came to be known by its Arabic title as the *Almagest*, and the more general *Planetary Hypotheses*. In them he used Aristotelian cosmology as the framework for a comprehensive account of solar, lunar, and planetary motions. His treatment of the sun's motion, distance and size, and position among the planets reveals a heavy debt to his predecessors and a passion for system.

Ptolemy adopted Hipparchus's theory of solar motion, making it easier to apply by adding appropriate tables (Pedersen and Pihl 1974). He determined the sun's distance and size by modifying a complex procedure devised by Hipparchus (Swerdlow 1968; Van Helden 1985). First Ptolemy computed the moon's greatest distance based on the discrepancy between its predicted and observed positions. Then, using this datum together with both a diagram for solar and lunar eclipses and the observed angular diameters of the sun and moon and of the earth's shadow cone where the moon passed through it during a full lunar eclipse, he calculated the moon's size and the sun's distance and size. The procedure, though geometrically sound, turned out to be highly sensitive to the empirical parameters. A mere 1 percent adjustment in their value—an adjustment that was far beneath his margin of error—would have more than doubled his estimate of the sun's distance. It was probably no coincidence, therefore, that Ptolemy obtained values of about 64 earth radii for the greatest lunar distance and 1210 earth radii for the sun's distance.

For one thing, these values indicated that the sun was nineteen times farther away than the moon—the same figure Aristarchus had reported earlier. These values had the added advantage of allowing ample, but not immoderate, room for a snugly nested model of the cosmos with the earth at its center and the fixed stars at its boundary. In constructing this model, Ptolemy started from the premise that there was no empty and hence useless space in the universe. The outer limit of each planet's orbit was the inner limit of the next one's orbit. First came the lunar sphere, including the mechanism guiding its motion, then those of Mercury, Venus, the sun, Mars, Jupiter, Saturn, and finally—at 20,000 earth radii—the celestial sphere. Thus the distance of the sun was intimately connected with the dimensions of the cosmos.

Ptolemy's astronomical system survived the collapse of the Roman Empire for at least four reasons—its technical quality, its coherent picture of the cosmos, its usefulness to astrology, and in the eyes of sophisticated Byzantine, Egyptian, and Syriac Christians, Sassanian Zoroastrians, Arabic Muslims, and eventually European Christians, its harmony with theology. As a consequence, until the development of the telescope and anti-Aristotelian worldviews in the early decades of the seventeenth century, serious astronomical treatises were—with a very few partial exceptions—essentially commentaries on the *Almagest* and *Planetary Hypotheses* (Van Helden 1985). Consider, for example, how the Italian churchman Campanus of Novara dealt with the sun in the *Theorica planetarum* that he completed in the late 1250s (Benjamin and Toomer 1971). He gave its distance as 1210 earth radii, its size as $5\frac{1}{2}$ times the earth's size, and its position as above the moon, Mercury, and Venus and below Mars, Jupiter, and Saturn. The numerical data attest the faithfulness of the scribal tradition during the preceding eleven centuries. They also illustrate Campanus's lack of originality. In fact, his only noteworthy departure from Ptolemy's account of the sun was to calculate the mean distance from its surface to the earth's surface in miles— $3,905,904\frac{9}{11}$ miles, to be precise!

The first European astronomer to develop an alternative to the Aristotelian-Ptolemaic system was Nicolaus Copernicus (1473–1543), a German-speaking canon in the Polish bishopric of Varmia (Rosen 1971; Swerdlow and Neugebauer 1984). It was about 1510, after studying in several universities and attending upon his uncle, the bishop of Varmia, that Copernicus (fig. 3) began work on a sun-centered system. The circumstances leading Copernicus to take this step are far from clear. In the late 1490s, while a student in Bologna, he had learned from Domenico Maria di Novara, an astronomer who was a vigorous critic of Ptolemy's geography, that ancient ideas were not beyond reproach. This lesson seems to have been reinforced at Padua between 1501 and 1503 through contact with some person or

Figure 3. Nicolaus Copernicus as visualized by the seventeenth-century astronomer and philosopher Pierre Gassendi; the heliocentric device portrays the earth with its moon circling the sun. (Source: Gassendi 1654 part 2: 2. Courtesy of Owen Gingerich.)

treatise advocating the use of compounded uniform circular motions to replace Ptolemy's uncomely geometric devices. Perhaps his inability to follow out this program was what induced Copernicus to cast about for an alternative to the orthodox cosmology. In any case, having once embraced heliocentrism he worked intermittently for some three decades on a large book expounding it. Finally on his deathbed, he received a printed copy of his *De revolutionibus orbium coelestium*.

Like most scientific revolutionaries, Copernicus joined old ideas with new in his heliocentric system. His conservatism and also his radicalism are manifest in his treatment of the sun. Copernicus was a conservative on the issue of the sun's distance and size (Swerdlow and Neugebauer 1984; Van Helden 1985). This problem did not depend, in the first instance, on whether it was the earth or the sun that was immobile. Using Ptolemy's method and Aristarchus's classic nineteen-to-one ratio for the distances of sun and moon, he eventually came up with values of 1179 earth radii and 5.45 earth radii for the sun's greatest distance and radius—not far from Ptolemy's corresponding values of 1260 and 5.5 earth radii. Though he was conservative here, Copernicus took a radical stance in accounting for the sun's spherical shape. Aristotle had explained the sphericity of the sun and other heavenly bodies as a consequence of their not needing an independent means of locomotion. Copernicus (1543/1976, 46) proposed instead that, like the earth, the sun, moon, and other planets were all centers of gravity: "I myself consider that gravity is merely a certain natural inclination with which parts are imbued by the architect of all things for gathering themselves together into unity and completeness by assembling into the form of a globe. It is easy to believe that . . . by [gravity's] agency [the sun, moon, and planets] retain the rounded shape in which they reveal themselves." Though Copernicus did nothing further with the idea, the cosmological revolution he initiated eventually gave rise to the principle of universal gravitation.

Copernicus's view of the sun's place in the cosmos was also radical. Since Aristarchus, no one of consequence had maintained that the sun sat stationary at the center of the universe. He went far beyond Aristarchus, however. He was able to present, for instance, detailed models for the revolution of Mercury around the sun in approximately eighty days, Venus in nine months, the earth with its moon in a year, Mars in two years, Jupiter in twelve years, and Saturn in thirty years. He was also able to determine the planets' least and greatest distances from the sun without relying on Ptolemy's hypothesis of nested spheres. The surprises here were that sizable gaps separated the planetary orbits and that Saturn's orbit was smaller in the heliocentric system than in the geocentric system.

Copernicus's *De revolutionibus* was above all a technical treatise designed

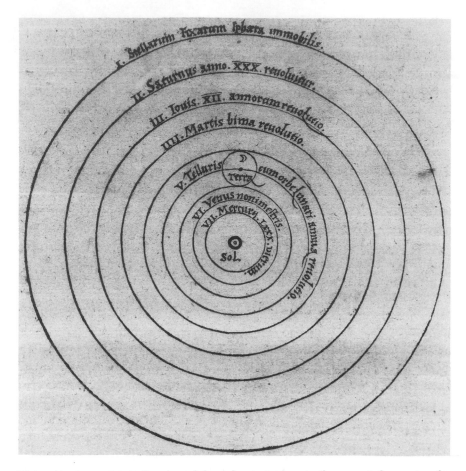

Figure 4. Copernicus's diagram of the solar system; note the sun at the center, the planetary spheres nested snugly in order of their periods of revolution, and the immovable sphere of fixed stars that would not be present if the diagram were to scale. (Source: Copernicus 1543, 10. Courtesy of the Biomedical Library, University of California, Los Angeles.)

to replace Ptolemy's *Almagest*. Nonetheless, after presenting a succinct account and simplified diagram (fig. 4) of his system, Copernicus (1543/1976, 50) waxed eloquent about heliocentrism:

> In the middle of all is the seat of the Sun. For who in this most beautiful of temples [the universe] would put this lamp in any other or better place than the one from which it can illuminate everything at the same time? Aptly indeed is he [the sun] named by some the lantern of the universe, by others the mind, by others the ruler. . . . Thus indeed the Sun as if seated on a royal throne governs his household of [planets] as they circle round him. . . . We

find, then, in this arrangement the marvellous symmetry of the universe, and a sure linking together in harmony of the motion and size of the spheres, such as could be perceived in no other way.

His panegyric expressed sentiments that would be echoed with increasing frequency over the next century.

Although ultimately rejected, the Aristotelian-Ptolemaic tradition still shaped most scientific thinking about the sun when the first telescopic studies of the heavens were undertaken about 1610. At this time natural philosophers and astronomers generally regarded the sun as a spherical body about $5\frac{1}{2}$ times larger than the earth. In their view, it and all the other heavenly bodies as well as the celestial spheres carrying them consisted of incorruptible ether. Its ethereal sphere, which was some 1200 earth radii distant, lay between the spheres of Venus and Mars. It illuminated the earth and moon, each of which occasionally intercepted the light destined for the other, causing an eclipse. Some of these ideas would withstand the scrutiny of the next century—the idea of a spherical sun larger than the earth and the explanation of eclipses. But most, as we shall see, would fall by the wayside as new instruments and worldviews gave birth to modern solar science.

PART ONE

Solar Science
Comes of Age

Part 1 opens about 1610 with Galileo's telescopic studies of the sun and closes about 1940 with Bethe's and Edlén's separate breakthroughs in identifying energy-generating nuclear reactions inside the sun and hot ions as the source of the solar corona's emission lines. At the beginning of this period, scientific discourse about the sun still focused on issues that were first raised nearly two millennia earlier by the Greeks—its distance, size, daily and annual motions, incorruptibility, and role in eclipses. However, Galileo with his telescope and his Copernican ideas soon made a persuasive case for bringing new questions into the realm of solar science. Others managed to do the same with increasing frequency in the ensuing thirty-three decades. By World War II solar physics, as the field had come to be called, comprised many specialized instruments and observing programs, an abundance of empirical relationships, and a promising set of interrelated theories dealing with the sun's energy generation, structure, and evolution.

In recounting the history of solar science down to 1940, part 1 emphasizes the debut of new lines of research concerning the sun. This narrative strategy has two related advantages over a more comprehensive telling of the story. It bypasses the technical minutiae that characterize so much of the ordinary scientific practice involved in refining novel instruments and techniques, following up on major discoveries, and exploiting interpretive breakthroughs. And in doing so, it lets the narration move through the decades at a brisk pace. One needs to guard, however, against the impression that solar science has advanced entirely by leaps and bounds.

Besides accenting major innovations, part 1 highlights changes in the self-images of those contributing to solar science and in the relations among them. It looks at the connections between the contributors' overall aspirations and their specific endeavors to enlarge knowledge of the sun. It examines how, as scientists who were ready to make career-long commitments to the field emerged in the mid-nineteenth century, solar physics became the site of a specialty-oriented community. It illuminates this community's subsequent organizational activities, which eventually gave rise to several commissions in the International Astronomical Union and to global networks for monitoring sunspots and solar flares. Finally, it considers the solar physics community's ongoing dependence for important advances on astronomers and physicists with little background in the field. By attending to such themes, part 1 contextualizes solar science's coming-of-age between 1610 and 1940.

CHAPTER ONE

The Beginnings of Modern Solar Science

1610–1810

During the seventeenth and eighteenth centuries, natural philosophers and astronomers set the stage for modern solar science. They recognized that the sun was the nearest star. They brought estimates of its distance, size, mass, rate of rotation, and direction of motion through space to within 10 percent of today's values. And they began trying to decipher its structure and behavior.

Their inauguration of modern solar science was made possible by both novel instruments and novel insights. The most important new tool was the telescope. This instrument—actually an evolving family of instruments, accessories, and observing techniques—enabled natural philosophers and astronomers to detect and study solar phenomena that had hitherto been beyond their ken. In itself, however, the telescope did not unequivocally determine how the phenomena it brought into view were understood. New ideas about matter and the universe played a crucial role in shaping interpretations. In fact, almost all the natural philosophers and astronomers who studied the sun in the seventeenth and eighteenth centuries were motivated by broad physical and cosmological concerns. They were not specialists whose primary interest was the sun itself. Only occasionally, if ever, did they concentrate their attention on solar phenomena. Despite the diverse motivations for the early solar studies and their intermittent character, each new generation believed its knowledge of the sun surpassed that of preceding generations.

This chapter recounts the chief developments in the study of the sun between 1610 and 1810, focusing first on Galileo and his classic study of sunspots. Next it considers how natural philosophers and astronomers came to regard the sun as a star and reviews a few of the inferences they drew from the solar-stellar analogy. Attention then shifts to the increasingly ambitious attempts between 1670 and 1780 to determine the sun's distance, size, and mass. Finally, William Herschel's efforts to ascertain the sun's place in the natural history of the heavens are scrutinized. The chapter closes with an assessment of the state of solar science about 1810.

Galileo's Study of Sunspots, 1610–1613

Sunspots are often large enough to be seen by persons with sharp vision when the sun's glare is reduced by haze, mist, or thin clouds (Eddy, Stephenson, and Yau 1989). We know from early Chinese, Middle Eastern, and European records that they were noticed from time to time for well over a millennium before the seventeenth century; but these sightings were too haphazard and fleeting to inspire systematic investigations. Such studies had to await the invention of the telescope (Van Helden 1977), which made it possible to see smaller spots and keep them under observation for extended periods. About 1610, four Europeans independently sighted spots with their telescopes (North 1974). The earliest recorded observation was made by Thomas Harriot near London in December 1610 (Shirley 1983). The first publication on sunspots, written by Johannes Fabricius, appeared in Germany about October 1611 (Berthold 1894). In the meantime Galileo was laying the groundwork for what would become the most penetrating and influential investigation of the phenomenon.

By his own testimony, Galileo Galilei (1564–1642) first noticed spots on the solar disk toward the end of 1610. During the preceding year he had discovered lunar mountains, Jupiter's moons, and Saturn's companions[1] with his twenty-power telescope (Galilei 1610/1989). These spectacular discoveries had transformed the Paduan professor of mathematics into a European celebrity and enabled him to secure a position in Florence as chief philosopher and mathematician to Cosimo II, the young grand duke of Tuscany (Westfall 1985; Biagioli 1990). They had also reinforced his belief in Copernicus's heliocentric cosmology. Galileo surely trained his telescope on the sun during his first year of telescopic observing, but the sun's brilliance seems to have prevented him from noting anything out of the ordinary. His eventual detection of sunspots was apparently the result of some simple improvement in technique that reduced the glare—perhaps observing near sunrise or sunset, as Harriot did, or placing tinted glass at one end or the other of his telescope.

During the next year and a half, Galileo was intermittently attentive to the

---

1. Christiaan Huygens (1629–95) later showed these features to be what we know as Saturn's rings.

spots (Drake 1978). In the spring of 1611, for instance, he showed them to various acquaintances while visiting Rome. Not long afterward, he concluded that the motion of the spots across the solar disk was a consequence of the sun's rotation. Although interested, he did not seize upon the discovery with the zeal he had brought to his earlier telescopic finds. He was preoccupied both with determining the periods of Jupiter's moons and with upholding his position in the Tuscan court. In addition, he seems not yet to have appreciated how the spots could be used in his campaign to replace the Aristotelian-Ptolemaic cosmology with the Copernican worldview.

Meanwhile, Christoph Scheiner (1575–1650), a professor of mathematics at the Jesuit university of Ingolstadt, was devoting himself wholeheartedly to the study of sunspots (Drake 1978; Shea 1972). He interpreted them as small planets passing in front of the solar disk, thereby conforming to the Aristotelian doctrine of celestial incorruptibility (Grant 1990). Soon he was sending reports about his work to Mark Welser, the Jesuits' banker in Augsburg and a prominent patron of learning. Welser secured permission to publish the reports on condition that the work appear under the pseudonym "Apelles" so as to protect the Jesuits' reputation in case of error. Upon the treatise's appearance in early January 1612, Welser sent Galileo a copy along with a request for his opinion of the author's theory that the spots were little planets.

Galileo was struck by the way Apelles's belief in celestial immutability led to the planetary interpretation of the spots. Welser had provided him with a marvelous opportunity for promoting the anti-Aristotelian cause. Galileo embarked at once on a systematic study of sunspots. His talented disciple Benedetto Castelli[2] devised the observing procedure. The sun's image was projected from the telescope onto a sheet of paper, where the spots were observed and registered on a standard inscribed circle. In May 1612 Galileo sent Welser a long letter expounding his own interpretation and criticizing that of Apelles. He followed this letter with a second in August and, after receiving the manuscript of Apelles's rebuttal, a third in December. Some four months later, friends in Rome published his letters on sunspots in a volume with a handsome frontispiece (fig. 1.1).

Galileo presented a powerful case that the spots, rather than being planets, were features of a rotating, spherical sun (fig. 1.2). Foreshortening on the sun's spherical surface, he argued, caused the spots to appear to broaden and accelerate as they traveled from one limb, or edge, toward the

2. Castelli, a Benedictine monk, had earlier given Galileo the idea of looking for the phases of Venus (Westfall 1985). He evidently devised the projection method of observing sunspots independently of Scheiner, who is ordinarily credited with its invention.

Figure 1.1. Galileo Galilei shortly before the publication of his letters on sunspots; the portrait's ornate frame reminds us that, for all his apparent modernity, Galileo's sensibilities differed immensely from our own. (Source: Galilei 1613, 5. Courtesy of the Bancroft Library, University of California, Berkeley.)

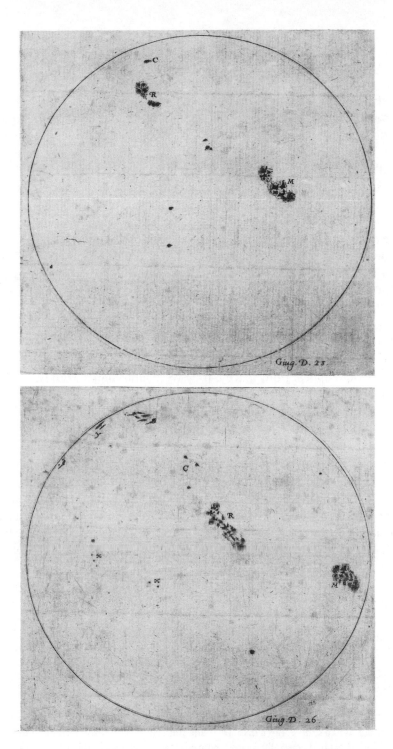

Figure 1.2. Galileo's observations of sunspots in 1612 on June 23 and June 26; note that during this three-day interval groups C, R, and M changed position and shape, a new group y rotated into view, and two new groups x and x made their appearance. (Source: Galilei 1613, 79, 82. Courtesy of the Bancroft Library, University of California, Berkeley.)

center of the solar disk. Likewise, foreshortening made the spots seem to get thinner and slower as they proceeded to the sun's other limb.[3] Galileo also pointed out that their motions had two characteristics that would be most improbable given the planetary hypothesis. The spots all moved across the solar disk at the same rate, making a complete passage in about fourteen days, and they all followed parallel paths.

Although Galileo was certain he was dealing with sunspots rather than planets, he was tentative when discussing their nature. He conceded that he did "not know, and [could] not know, what the material of the solar spots may be" (Galilei 1613/1957, 98). Even so, he thought it possible to determine many of their properties and, on the basis of this knowledge, to speculate on their closest terrestrial analogue. In particular, his observations indicated that the lifetimes, shapes, shadings, and numbers of sunspots were highly variable. All these characteristics reminded Galileo of clouds, which also "are vast and immense, are produced and dissolved in brief times, endure for long or short periods, expand and contract, easily change shape, and are more dense and opaque in some places and less so in others" (99). In supporting the analogy, he suggested that readers consider how the earth's clouds would appear from space if they were illuminated from below. Notwithstanding their resemblance to clouds, he granted that sunspots could be "any of a thousand other things not perceived by us" (100).

Besides being uncertain about the nature of sunspots, Galileo was puzzled that they appeared only in a band extending some 30° on each side of the solar equator. At first he thought this fact might illuminate their origin. Thus he suggested in his second letter to Welser that since the ecliptic, or plane of the planetary orbits, passed through this same band, the planets might somehow cause the sunspots (Shea 1972). However, he deleted this speculation from the published version of his letters (Galilei 1613/1957, 125), leaving it for others to explain why sunspots were confined to the lower solar latitudes.

Although Galileo could not be certain about the nature of sunspots or explain their concentration along the solar equator, he was confident that they demonstrated the mutability of the sun. He declared, for instance, "that the modern observations deprive all former writers of any authority, since if they had seen what we see, they would have judged as we judge. As a matter of fact those authorities who did not believe the sun could be yielding and changeable were still farther from believing that it is sprinkled with dark

3. Galileo's argument provided the first observational grounding for the proposition that the sun was a sphere, not a disk. He made nothing of this, since natural philosophers and astronomers had generally accepted the sun's sphericity since the fourth century B.C. (see Prologue).

spots. And now that its supposed immaculacy must yield to observation, it is vain to run to such men asking for support of the opinion that the sun is hard and unchangeable" (Galilei 1613/1957, 135). He implied, indeed, that it was no longer moral to believe in celestial immutability because nature, "in order to aid our understanding of her great works, has given us two thousand more years of observations, and sight twenty times as acute as that which she gave Aristotle" (143).

Galileo certainly had no qualms about drawing large conclusions from his observations. Small wonder, for it was his enthusiasm for the Copernican worldview, not his curiosity about the sun per se, that motivated his study of sunspots. Even though Galileo's interest in sunspots was fleeting, his little treatise was seminal. Because of the strength of the Aristotelian doctrine of celestial immutability, it did not immediately quash the idea that they were small intra-Mercurial satellites (Grant 1990; Baumgartner 1987). But his interpretation of them as spots on a rotating, spherical sun served the growing number of Copernicans as a starting point for thinking about the sun's nature. Galileo had inaugurated modern solar science.

The Sun as a Star, 1610–1700

In the early seventeenth century, on the eve of the first telescopic observations of the heavens, almost all natural philosophers and astronomers embraced one of three worldviews (Koyré 1957; Donahue 1972). A very large majority still believed that a stationary earth was at the center of the cosmos and that a rotating celestial sphere bearing the fixed stars was its outer boundary. A distinct minority followed the heliocentric cosmology of Copernicus. They presumed that the sun and its family of planets were at the center of a stationary celestial sphere of such vastness that no change could be detected in the relative positions of the stars as the earth orbited the sun. A few renegades followed Giordano Bruno[4] in denying the existence of a finite stellar sphere. They maintained instead that the stars were so many suns, each with its own planetary system, scattered through infinite space.

Although the Copernicans made immediate and effective use of the telescope in combating Aristotelianism, they did not feel that the new observations obliged them to adopt Bruno's unorthodox worldview. In fact, Galileo and the German astronomer Johannes Kepler (1571–1630), the two most

4. Bruno (1548–1600) was executed as a religious heretic by the Roman Inquisition. The influence of Copernicanism on his unorthodox cosmology has been much debated (McMullin 1987).

prominent Copernicans in the early decades of the seventeenth century, both rejected Bruno's prescient idea that the stars were suns. They did so for very different reasons, however, reflecting the state of flux in that era's cosmological thinking. Kepler argued, primarily on Pythagorean grounds, that the stars were situated on a sphere having a radius of 4 million and a thickness of $\frac{1}{6000}$ solar radii (Van Helden 1985). He supported this vision's startling consequence that the stars were but a few miles in radius by maintaining—many decades before most astronomers—that the telescope had revealed the impossibility of direct measurements of stellar angular diameters. Unlike Kepler, Galileo refused to make any definitive assertions about the stars. He thought they could be suns scattered through an infinite universe. But he contended that the shape and dimensions of the cosmos, and hence the nature of the stars, were and would forever remain beyond the reach of human reason.

More than a generation younger than Kepler and Galileo, René Descartes (1596–1650)—mathematician, philosopher, and Copernican—had no trouble believing that the stars, rather than being attached to a finite celestial sphere, were suns dispersed through an infinite universe (Aiton 1972; Dick 1982). In 1628 he moved from France to Holland so that he could more freely pursue his goal of deducing a mechanistic alternative to the Aristotelian worldview. His first cosmological work, completed four years later, portrayed the universe as unbounded space full of matter in motion. Descartes (1632/1967, 29–30) maintained that there were three main categories of matter: "First the sun & fixed stars, second the heavens, & third the earth with the planets & comets. This is why we have strong grounds for thinking that the sun & fixed stars [consist of] nothing other than the first element in completely pure form; the heavens, of the second; & the earth, with the planets & comets, of the third. . . . I unite therefore the sun with the fixed stars, & attribute to them a nature totally contrary to that of the earth, for the action of their light alone assures me that their bodies consist of a very subtle & very agitated substance." The sun and stars must, in short, be similar because they were all celestial bodies that shone by their own light. He went on to propose that a universe that began its existence simply as matter in motion would eventually partition itself into ethereal vortices whirling around stars composed of the first element. His single concession to orthodoxy was that he pictured the evolution of planets as occurring around only one star—our sun. Not long after finishing his manuscript, Descartes concluded that this lone concession was not sufficient to keep him out of trouble. He was persuaded by the Catholic church's heavy-handed response to Galileo's *Dialogo sopra i due massimi sistemi del mondo* (1632) that it would be prudent, even in Holland, not to publish (Russell 1989).

A decade later, after deciding he could defend the earth's immobility by saying that it, like a boat floating down a river, was not moving with respect to its immediate surroundings, Descartes gave a full account of his mechanistic cosmology in his *Principia philosophiae* (1644/1982). As in his earlier treatise, he discussed the sun and stars (147–52). He numbered the sun among the fixed stars on the grounds that their shared property of luminosity revealed their similar natures. Likewise, he envisioned the sun and stars as rotating spheres composed of extremely fine corpuscles at the centers of ethereal vortices distributed through an unbounded universe (fig. 1.3). On this occasion, however, he also sought to explain why spots appeared on the sun and stars. He postulated that the formation of spots was analogous to the appearance of scum atop a boiling liquid—irregular corpuscles inside the sun or star became entangled and then rose to its surface. And just as continued boiling would eventually dissolve the scum, so the constant agitation of the subtle first element sooner or later overwhelmed the spots. The stars, of course, were too distant for their ordinary spots to be seen. But Descartes was confident that the formation and dissolution of exceptionally large spots lay behind such reported events as the dimming and disappearance of some stars and the sudden appearance of others. His interpretation of sunspots was a crude yet pioneering attempt to forge a connection between solar and stellar phenomena with the aid of terrestrial experience.

Inspired by Descartes, the next generations of Copernicans went beyond him to embrace the idea of an unbounded universe occupied by an infinity of suns, each with its own retinue of inhabited planets (Dick 1982). The first to do so was the young Cambridge philosopher Henry More. A few lines from a long poem of 1646 indicate More's fascination with the theme:

> The skirts of his [the sun's] large Kingdome surely lie
> Near to the confines of some other worlds
> Whose Centres are the fixed starres on high,
> 'Bout which as their own proper Suns are hurld
> Joves, Earths and Saturns; round on their own axes twurld.
> And as the Planets in our world (of which
> The sun's the heart and kernell) do receive
> Their nightly light from [distant] suns . . .
> . . . so our worlds sunne
> Becomes a starre elsewhere, and doth derive
> Joynt light with others, [to] cheareth all that won [dwell]
> In those dim duskish Orbs round other suns that run.
> (More 1647, stanzas 22, 24)

Not surprisingly, given such enthusiasm, More took Descartes to task for making our solar system unique.

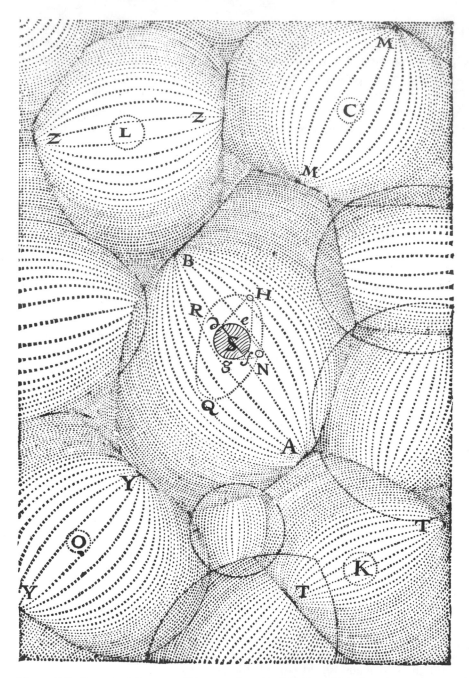

Figure 1.3. Descartes' diagram of his cosmology: L, C, K, O, and S—the sun—represent stars at vortex centers and RHNQ, the solar system's ecliptic plane. (Source: Descartes 1644, 114. Courtesy of the Bancroft Library, University of California, Berkeley.)

During the next four decades, such Cartesian speculations about the nature of the universe spread rapidly throughout northwestern Europe (Jacob 1988). This diffusion was aided by the eagerness of professors at several universities to step forward as champions of Descartes's cosmology. It was also promoted by the establishment of societies—especially the Royal Society of London and the Académie des Sciences in Paris in the early 1660s—where natural philosophers, mathematicians, astronomers, naturalists, and their patrons could meet to discuss the new science. In the mid-1680s, the young French writer Bernard de Fontenelle gave the Cartesian cosmology greater currency with his *Entretiens sur la pluralité des mondes*. This book, which espoused the new worldview in a series of imaginary conversations, went through nine French printings and five English ones by 1700 (Fontenelle 1686–1742/1966). To judge from the exuberant translation of Afra Behn, the first Englishwoman to live by her pen, readers thrilled when Fontenelle wrote that the immensity of the universe

> pleases and rejoices me; when I believ'd the Universe to be nothing, but this great Azure Vault of the Heavens, wherein the Stars were placed, as it were so many golden Nails or Studs, the Universe seem'd to me too little and strait; I fansied my self to be confin'd and oppress'd: But now when I am perswaded, that this Azure Vault has a greater depth and vaster Extent, and that 'tis divided into a thousand and a thousand different *Tourbillions* or Whirlings, I imagine I am at more Liberty, and breath a freer Air; and the Universe appears to me to be infinitely more Magnificent. (Fontenelle 1688, 133)

Fontenelle's readers also seem to have been quick to accept the view that the stars "are enlightened of themselves; and are by consequence so many Suns" (131–32). But some may have balked when he used the "rule of resemblance . . . betwixt our Sun and the fix'd Stars" to argue for a universe full of inhabited solar systems (138). At least, Behn criticized Fontenelle for pushing "his wild Notion of the Plurality of Worlds to that height of Extravagancy, that he most certainly will confound those Readers, who have not Judgment and Wit to distinguish between what is truly solid (or, at least probable) and what is trifling and airy" ([xiii]).

Sizing up the Sun, 1670–1780

Even as the intelligentsia were embracing the Cartesian cosmology with its infinitude of solar systems scattered through unbounded space, the astronomers were increasing the accuracy of their instruments and techniques (Van Helden 1985). First they adopted the "astronomical" telescope, which, un-

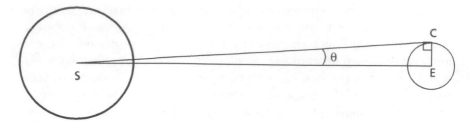

Figure 1.4. The solar parallax, that is, the mean value of the angle (θ) between the line connecting the sun's center (S) to the earth's center (E) and a tangent line connecting the sun's center to the earth's circumference (C); the sun's mean distance from the earth, SE = EC/sin θ, where EC is the earth's radius. Since sin θ approaches zero as θ decreases, the smaller the solar parallax, the greater the sun's distance. (Courtesy of NASA History Office.)

like the "Dutch" telescopes used by Galileo's generation, consisted entirely of convex lenses and hence provided a sizable field of view at magnifications of fifty and above. Then they began transforming this instrument into a tool for precision work by adding the micrometer for measuring small angles and cross hairs for fine alignment. By 1670, largely as a consequence of the trend toward higher accuracy, Copernican astronomers were eager to reduce their uncertainty concerning the sun's distance. Their figures for this quantity ranged from 7300 earth radii upward to 25,000 earth radii. This range corresponded to a range in their estimates of the solar parallax—the angle that would be subtended by the earth's radius if it were viewed from the distance of the sun (fig. 1.4)—from 30″ (second of arc) down to 8.2″. Only by narrowing this spread could they satisfy their curiosity about the solar system's true dimensions. The trouble was that the methods the Greeks devised for ascertaining the solar parallax were not at all suited—even with the best telescopes—for direct measurements of such small angles.

Cassini, Flamsteed, and the Sun's Distance, 1670–1700

In 1670 the director designate of the Paris Observatory—the Italian astronomer Giovanni Domenico Cassini—proposed a fresh means of attacking the solar-parallax problem (Olmsted 1942; Van Helden 1985). He could use the geometry of the Copernican system to determine the sun's parallax from any planet's parallax. During August–October 1672, when the earth overtook Mars, the red planet's parallax would be at a maximum because of its proximity. Cassini realized that Mars's parallax could be ascertained at this time by making simultaneous observations from Paris and from Cayenne, a French colony near the equator to which the Académie des Sciences would soon be sending an expedition. His proposal for such observations won the assent of his fellow academicians and the government. In the fall of 1671, therefore, he and the expedition's leader, Jean Richer, agreed on the observing procedures for this project and related ones. Everything worked according to plan a year later. While Cassini and his colleagues observed Mars's celestial position from Paris and other sites in France, Richer and his assistant were doing the same from Cayenne.

Besides seeking data to correlate with those from Cayenne, Cassini followed an observing program that had recently been promoted by the young English astronomer John Flamsteed (Van Helden 1985). Measuring Mars's parallax required observing the planet from two widely separated points, such as Paris and Cayenne. On account of the earth's rotation, however, comparable separation could be achieved by making two sightings from the same observatory a few hours apart. This method had the disadvantage of requiring corrections for the orbital motions of the earth and Mars during the interval between the observations, but it had the great advantage of ensuring that the requisite observations were obtained with the same instruments and procedures. Accordingly, Cassini and his Danish assistant Ole Roemer[5] observed Mars twice nightly on several occasions in September 1672. Flamsteed, whose plans for observing during this month were thwarted by the demands of family business, finally got a pair of observations one night in early October.

Flamsteed was the first to reduce his data to publishable results (Van Helden 1985). In an open letter to Cassini, which appeared in the Royal Society's *Philosophical Transactions* in 1673, he reported that Mars's parallax during the opposition was 25″, corresponding to a solar parallax of 10″ or a solar distance of about 21,000 earth radii. Cassini responded that his observations in Paris had yielded similar results. But contrary to his usual practice, the Italian postponed publishing the details of his investigation. He evidently suspected that the wide range in values for Mars's parallax emerging from his own data meant the true value was being swamped by observational errors. If so, his suspicions must have increased when his reduction of the simultaneous observations made from Paris and Cayenne also yielded a wide range of results. Nevertheless, convinced on other, rather subtle grounds that the solar parallax must be less than 12″,[6] Cassini eventually overcame most of his doubts. In his detailed report of 1684, he excluded those observations from his analysis that yielded values he regarded as

5. Roemer later achieved renown with his determination of the velocity of light.
6. Cassini had found that the value of the obliquity of the ecliptic at Cayenne agreed with the value found in Europe only if it was assumed that the difference between the measured values was due mainly to atmospheric refraction, not solar parallax. This finding indicated that, at most, the solar parallax was 12″ (Van Helden 1985, 139).

unreasonable. This selective use of the data enabled him to conclude that Mars's maximum parallax was slightly larger than 25", giving a solar parallax of 9.5" and a distance of 21,600 earth radii. These figures, which thanks to no little luck were within 10 percent of modern values, were slow to find acceptance. Astronomers seem to have realized that it was largely fortuitous that the solar parallaxes of Cassini and of Flamsteed, who had meanwhile become the first astronomer royal, agreed so closely. Nonetheless, by the early 1700s most astronomers subscribed to a solar distance of about 21,000 earth radii and, by simple geometry, a solar radius of about 97 earth radii.

Newton and the Sun's Mass, 1685–1725

In the mid-1680s, just after Cassini finally reported his solar-parallax results, the Lucasian Professor of Mathematics at Cambridge University—Isaac Newton (1642–1727)—turned his attention briefly to the sun. At the time, Newton (fig. 1.5) was hard at work on his *Philosophiae naturalis principia mathematica,* a treatise that would put mechanics on a firm footing and earn him a preeminent place in the pantheon of science (Westfall 1980). Among the many examples he used to demonstrate the power of his mathematical principles were determinations of two new solar parameters—the sun's mass and density relative to the earth's mass and density.

His procedure for determining these parameters was astute (Newton 1687–1726/1972, 577–82). According to his law of centripetal force, the acceleration of any satellite toward a central body varied directly with its radius of revolution and inversely with the square of its period. Moreover, according to his law of gravitation, its centripetal acceleration in any other orbit would equal its actual acceleration multiplied by the square of the ratio of the original and new radii. Consequently one can compare the centripetal accelerations that, say, Venus and the moon would have were they at equal distances from the sun and the earth, respectively. In making this comparison Newton could rely on generally accepted values for the distance of the moon and the periods of the moon and Venus. Unpersuaded by Flamsteed's and Cassini's arguments for a solar parallax of about 10", however, he used the value of 20" for determining the distance of Venus from the sun. On this assumption, the centripetal acceleration toward the sun for any given distance ended up 28,700 times greater than that toward the earth from the same distance. This result was equivalent, once again according to his law of gravitation, to the claim that the sun is 28,700 times more massive than the earth. Newton went on to compare the solar and terrestrial densities, finding that the sun has slightly more than a quarter of the earth's density. He was especially confident in this result because it did not depend on his assumed solar parallax of 20"—and well might he have been. While

Figure 1.5. Isaac Newton in 1689, two years after the publication of his *Principia;* his strong sense of self-esteem was manifest in his choice of Sir Godfrey Kneller, the leading portraitist of the day. (Courtesy of Richard S. Westfall.)

his initial estimate of the sun's mass turned out to be too small by an order of magnitude, his result for the sun's relative density was within 2 percent of the modern value.

In the second (1713) and third (1726) editions of his *Principia,* Newton

Table 1.1
The Sun's Parameters in Three Editions of Newton's *Principia*, 1687–1726

Parameter	1687 Edition	1713 Edition	1726 Edition	Modern Values
Solar parallax	20″	10″	10.5″	8.7942″
Solar distance[a]	10,000 r_e	21,000 r_e	19,600 r_e	23,455 r_e
Solar radius[a]	48 r_e	96 r_e	92 r_e	109 r_e
Solar mass	28,700 m_e	227,512 m_e	169,282 m_e[b]	332,945 m_e
Relative solar density[a]	.258	.253	.250	.255

Sources: Newton 1687–1726/1972, 580–82; Allen 1973, 16, 161.

Note: r_e and m_e = the earth's radius and mass.

[a]Calculated from related numbers given by Newton.
[b]On the basis of Newton's values for the observational parameters, this figure should be 196,282 m_e. Encke (1842) pointed out that something was amiss here.

remained true to his original method for calculating the sun's relative mass and density; but on both occasions he revised some of the observational parameters, especially the value for the solar parallax, to bring them into accord with the trend of astronomical opinion. These changes influenced his results for the sun's relative mass and density, particularly the mass (table 1.1). Newton's new figures for the mass illuminate his own and his age's scientific practice. The values, with their six significant figures, attest to a certain enthusiasm for calculation. They also attest to a continuing insensitivity to the limits that imprecise data impose on the accuracy of end results. In offering these revised values, Newton started this parameter's journey along a damped oscillatory path to its present value (table 1.2).

Halley, the Transits of Venus, and the Sun's Distance, 1715–1780

Between the 1713 and 1726 editions of his *Principia*, Newton considered adjusting the solar parallax to 13″, then to 12″, taking this figure seriously enough to compute the corresponding mass, and then to 11″ (Newton 1687–1726/1972). Finally, as table 1.1 indicates, he took the average between 11″ and his 1713 figure of 10″. Newton's uncertainty on this point reflected the growing uneasiness among astronomers about the accuracy of Flamsteed's and Cassini's determinations of the solar parallax. Flamsteed's successor as astronomer royal, the versatile investigator of cometary orbits Edmond Halley (1656–1743), was the chief critic of the usefulness of Mars as a bridge to the sun (Woolf 1959). His alternative, proposed first in 1678, again in 1691, and most forcefully in 1716, was for astronomers to use a passage of Venus in front of the solar disk to find the parallax of that planet

Table 1.2
Determinations of the Sun's Mass, 1687–1976
(Values in Earth Masses)

Year	Value	Percentage Difference from 1976 Value	Source
1687	28,700	−91	Newton 1687–1726/1972, 581–82
1713	227,512	−32	Newton 1687–1726/1972, 581–82
1726	196,282	−41	Note to table 1.1
1774	365,412	+ 9.8	Lalande 1774, 503
1796	329,809	− 1.0	Laplace 1796, 32–34
1813	337,102	+ 1.2	Laplace 1813, 217
1831	357,500	+ 7.5	Encke 1831, 342
1842	359,551	+ 8.0	Encke 1842, 187–88
1863	319,455	− 4.1	Hansen 1863, 10–11
1895	332,040	− 0.27	Newcomb 1895, 100
1926	331,950	− 0.30	Russell, Dugan and Stewart 1926, app. 2
1955	333,100	+ 0.046	Allen 1955, 16
1964	332,958	+ 0.0036	Clemence 1964, 102
1967	332,945	− 0.00030	Ash, Shapiro, and Smith 1967, 338
1976	332,946.0 ± 0.3	—	Duncombe et al. 1976, 63

Note: The 1967 value has been calculated from values given for the ratios of the solar mass to the terrestrial-lunar mass and of the terrestrial mass to the lunar mass.

and thence of the sun. Halley argued that the differences in timing and path of Venus's transit, when viewed from different sites around the earth, would enable a precise calculation of the planet's distance. The only difficulty appeared to be that transits of Venus were rare, coming in pairs after intervals longer than a century. His own death, Halley realized, would precede the next two transits, which were predicted for 1761 and 1769.

As the 1761 transit of Venus approached, scientific interest in the solar-parallax problem intensified (Woolf 1959). Indeed, having no other outlet, this interest stimulated fresh applications of methods that Halley had judged inferior. In the fall of 1751, astronomers in Greenwich, Berlin, and Saint Petersburg responded to the appeal of the French academician Nicolas de Lacaille to observe Mars and Venus at the same times and with the same methods he would be using at Cape Town in southern Africa. This cooperative venture, which involved more stations than the Paris-Cayenne observing program of 1672, yielded data that enabled Lacaille to report a solar parallax of 9.5" several years later. Meanwhile in 1753, largely at the instigation of the veteran French astronomer Joseph-Nicolas Delisle, many observers in Europe and North America monitored a transit of Mercury. As Halley had prophesied, their efforts to obtain reliable parallax data were thwarted by this planet's proximity to the sun. Still, Mercury gave astronomers valuable practice in observing transits and heightened their appreciation of the opportunity presented by Venus's forthcoming passages across the sun's disk.

In 1756, five years before the first transit of Venus, the European powers and their colonies in North America were at one another's throats in what would come to be called the Seven Years' War in Europe and the French and Indian War in America. Despite the hostilities, the Paris Academy arranged royal patronage for three major expeditions to various parts of the globe (Woolf 1959). Delisle also convinced many astronomers throughout central Europe to ready themselves for the transit. Meanwhile in Scandinavia, the astronomer Pehr Wilhelm Wargentin led the Swedish Academy's successful campaign for support of several observing parties. In Britain, no astronomer stepped forward in the manner of Delisle or Wargentin to promote observations of the transit. But in mid-1760, upon receiving an appeal from Delisle, the Royal Society decided it would be shameful to let the opportunity pass. Stressing that "the Improvement of Astronomy and the Honour of this Nation" were at stake (Woolf 1959, 83), the society soon persuaded the government to send out two observing expeditions.

On June 6, 1761, 120 scientists at sixty-two stations observed the transit of Venus with sufficient attention to detail that their data could be given a place in the astronomical literature (Woolf 1959). To everyone's dismay, however, the data did not yield a definitive value for the solar parallax. Results ranged from 8.28" to 10.60", corresponding to a range of distances from 24,900 to 19,500 earth radii. This discordance was traced to two main sources. First, it was difficult for observers to furnish exact times of Venus's entry onto and exit from the sun's disk because the planetary and solar circumferences often failed to separate or merge cleanly. Second, and more important, it was difficult for the analysts to make effective use of the transit data provided because the coordinates of many observing stations were poorly known.

Meanwhile, spurred on in part by the quarrels emerging from the analysis of the 1761 data, astronomers impatiently prepared for the next transit, which would also be visible from around the globe (fig. 1.6). With the war's end, patronage was more abundant and travel less risky. Some 150 astronomers at seventy-seven stations observed the transit on June 3, 1769 (Woolf 1959). Their observations enabled analysts to narrow the range for the solar parallax to between 8.43" and 8.80". Although not the definitive figure Halley had anticipated, this range of values may be regarded—unlike the figures reported by Flamsteed and Cassini—as genuine measurements (Van Helden 1985).

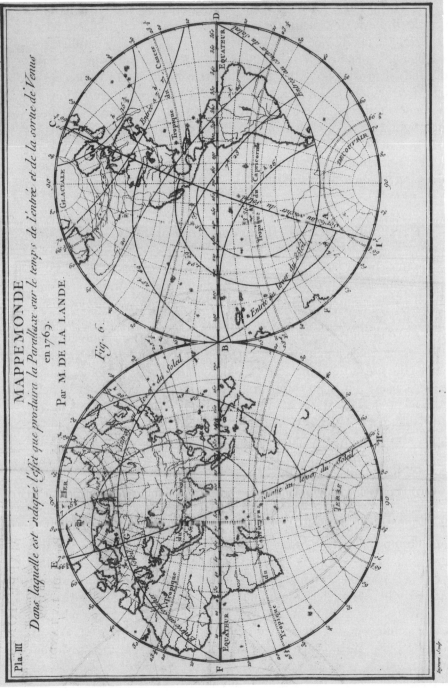

Figure 1.6. Joseph-Jérôme de Lalande's *mappemonde* predicting when Venus's entry and exit from the solar disk would be observed in 1769 from around the world; this map, which was presented to the Paris Academy in 1760, indicates the level of sophistication attained by celestial mechanicians within three generations of Newton's *Principia*. (Source: Lalande 1762, plate 15 facing p. 250. Courtesy of the Bancroft Library, University of California, Berkeley.)

By 1780, after a century of increasingly sophisticated attempts to determine the scale of the solar system, the astronomers had a fairly reliable set of figures. In fact, taking 8.6″ as a reasonable mean for their solar parallax, the corresponding solar distance of 24,000 earth radii, solar radius of 112 earth radii, and solar mass of 357,000 earth masses were all within 10 percent of modern values. In the two ensuing centuries, succeeding generations of astronomers returned time and again to the problem of refining all these related figures. The story has many interesting twists and turns (e.g., Clerke 1902; Dyson 1912; Shapiro 1968). Nonetheless, the emphasis henceforth will be on scientific endeavors to go beyond the sun's dimensions and understand its physical constitution.

Herschel's Natural History of the Sun, 1780–1810

William Herschel (1738–1822) is best known today for his large reflecting telescopes, his discovery of Uranus, and his studies of the distribution of stars through space. As this section establishes, however, he also gave the sun considerable attention. His approach to the heavens was that of a naturalist who wanted to classify all the objects in the celestial realm and to discover the stages of development within each class (Schaffer 1980b). Since he regarded the sun as the prime exemplar of the stellar class, he took up solar questions again and again throughout his astronomical career. Herschel's involvement with the sun is of interest partly for what it reveals about the state of solar science between 1780 and 1810. But the evolution of his solar ideas is also of interest because, despite the extravagance of some of his specific conclusions, he had a talent for raising important questions and proposing useful lines of inquiry.

Even in his own day, when amateurs were common in science, Herschel's career was regarded as marvelous (Lubbock 1933; Hoskin 1963; Turner 1977, 1988; Krafft 1986). The third child of an oboist in the Hanoverian foot guard, he joined his father's band at the age of fourteen. He remained in the guard until 1757, when his battalion became so heavily engaged in the Seven Years' War that it had no time for music. Then, on his father's advice, young Herschel deserted. His destination was England, where he had earlier spent a few happy months on tour with the band. Reaching London before his nineteenth birthday, he earned his living there for a year or so as a musical copyist. Then, eager to perform, he went out into the provinces. Over the next few years, he worked as conductor, composer, music teacher, violinist, and organist. In 1766 he finally landed a secure position as the organist for the Octagon Chapel in Bath.

Fifteen years later, Herschel (fig. 1.7) won fame by detecting what was soon recognized as a new planet. Not only Uranus, as the planet came to be known, but also its discoverer—a middle-aged German musician residing in an English resort town—astounded the European intelligentsia. By 1781, however, Herschel was already a seasoned observer of the heavens. During the preceding seven years, evidently bored with his music, he had managed to construct telescopes far more powerful than any then in existence. And during the preceding eighteen months, he had systematically used one of his medium-sized instruments—a reflector of 227 power—in his *second* celestial survey. It was in the course of this survey that he spotted a "comet" that, after much debate, was identified as the first planet to be discovered since antiquity (Schaffer 1981). The discovery gave Herschel and his unorthodox approach to the heavens a passport into the world of science. Before the debate over the object's nature was settled, he was awarded the Royal Society's Copley Medal and elected to its membership. Then in 1782, after demonstrating the power of his instruments to the astronomer royal and their entertainment value to the royal family, he was provided with a pension so he could establish himself near the palace at Windsor. Abandoning music, Herschel devoted his remaining forty years to astronomy and natural philosophy.

Herschel and the Solar-Stellar Analogy, 1780s

Just before achieving fame, Herschel revealed his conception of the sun in two papers he read at Bath's Philosophical Society (Schaffer 1980a). It blended recent ideas, especially the theory of Alexander Wilson (1774) that sunspots gave glimpses of a solid and dark core below the sun's brilliant exterior, with older views, particularly the insight that the stars were suns. In the first paper—a discourse on the theory of light—Herschel (1780, xcvi) disputed the idea that "the whole mass of the Sun should consist of one intire clear flame. Nay, the contrary is proved by the observation of so many spots, which indicate plainly that under the outward flaming Surface is most probably contained a solid globe of unignited Matter. The gravitation of all the planets to the Sun indeed is alone a sufficient demonstration of [its] solidity."[7] In his second paper he dramatized the variability of the star Collo Ceti by describing it as "a *Sun* . . . perhaps surrounded with a system of Planets [that] undergoes a change, which, were it to happen to *our* Sun would probably be the total destruction of every living creature!" He went on to interpret this variable star as a rotating body with a superabundance of

---

7. Herschel's gravitational argument for the sun's solidity reveals that, like many of his contemporaries, he did not regard gravity as a universal property of matter.

Figure 1.7. William Herschel, "the Celebrated Discoverer of the New Planet." (Source: Herschel 1785, facing p. 1.)

spots, some of which were "occasionally consumed, as . . . on the Sun" (Herschel 1781, civ–cv).

During the 1780s, Herschel frequently returned to the solar-stellar analogy. He used it in two main ways. Hoping to ascertain how the stars were distributed through space, he presumed that they were all about the same size as the sun and hence that their distances varied inversely with the square of their apparent brightnesses. His warrant for this approximation was the natural historian's concept of a species:

> When I say "Let the stars be supposed *one with another* to be *about* the size of the Sun," I only mean this in the same extensive signification in which we affirm that *one with another* Men are of such or such a particular height. This does neither exclude the Dwarf, nor the Giant. An Oak-tree also is of a certain general size tho' it admits of very great variety. . . . If we see such conformity in the whole animal and vegetable kingdoms, that we can without injury to truth affix a certain general Idea to the sizes of the species, it appears to me highly probable and analogous to Nature, that the same regularity will hold good with respect to the fixt stars. (Herschel 1782)

This was an appealing argument. As Herschel himself would demonstrate later in his life, however, the range in intrinsic brightness of the stars was far greater than the analogy allowed.[8]

While Herschel used the sun to investigate the distances of the stars, he used the stars to explore the motion of the sun (Hoskin 1980; Hendry 1983). For instance, he made a detailed case for the universality of stellar motions and then asked rhetorically, "Who can refuse to allow that our sun, with all its planets and comets, that is, the solar system is no less liable to such a general agitation as we find to obtain among all the rest of the celestial bodies?" (1783, 260). He developed this idea by arguing that, since the stars whose motions could be measured tended to move away from the star L Herculis, the sun must in fact be moving toward this star. In making this argument, Herschel showed that the solar-stellar analogy could be used to enrich knowledge of the sun as well as of the stars.

Herschel on the Sun's Structure and Evolution, 1790s

In the early 1790s, Herschel's interest in the structure of the sun and stars was rekindled by two observations. After several years of puzzling over the

8. Herschel showed in 1803 that many double stars, rather than lying by coincidence along nearly the same line of sight, were revolving around one another as binary systems. Since both stars in a binary system were about the same distance from earth, a large difference in their apparent brightness revealed an equally large difference in their intrinsic brightness.

nature of planetary nebulae,[9] he discovered one whose luminous halo surrounded a star. He saw at once that the intimate association of the halo and star subverted his presumption that powerful telescopes would eventually resolve such nebulae into stars. The halo must, he announced (1791, 83–84), consist of "a shining fluid, of a nature totally unknown to us." He went a step further to conjecture that the "condensation [of] luminous matter" in planetary nebulae regenerated stars that had been destroyed in collisions. Not long after this study, the appearance of a particularly large sunspot drew Herschel's attention directly to the sun. He had observed giant spots in 1779 and again in 1783, inferring along with Wilson that they were depressions in the sun's surface. In the context of his fresh interest in luminous matter, his observations of the big spot of 1791 set him to thinking about the role of such matter in solar and stellar structure. During the next three years, accordingly, he familiarized himself with typical sunspots and the ordinary appearance of the sun's surface.

Herschel (1795) reported his conclusions to the Royal Society in a paper, "On the Nature and Construction of the Sun and fixed Stars." He opened this study with a spirited justification of his undertaking: "Among the celestial bodies the sun is certainly the first which should attract our notice. It is a fountain of light that illuminates the world! it is the cause of that heat which maintains the productive power of nature, and makes the earth a fit habitation for man! it is the central body of the planetary system; and what renders a knowledge of its nature still more interesting to us is, that the numberless stars which compose the universe, appear, by the strictest analogy, to be similar bodies" (46). He proceeded to laud Newton for his analysis of the sun's gravity and mass, Galileo, Scheiner, and others for determining the rate of the sun's rotation and the inclination of its axis, and the mathematicians for using data from the transits of Venus to ascertain the sun's distance—given a few pages later as 95,000,000 miles (equivalent to a solar parallax of 8.7")—and thence its true dimensions and surface gravity. His predecessors' achievements had, however, shed little light on the "internal [or] physical construction of the sun" (47–48). Hoping to clarify this issue, he adduced his observations of sunspots in support of Wilson's idea that they were depressions penetrating to "the real solid body of the sun itself" (51). Then, taking full advantage of the license provided by his renown, Herschel went on to pile conjecture upon conjecture.

The sun, he surmised, possessed "a very extensive atmosphere"

9. Any diffuse source of light in the heavens was called a nebula. Some nebulae turned out to be gas clouds and star clusters within our galaxy; others, galaxies of stars at great distances from our galaxy. A ring of diffuse light was the identifying characteristic of a planetary nebula.

(Herschel 1795, 58). Chemical interactions in this atmosphere generated a stratum of "lucid fluid" that emitted the sun's light (59). Beneath the resulting "luminous solar clouds" was the sun's body, which, to judge from the uneven contours of the spots, was probably covered with "mountains and vallies" (62). The sun, in other words, was "nothing else than a very eminent, large, and lucid planet [which] is most probably also inhabited, like the rest of the planets, by beings whose organs are adapted to the peculiar circumstances of that vast globe" (63). The sun's inhabitants, he suggested, were not scorched by its rays because these rays, rather than conveying heat themselves, could only liberate heat by interacting with substances that easily released their "matter of fire."[10] He deemed it likely that the gases composing the sun's "atmosphere, and the matter on its surface, are of such a nature as not to be capable of an excessive affection from its own rays" (64). Herschel saw no reason to stop there. Certain that the sun could be numbered among the fixed stars, he argued that "if stars are suns, and suns are inhabitable, we see at once what an extensive field for animation opens itself to our view" (68). Thus, following an analogical chain, Herschel had progressed from sunspots to a solid sun to a universe replete with inhabited stars.[11]

A year later, Herschel (1796) returned to the sun in a paper proposing means of detecting changes in the apparent brightness of the stars. His interest in this issue was motivated by his desire to solve a problem of general concern: "[W]hat degree of permanency we ought to ascribe to the lustre of our sun? Not only the stability of our climates, but the very existence of the whole animal and vegetable creation itself is involved in the question. Where can we hope to receive information upon this subject but from astronomical observations. If it be allowed to admit the similarity of stars with our sun as a point established, how necessary will it be to take notice of the fate of our neighboring *suns*, in order to guess at that of [the] *star* which among the multitude we have dignified by the name of *sun*" (541). His own opinion, based on the evidence of natural history (presumably fossils), was that the earth's climate had changed through the ages. He supposed that these changes were caused by changes in the sun's brightness. In any case, and here he may have been influenced by the turmoil in revolutionary France, he regarded it as "highly presumptuous to lay any

10. Herschel revealed himself here as a supporter of the chemical theory that combustibles contained phlogiston, the principle of combustion. At the time, this theory had been vanquished in most of Europe by Lavoisier's chemical system (Hufbauer 1982).

11. Although preposterous to today's reader, Herschel's conjecture that the sun was inhabited was readily accepted by many of his contemporaries and immediate successors as a reasonable extension of the doctrine of the plurality of worlds (Crowe 1986).

great stress upon the stability of the present order" (541). He thought it might be possible to devise "some photometer" that would reveal the "instability of the sun's lustre" (542). Presciently, he suggested that such observations would best be made on "some high and insulated mountain" (542).

Herschel's Last Solar Discoveries and Speculations, Early 1800s

In 1800 Herschel's interest in the sun's structure and variability led him to discover the phenomenon that would later be called infrared radiation (Herschel 1800a; Lovell 1968). The year before, he had embarked on his first prolonged set of solar observations. As the work progressed, he tried to minimize the painful glare by viewing the sun through colored glass plates. He soon noted that red glass reduced the intensity of the light much more than that of the heat and that green glass had the opposite effect. By March 1800 he had concluded that—contrary to his view of 1795—solar rays conveyed not only light but heat. Moreover, like light, radiant heat was dispersed into a spectrum when it passed through a prism. His probing of such a spectrum with thermometers indicated that the intensity of the solar heat rays increased from the violet through the red, reaching a maximum beyond the visible spectrum. Intrigued, Herschel set aside his solar observations to inquire further into the properties of radiant heat. Although he found it could be reflected as well as refracted, he was unable to abandon the idea that the rays giving rise to the sensations of light and heat were essentially different. Before long, however, his young contemporary Thomas Young, an ardent advocate of the new wave theory of light, proposed that light and heat rays differed only in the frequency of their vibrations.

Herschel soon returned to his study of the sun. His discovery of solar heat rays obliged him to rework his theory of the sun's structure (Herschel 1801; Kawaler and Veverka 1981). He had, in particular, to devise a new explanation of the solar inhabitants' exemption from scorching. He ended up arguing that the sun's atmosphere contained two layers of clouds, a self-luminous stratum and beneath it a reflecting stratum that shielded the solarians from most of the luminous clouds' light and heat. Sunspots, he went on to hypothesize, were openings in the cloud strata (fig. 1.8) caused by the upwelling of a buoyant gas generated by the body of the sun. The interaction of this rising gas with the gases making up the solar atmosphere produced the self-luminous clouds. Consequently, the greater the abundance of sunspots, the greater should be the number of self-luminous clouds and hence the sun's luminosity.

Herschel realized that a direct test of this hypothetical relation was beyond his means. But knowing from his own experience that the sun was sometimes spotless, he went over observational records since 1610 and

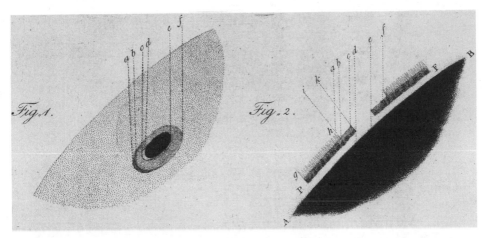

Figure 1.8. Herschel's conception of sunspots as openings in the solar atmosphere: his first figure portrays the observer's view of a sunspot shortly after it has come over the sun's limb, and the second offers a cross section through this sunspot. (Source: Herschel 1801, plate 18 at end of volume.)

identified five periods longer than two years in which no spots had been recorded. During these periods, according to his hypothesis, the sun should have had a lower luminosity than usual. Presuming that lower luminosity would result in colder weather, reduced wheat harvests, and higher wheat prices, Herschel then compared the price series for grain appearing in Adam Smith's *Wealth of Nations* with his own historical data on sunspots. The outcome, he judged, was favorable to his theory of solar variability. Although many of his contemporaries regarded his attempt to correlate wheat prices and the sun's luminosity as laughable, Herschel's idea that the sun's historical behavior might be traced with terrestrial proxies eventually yielded significant results (see chap. 7).

After laying out his revised solar theory in the spring of 1801, Herschel seems to have concluded that his inquiries into the sun's structure and variability had reached the point of diminishing returns. The problem was that he could not observe the sun at the very high magnifications he routinely used for studying the stars and nebulae. He soon established that the sun's heat rays were distorting his mirrors (Herschel 1803). At this point he could have set himself the task of developing a special solar telescope. Preferring to continue his stellar investigations, he limited his future discussions of the sun to issues not requiring observations. He used the data on stellar proper motions that had been accumulated since his 1783 paper to reexamine the sun's motion through space (Herschel 1805, 1806). Five years later, he reconsidered his ideas of the early 1790s on celestial nebulosity. He now took a stance similar to that of the French mathematical astronomer Pierre-Simon Laplace (1749–1827). The two had met in the summer of 1802 when Herschel visited Paris. During an audience with Napoleon, they had discussed Laplace's theory that the sun and planets condensed out of a rotating cloud of matter (Lubbock 1933). Reflecting on Laplace's views as well as his own observations, Herschel was no longer content to limit the role of "nebulous matter" to the occasional regeneration of stars. He had come to regard its condensation as the birth process of all stars including our sun (Herschel 1811). His paper added important weight to Laplace's case for what came to be called the nebular hypothesis or nebular cosmogony (Numbers 1977).

From 1780 through 1810, therefore, Herschel's thinking about the sun was guided by two ideas he had adopted from his predecessors—that sunspots gave glimpses through the sun's luminous atmosphere to its dark, solid body and that the sun was a typical star. Approaching the heavens with a natural historian's perspective, he took these ideas in very interesting directions. Some of these directions would, with the further development of physics in the nineteenth century, come to be regarded as utterly wild—particularly his attempts to populate the sun and stars. But a remarkable number of Herschel's initiatives adumbrated future lines of research—especially his case for the sun's motion, his emphasis on its physical structure, his arguments for its variability, his discovery of its infrared radiation, and his speculations about its origins.

The State of Solar Science about 1810

By 1810 astronomers possessed a fair knowledge of several of the sun's properties. They could determine its approximate distance from the data that had been acquired during the transits of Venus. They could compute its dimensions from their knowledge of its distance and angular diameter. They could calculate its relative mass and density from their knowledge of the distances and periods of Venus and the moon. They could ascertain its rate of rotation by tracking sunspots across the solar disk. And they could estimate its velocity through space from their slowly increasing knowledge of the motions of nearby stars.

The astronomers of 1810 could make these determinations, which were for the most part beyond the imaginings of their pretelescopic forerunners, partly because their observational capabilities were greater. With their telescopes, they could observe phenomena such as ordinary sunspots and planetary transits that had hitherto been invisible. With the accessories to their telescopes, they could measure celestial positions within seconds rather than minutes of arc. With their growing numbers and better access to patronage, they could use techniques involving the cooperation of observers around the globe. Besides their greater observational capabilities, the astronomers of the early nineteenth century had more potent theoretical tools. With their mechanistic worldview, they could presume that the sun obeyed the same laws as all other matter in motion. And with their celestial mechanics based on Newton's laws, they could not only devise more sophisticated observing techniques but also discern more subtle ways of using familiar data to determine the sun's properties. Indeed, as Laplace had recently demonstrated, they could use celestial mechanics to deduce a reasonable value for the sun's distance from its perturbing effect on the moon's orbit.

Although the astronomers of 1810 could deal fairly confidently with the sun as an object of celestial mechanics, they did not yet have many instrumental or interpretive resources for studying its structure, workings, or evolution. The inadequacy of their observational and theoretical tools for such inquiries was manifest in Herschel's solar studies. Although he possessed the age's best telescopes, he found that the heat of the sun's radiation limited their utility. Although he used analogical insights to raise important questions about the sun and to initiate novel lines of research, he was also beguiled by his analogies into extravagant conjectures about the sun's landscape and inhabitants. During the nineteenth century, astronomers would acquire much better tools for investigating the physical constitution of the sun than Herschel's telescopes and analogies. The new tools were so powerful, in fact, that it was possible for solar science to emerge as the focus of a specialized research community.

CHAPTER TWO

The Rise of Solar Physics

1810–1910

During the nineteenth century, scientists developed a fundamentally new picture of the sun's physical constitution. They abandoned William Herschel's planetary analogy for what may be called the heat-engine model of the sun. They came, that is, to regard the sun as composed of dense gases that through progressive contraction or some other process generated heat that, as it radiated outward, sustained incessant and manifold activities in the visible atmospheric layers. Their new picture of the sun was the fruit of new research tools, chiefly from physics. In the 1860s, accordingly, those studying the sun's structure and behavior started calling their field "solar physics." Heretofore, most contributors to solar science had been men who, like Galileo and Herschel, gave only intermittent attention to the sun. By 1870 some scientists were beginning to devote their careers to solar studies. Bound together by shared experiences, outlooks, and interests, the solar physicists came to constitute a specialty-oriented community within the discipline of astronomy.

This chapter describes the rise of solar physics. It opens by examining how, between 1810 and 1860, astronomers and physicists developed what would become essential tools for the new specialty. The focus then shifts to the formation of the solar physics community in the 1860s and its subsequent, not altogether successful, endeavors to consolidate the field's position. Next the chapter considers how the American solar physicist George Ellery Hale revitalized the field at the turn of the century. It concludes with a survey of the state of solar physics about 1910.

New Solar Observing Programs, 1810–1860

During the early decades of the nineteenth century, astronomers devoted themselves primarily to observing the positions and motions of the heavenly bodies and preparing reliable ephemerides and star catalogs (Grosser 1962; Herrmann 1985; Krafft 1981). A traditional objective was to provide a more accurate basis for navigation and geodesy. Another, dearer to the heart of most astronomers, was to discern order within the solar system and the stellar regions beyond. Some, indeed, placed such a high value on this goal that they excluded all other concerns from astronomy's proper domain.

One such purist was Germany's foremost astronomer, Friedrich Wilhelm Bessel.[1] In a lecture of 1832 he declared that

> what astronomy must do has always been clear—it must lay down the rules for determining the motions of the heavenly bodies as they appear to us from the earth. Everything else that can be learned about the heavenly bodies, e.g. their appearance and the constitution of their surfaces, is certainly not unworthy of attention; but it is not properly of astronomical interest. Whether the lunar mountains have this or that shape is no more interesting for the astronomer than is such knowledge of terrestrial mountains for the non-astronomer; whether Jupiter displays dark streaks on its surface or appears evenly illuminated arouses just as little of the astronomer's curiosity; even its four moons interest him only by their motions. (Bessel 1848, 5–6)

Not all of the age's leading astronomers viewed their enterprise so narrowly. But aware of the limitations of their research tools, they were pessimistic about the prospects for ascertaining the constitution of the heavenly bodies. In particular, they agreed with Bessel (1848, 84) in thinking that the very brightness of the sun's atmospheric envelope prevented them from learning how the dark body of "the sun itself" was constituted. Little did they imagine that it would soon prove possible to study many properties of the sun besides its distance, size, mass, and rate of rotation.

Solar Eclipses

The astronomers' orientation to positional work was manifest in their routines for observing solar eclipses before the 1840s (e.g., Arago 1836). They typically recorded the moon's first and last contact with the solar disk and the beginning and end of totality. Such information, if taken at a place of known geographical position, could be used for refining the empirical parameters entering into lunar orbital theory. And if taken at a place of uncertain location, it could be used to calculate precise geographical coordinates. Timing the stages of a solar eclipse was such a demanding task that astronomers rarely did more than take notice of such attendant phenomena as unusual winds, the moon's outline, or the white corona that sometimes appeared around the sun at totality.

1. Bessel (1784–1846) made numerous contributions to positional astronomy, including one of the first convincing determinations of a stellar distance in 1837.

The astronomer who did the most to stimulate interest in the last of these phenomena was Francis Baily (1774–1844), a wealthy amateur who had helped found the Royal Astronomical Society[2] in London during the 1820s (Clerke 1902). His forte was star catalogs. Still, being a man of broad curiosity and ample means, he did not hesitate to journey to Scotland in 1836 just to witness the spectacular effects that, according to earlier accounts, sometimes accompanied annular eclipses. The eclipse exceeded all his expectations. In his first report to the Royal Astronomical Society, Baily (1836, 16) recounted his delight that "a row of lucid points, like a string of beads, irregular in size, and distance from each other, *suddenly* formed round that part of the circumference of the moon that was about to enter on the sun's disc." He was also struck that the dark regions separating the beads stretched out into lines as the moon's trailing edge receded from the sun's limb and that "all at once, they [the lines] *suddenly* gave way." In concluding his report, Baily urged astronomers to pay particular heed to these phenomena at future eclipses.

As it happened, the next total eclipse was to be visible along a path through southern and central Europe on July 8, 1842. Over one hundred observers took up stations along the path. Weather conditions were generally excellent. Although many observers gave first priority to the timing of the eclipse, a good number focused on the phenomena that were visible near and during totality. One of these was Baily, who observed the eclipse from Pavia, Italy. Despite reading all available accounts of earlier eclipses, he was surprised by the size and brilliance of the corona, and he was astounded by the presence of four large rose-colored prominences that extended outward from the eclipsed sun (fig. 2.1). Although other observers shared Baily's wonder at the prominences and corona, they could not reach consensus on their interpretations. Some thought the prominences shone by reflected light, others that they were luminous. Some thought the prominences and corona were solar features, others that one or both evidenced a lunar atmosphere (Ranyard 1879; Clerke 1902). Such disagreements led to general assent to Baily's recommendation (1842, 211) that future eclipses be studied by teams in which "a *division of labour*" would enable each observer to attend "solely to the part which he has selected for his particular object."

The debates engendered by the 1842 eclipse nurtured intense interest in the next total eclipse in Europe, which would be visible in Scandinavia and along the Baltic seaboard on July 18, 1851. When the long-awaited day arrived, almost a hundred astronomers—mainly Britons, Scandinavians,

2. Founded in 1820, the Royal Astronomical Society was astronomy's first discipline-oriented association.

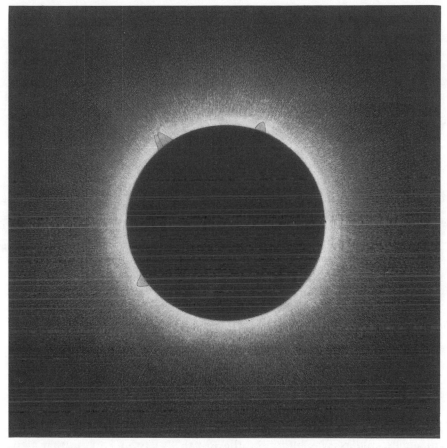

Figure 2.1. Francis Baily's illustration of the 1842 eclipse; note his stylized representation of the four prominences that were visible during totality. (Source: Baily 1842, facing p. 212.)

and Germans—were stationed along the path of totality. A good many parties were unable to observe the eclipse on account of bad weather, but a rich set of observations was secured at the sites with clear skies. After comparing many of the observational accounts, the Royal Astronomical Society Council (1852, 105–7) expressed its satisfaction with the results. The corona appeared "concentric with the sun," confirming its solar nature. Likewise, the red prominences "were respectively covered and revealed on the eastern and western limbs of the sun by the advancing moon, [which] most decisively proved . . . that these wonderful phenomena belong to the sun." Last, many observers had detected "deep red *sierras*, or jagged mountain chains [joined] at the base by a continuous red band" between the prominences and the sun's luminous surface, or "photosphere."

The Sunspot Cycle

Even as observations of the 1851 eclipse were being written up, an independent set of developments was giving prominence to another kind of solar observing program—sustained sunspot monitoring. The pathfinder here was Heinrich Schwabe (1789–1875) of Dessau, Germany (Johnson 1857; Wolf 1876). In 1826 this prosperous pharmacist started a daily record of the number of sunspots and sunspot groups as an aid to his search for an intra-Mercurial planet. Three years later he sold his shop so that he could devote himself to his inquiries. As his data accumulated, he began to suspect that the number of sunspots followed a slow cyclical pattern. At the end of 1843, after observing two maxima and two minima in what appeared to be a ten-year cycle, Schwabe sent a report of his discovery to the *Astronomische Nachrichten*.[3] His brief announcement was passed over by most of his contemporaries. Evidently he was too far outside the astronomical mainstream to receive the attention that, in retrospect, he deserved.

It was the aged German polymath Alexander von Humboldt[4] who thrust Schwabe's discovery into the limelight. He did so by including an account of Schwabe's observations between 1826 and 1850 in his majestic and immensely popular *Kosmos* (Humboldt 1851, 401–4). His report received a warm welcome, partly because his readers respected his judgment and partly because the volume appeared just as observations of terrestrial magnetism, which had been collected since the mid-1830s (Jungnickel and McCormmach 1986), were being subjected to systematic analysis (Meadows and Kennedy 1981). Soon Edward Sabine, the chief British promoter of magnetic studies, was informing the Royal Society in London that "it is certainly a most striking coincidence, that the period, and the epochs of minima and maxima, which M. Schwabe has assigned to the variation of the solar spots are absolutely identical with those which have been [found for] magnetic variations" (Sabine 1852, 121). Not long afterward, J. Rudolf Wolf (1816–93), a lecturer in mathematics and astronomy at Bern University, reported the same correlation between the sunspot and magnetic cycles (fig. 2.2). He also reported that his review of sunspot observations since Galileo's day indicated that the spot cycle's typical duration was eleven years, not ten (Schröder 1984). Delighted with the sudden interest in his work, Schwabe (1852) wrote Wolf of his hope that additional insights "on the mysterious nature of the sun" would soon be forthcoming.

3. This journal was founded in 1821 by H. C. Schumacher of Copenhagen (Herrmann 1985).
4. Humboldt (1769–1859) made his name as a scientific explorer at the beginning of the nineteenth century.

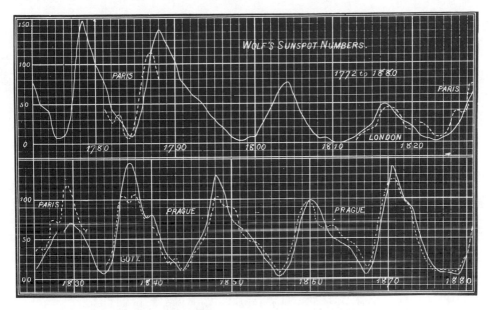

Figure 2.2. Numbers of sunspots (*solid line*) and intensity of terrestrial magnetic activity (*dashed line*) through five eleven-year cycles; the American solar physicist Charles Young, who was one of the first to make extensive use of graphs for characterizing solar phenomena, prepared this diagram for his monograph *The Sun* on the basis of data assembled by the Swiss solar physicist Rudolf Wolf. (Source: Young 1884, 146.)

The astronomer who secured the most interesting results from sunspot monitoring during the remainder of the decade was Richard C. Carrington (1826–75) (Clerke 1902). Supported by his father, a wealthy brewer, he was establishing a private observatory at Redhill in Surrey just at the time when Sabine and Wolf were announcing the correlation between the sunspot and magnetic cycles. His original objective had been to prepare a detailed catalog of polar stars. Intrigued by their announcements, he soon decided to expand his observing program. By night he would measure stellar positions; by day he would map the sunspots. Before long, his solar program was yielding important results. Carrington (1858) first reported that the latitudinal distribution of spots changed in a regular way through the sunspot cycle. After a minimum, spots appeared on both sides of the solar equator in zones between 20° and 40° latitude. As the cycle progressed, the spot zones contracted toward the equator, eventually disappearing there at the next minimum (fig. 2.3). Carrington's second finding (1859a) was that spots near the equator traversed the solar disk more rapidly than those toward either pole. In interpreting this phenomenon, which came to be known as the differential rotation of the sunspots, he accepted Herschel's hypothesis that spots

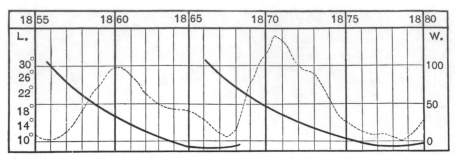

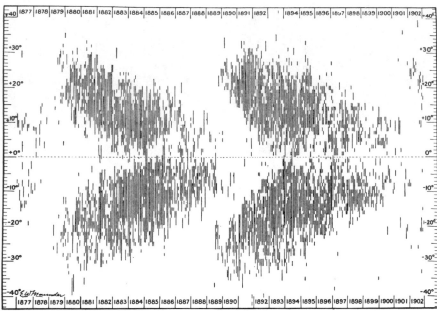

Figure 2.3. Sunspot latitudes during the solar cycle: *Top,* average spot latitude (*solid line*) during two solar cycles (*dashed line*); another graph devised by Young, this time using data reported by the German solar physicist Gustav Spörer; note that between 1866 and 1868 spots from the old and new cycles were simultaneously present at low and high latitudes respectively. (Source: Young 1895, 157.) *Bottom,* the same phenomenon, more vividly portrayed with "butterfly" graphs by the British solar physicist E. Walter Maunder a decade later. (Source: Maunder 1904a, facing p. 760).

were openings in the sun's atmosphere that gave glimpses of its solid core. Accordingly, he attributed the differential rotation to winds—an equatorial current in the direction of the sun's rotation and countercurrents at higher latitudes.

Carrington's third discovery was serendipitous. While mapping sunspots on the morning of September 1, 1859, he was startled by the sudden appearance of "two patches of intensely bright and white light." Over the next five minutes, as the patches swiftly faded away, they "traversed a space of about 35,000 miles" on the solar disk. He immediately compared the configuration of the local spots with the drawing that, by good luck, he had completed just before the appearance of the bright patches. They were unchanged. Carrington (1859b, 181–83) inferred that "the phenomenon took place . . . altogether above and over the great [sunspot] group in which it was seen projected."[5] His subsequent inquiries revealed that a London observer had also seen the patches and that magnetic instruments had registered a significant disturbance just three minutes before he noticed the phenomenon. Carrington rightly suspected that his solar conflagration had somehow affected terrestrial magnetism. But, he cautioned, "One swallow does not make a summer."

Although Carrington made his discoveries with traditional techniques, some scientists believed the future of solar monitoring lay with photography.[6] Regular photographs of the sun would be—if only the practical problems could be solved—a quick means of building up an impersonal record of sunspot numbers and distribution. These problems were first solved at the Kew Observatory, which the British Association for the Advancement of Science supported near London. The project began in 1854 when the Association's Kew Committee gave the job of constructing a "photoheliograph" to Warren de la Rue (1815–89), an amateur astronomer who was introducing the wet collodion process[7] into lunar photography (Gassiot 1854; De la Rue 1859; Smith 1981). Over the next three years, he supervised the construction of a shuttered refracting telescope that could produce 10-cm images of the solar disk. The main difficulty was devising a shutter fast enough to keep the sensitive collodion plates from being overexposed by the sun's glare. In March 1858, De la Rue's group began getting solar photographs that showed spots (Gassiot 1858). Although many technical and funding details had yet to be worked out, the Kew photoheliograph would soon be yielding a record of sunspots that surpassed the sketches of Schwabe, Carrington, and others in quickness and objectivity.

5. Nearly a century later, Carrington's phenomenon would be recognized as a rare kind of flare.

6. The first practical photographic process to yield permanent pictures with good definition was developed in 1837 by Louis Daguerre, an artist and inventor who resided in Paris (Gernsheim 1982).

7. Invented in 1851, the wet collodion process involved coating a glass plate with gun cotton and potassium iodide dissolved in alcohol or ether, letting this collodion dry, and then, just before use, dipping the prepared plate into a silver nitrate solution (Lankford 1984).

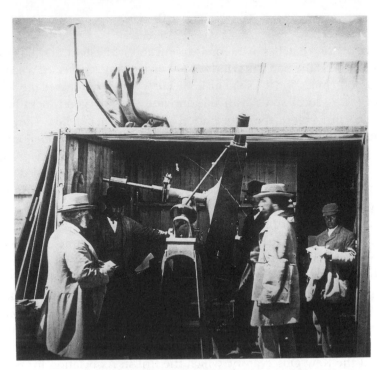

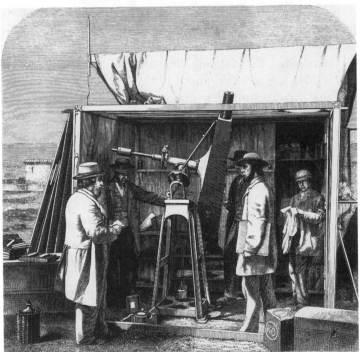

Figure 2.4. Warren de la Rue (*right*) and his observing team at Rivabellosa for the 1860 eclipse; their principal instrument was the Kew Observatory's photoheliograph: *Top*, original photograph. (Courtesy of Robert W. Smith.) *Bottom*, engraving from the photograph done for de la Rue's Bakerian Lecture to the Royal Society. (Source: De la Rue 1862, 363.)

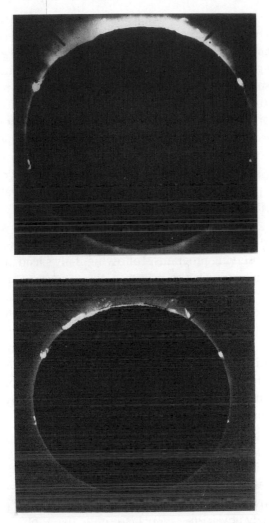

Figure 2.6. The 1860 eclipse during totality. *Top*, original photograph. (Courtesy of Robert W. Smith.) *Bottom*, an engraving from this photograph, which reveals the artist's ability to discern detail at various levels of exposure. (Source: De la Rue 1862, plate 8 at end of volume.)

The 1860 Eclipse

On July 18, 1860, after an eight-year waiting period during which scientific interest in the sun had greatly intensified, European and also American astronomers had an opportunity to observe another total eclipse. Over 25 astronomers and physicists published advisory brochures and articles before the event. And over 150 observers took up stations along the path of totality, which stretched from Oregon through Canada to Spain and North Africa

(Ranyard 1879; Bracher 1981). The most celebrated observations were made in Spain by photographic means (De la Rue 1862; Clerke 1902; Smith 1981). At Rivabellosa, De la Rue's group (fig. 2.4) followed a tightly coordinated observing program with the Kew photoheliograph. They succeeded in getting a fine series of eclipse photographs (fig. 2.5). Meanwhile, some 400 km to the southeast at Desierto de la Palmas, Father Angelo Secchi (1818–78) and his party from the Collegio Romano's observatory were also photographing the eclipse. Their images agreed point for point with those of the English party. This agreement clinched the case that prominences were features of the sun, thereby demonstrating the desirability of equipping all future eclipse parties with photographic apparatus (Royal Astronomical Society Council 1861, 115–7).

Between the mid-1820s and 1860, therefore, astronomers established two kinds of solar observing programs. Following Baily's lead, a growing number of astronomers traveled to eclipse paths to observe the sun's prominences and corona during the brief moments of totality. Meanwhile, following Schwabe's lead, a growing number monitored the sun with the goal of determining patterns in the frequency and distribution of sunspots. Not surprisingly, the most renowned astronomers and observatories played only a modest role in establishing these observing programs. They were already busy meeting the exacting demands of positional astronomy. Rather, the astronomers setting the pace in solar observing were mainly amateurs who had the freedom to set their own research agendas.

New Interpretive Tools, 1810–1860

While some astronomers were starting new solar observing programs, other scientists—chiefly physicists—were fashioning two tools that would be invaluable for interpreting the sun. One was spectroscopy, the science of spectra, the other was thermodynamics, the science of heat. The creators of these sciences did not evince a sustained interest in the sun as did Schwabe, Wolf, Carrington, or De la Rue. Nonetheless, their work was if anything more important in preparing the way for the emergence of solar physics during the 1860s.

The Fraunhofer Spectrum

As of 1810, knowledge of the solar spectrum was still very limited. Most physicists followed Isaac Newton in presuming that sunlight was spread by a prism into a spectrum because the light's constituent colors had different indices of refraction. Most also knew, thanks to recent investigations by William Herschel and Johann Wilhelm Ritter[8] respectively, that the sun's visible spectrum was bordered on its red and blue ends by invisible infrared and ultraviolet spectra. But they were uncertain about how the sun, or any other light source, produced colored, infrared, and ultraviolet rays in the first place.

The puzzle of the origin of the solar spectrum was deepened in 1817 by the optician Joseph Fraunhofer (1787–1826), a partner in a Munich instrument-making firm (Roth 1976; James 1985; Jungnickel and Mc-Cormmach 1986; Sang 1987). That year he reported his method of determining the refractive properties of the different kinds of glass used in compound telescope lenses. His goal was to perfect the achromatic lens[9] by minimizing the dispersion, or spectral spreading, of stellar images. Fraunhofer had realized that success in figuring lens components depended upon finding homogeneous colors with which to measure the refractive properties of each kind of glass. His quest led him to the surprising discovery that the spectrum of sunlight was crossed by hundreds of dark lines. Each line had a distinctive index of refraction and was therefore homogeneous. Fraunhofer proceeded to map the positions of more than five hundred of the solar lines, labeling with letters ten prominent lines that were useful as reference lines in his lens work (fig. 2.6). In reporting his discovery, he pointed out that the bright R line in a candle flame coincided with the dark D line in the solar spectrum. He did not, however, recognize this fact as a clue bearing on the origin of the dark solar lines.

During the 1820s and 1830s, British scientists did the most to follow up on Fraunhofer's discovery (James 1985). Their attention was called to the lines by the Scottish natural philosopher David Brewster, who published a translation of Fraunhofer's article in 1823. Not long afterward, William Herschel's versatile son John Herschel advanced an explanation of the Fraunhofer lines. He suggested that the molecules in the sun's atmosphere possessed "an intense absorbent power" for certain colors. This power, he supposed, caused "the deficient rays [dark lines] in the light of the sun and stars [to] be absorbed in passing through their own atmospheres" (James 1985, 63). A few years later, Brewster thought he had found evidence that the absorption spectrum produced by nitrous oxide matched some of the dark lines in the solar spectrum (McGucken 1969; James 1981). Hoping to

8. Working in Jena, the German chemist Ritter noticed in 1801 that the sun darkened damp silver chloride on a piece of paper. His exploration of the phenomenon indicated that invisible rays adjacent to the violet part of the visible spectrum had the greatest effect.

9. Invented by John Dollond in 1758, the achromatic lens corrected the dispersion of colors by combining lens components with different refractive indices.

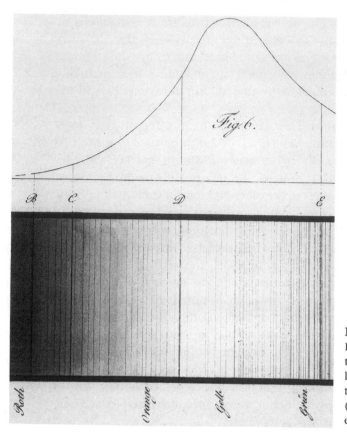

Figure 2.6. A section from Joseph Fraunhofer's map of the solar spectrum; the D line in the yellow would later figure prominently in the establishment of spectral analysis. (Source: Fraunhofer 1817, table 2 at end of volume.)

prove that this gas was responsible for the dark-line spectrum, he proceeded to map about two thousand Fraunhofer lines. He noticed during this project that some lines were unvarying in their intensity while others got darker as the length of the sunlight's path through the terrestrial atmosphere increased. He inferred that the unvarying lines had a solar origin and the variable lines a terrestrial one. Brewster's conclusion was that nitrous oxide must be present in the atmosphere of both the sun and the earth.

James D. Forbes, another Scottish natural philosopher, soon raised doubts about the solar origin of the unvarying lines. Like Baily, he observed the 1836 annular eclipse in Scotland. His focus was on the spectral lines from the sun's limb, which he reasoned should be more numerous and broader if they were caused by absorption in the sun's atmosphere. He found that, despite the sunlight's longer path through the solar atmosphere, the lines had the same appearance as those from the center of the sun's disk. This result struck Forbes (1836) as strong evidence against the hypothesis that the lines were caused by solar absorption. His conclusion was that all the lines in the solar spectrum must be terrestrial in origin. As of 1840, therefore, it was still very much an open question whether the Fraunhofer lines were intrinsic to sunlight and, if so, how they were produced.

Solar Thermodynamics

Meanwhile, scientists were wondering how the sun continued shining through the ages. This question had inspired a good deal of speculation since Newton raised it in the early eighteenth century (Schaffer 1978; Kidwell 1979). Updating this speculative tradition, John Herschel (1833, 212) remarked that the "great mystery . . . is to conceive how so enormous a conflagration (if such it be) can be kept up. Every discovery in chemical science here leaves us completely at a loss, or rather, seems to remove farther the prospect of probable explanation. If conjecture might be hazarded, we should look rather to the known possibility of an indefinite generation of heat by friction, or to its excitement by the electric discharge . . . for the origin of the solar radiation." It was one thing to recognize the mystery of the sun's perdurance and quite another, to judge from Herschel's feeble conjectures, to come up with a plausible explanation.

Although baffled by the sun's durability, Herschel and his contemporaries established the magnitude of the problem by making the first measurements of the sun's heat output (Kidwell 1981). The most influential study was carried out by the Parisian physicist Claude Pouillet. He defined the "solar constant" as the amount of heat from the sun that impinged on a square centimeter of the earth's atmosphere each minute. This quantity, he pointed out, could not be directly measured at the earth's surface because some of this heat must be absorbed during its passage through the atmosphere. Accordingly, he supplemented his measurement of the incident heat with an estimate of the amount absorbed. The upshot of his investigation was a value for the solar constant of about 1.0 cal/cm^2/min. This was no small figure. Even at its great distance from earth, the sun could melt an ice blanket 31 m thick in a year.

The first persons to recognize the insufficiency of the sort of mechanisms proposed by Herschel to generate this amount of heat were Julius Robert Mayer, a young German physician, and John James Waterston, a Scottish engineer in Bombay (Kidwell 1979; James 1982). In the early 1840s these aspiring natural philosophers were thinking about the relation of heat to other forms of energy. Independently of one another both realized—Mayer in 1841 and Waterston in 1843—that the problem of the sun's heat provided a good test for their views. Both later went on to propose that solar radiation originated in the conversion of gravitational energy into heat. In a manuscript that was denied publication by the Paris Academy, Mayer speculated that meteors falling into the sun engendered the solar heat. Waterston, whose manuscript was turned down by the Royal Society of London, developed an alternative scenario. Working within the framework of the increasingly popular nebular theory of the solar system's origin (Numbers 1977; Brush 1987), he hypothesized that the gradual contraction of the sun's mass was the source of solar heat. He went an interesting step further. He calculated how long the sun could sustain its prodigious output of energy. First he determined the rate at which the sun produced heat from the published values of its distance and the solar constant. Then, estimating the conversion factor for heat and gravitational energy, he translated the rate of heat generation into the rate of contraction. His calculation indicated that a shrinking sun would continue shining for another nine thousand years.

Despite being rebuffed, both Mayer and Waterston attempted over the next few years to gain a hearing. Their scientific styles were, however, too loose and speculative for them to have much success. In 1853, Waterston finally stirred up some interest at the annual meeting of the British Association for the Advancement of Science. Still working within the context of the nebular cosmogony, he developed both the meteoric and the contraction hypotheses for the origin of the sun's heat (Waterston 1853). If meteors supplied all the energy, the sun's radius would be increasing by about 5 m a year. On the other hand, if contraction alone was the source, its radius would be decreasing at an annual rate of some 140 m. Since he did not indicate a definite preference for either alternative, Waterston may well have believed that both processes were at work.

Waterston's ideas fell on fertile ground. At the time, several physicists were aggressively developing the new science of thermodynamics using the principles of the conservation of energy and the increase of entropy, a measure of disorder, as their starting points. Two of the most successful thermodynamicists—William Thomson of Glasgow (later known as Lord Kelvin) and Hermann Helmholtz of Königsberg—were smitten by Waterston's ideas. Within a year, they gave solar thermodynamics a place in the scientific mainstream.

Upon reading Waterston's paper, Thomson (1824–1907), who had been pondering the origin of the sun's heat for two years, latched onto the meteoric hypothesis (James 1982; Smith and Wise 1989). Soon he was corresponding with Gabriel G. Stokes[10] about the theory. He wanted Stokes's opinion not only of the general approach but also of the possibility of using the Fraunhofer lines to confirm the theory's prediction, based on the com-

10. Stokes, a mathematical physicist in Cambridge, was then the foremost expert on the wave theory of light.

position of meteorites, that iron was present in the sun's atmosphere. Although generally positive, Stokes reported that research on the solar spectrum was not yet sufficiently advanced to warrant any inference about iron in the sun. Undaunted, Thomson propounded the meteoric theory of solar energy before the Royal Society of Edinburgh in 1854. He anticipated that the sun's remaining life must be much shorter than the millions of years some geologists were postulating for the age of the earth. In particular, convinced by the orbital data on Venus and Mercury that meteoric material could not be abundant, he estimated that the meteors remaining would sustain the sun for only another 300,000 years.

Shortly before Thomson gave his paper, Helmholtz (1821–94) included a discussion of solar heat in a popular lecture on the interrelations of the various forms of energy. Since the German physicist had attended the British Association's meeting and since he later published an abstract of Waterston's paper, it seems fair to assume that this paper started him thinking about the contraction hypothesis (James 1982). Helmholtz (1854/1962) was clearly captivated by the ease with which it could be integrated into the nebular cosmogony. Working on a large canvas, he told his audience in Königsberg how gravity had first shaped the solar system and then provided, through the steady contraction of the sun, the radiant heat that in its turn was so important in the geological, meteorological, and biological history of the earth. Eventually, he was sure, the sun's store of gravitational energy would be exhausted. But this would be far in the future, for there had been no sensible decline in the earth's temperature in the past four millennia. Indeed, his calculations after the lecture indicated that a reduction in the sun's radius by a mere $\frac{1}{10000}$ would supply energy at the rate measured by Pouillet for 2289 years.

Solar Spectroscopy

Some five years after Thomson and Helmholtz gave currency to the meteoric and contraction theories of the sun's heat, there was a breakthrough in the interpretation of the solar spectrum. In the early 1850s, as Stokes's response to Thomson shows, physicists still had no convincing explanation of the Fraunhofer lines. Later in the decade a few chemists, all of them apparently unaware of the interpretive problems posed by the solar lines, explored the possibility of using spectra in chemical analysis (James 1983). Their presumption was that each element emitted a characteristic set of spectral lines at a high temperature. Their efforts to use bright emission lines as signatures for the elements were bedeviled, however, by their failure to appreciate that trace impurities were adding many unwanted lines to their flame spectra.

Among the chemists trying to develop spectral analysis in the late 1850s

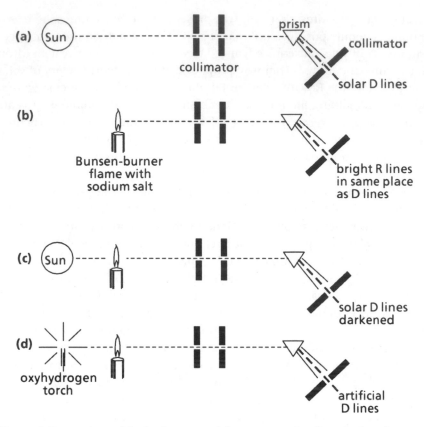

Figure 2.7. Gustav Kirchhoff's discovery of the presence of sodium in the solar atmosphere: the revelation came in the third experiment (c) when the sodium in the flame darkened the solar D lines, and the case was clinched by the fourth experiment (d) when the passage of homogeneous light through the sodium flame produced artificial D lines. (Courtesy of NASA History Office.)

was Robert W. Bunsen of Heidelberg University. In mid-1859, after a year of frustration, he sought advice from Gustav R. Kirchhoff (1824–87), a physicist he had recruited to Heidelberg a few years before. Soon they were collaborating (Kirchhoff 1859–62/1972; James 1983; Jungnickel and McCormmach 1986). Kirchhoff evidently suggested they compare the flame spectra they were trying to decipher with the solar spectrum. As others had noted before them, they found that the bright R lines emitted from flames impregnated with sodium salts coincided with the dark solar D lines. Then, to their surprise, they found that when sunlight was passed through a sodium flame, the D lines, rather than being weakened by the superposition of the R lines, got darker. Kirchhoff reasoned that the sodium in the flame must be intensifying a process of absorption begun by sodium in the solar atmosphere. To check this interpretation, he passed the intense light produced by putting chalk at the focus of an oxyhydrogen torch through a sodium flame. As expected, he produced artificial D lines (fig. 2.7). Next Kirchhoff looked for evidence of iron in the solar spectrum. He may have chosen this element at the suggestion of their new colleague Helmholtz, who, believing in a common material origin for every member of the solar system, would have welcomed proof of iron's presence in the sun. In any case, Kirchhoff found that the bright lines in the flame spectrum of iron matched the solar E lines and many of the b lines. He was satisfied that he had broken the Fraunhofer code.

Kirchhoff announced his discovery in late October 1859. His colleagues helped spread the exciting news. Bunsen (1859) soon wrote to a former English student about Kirchhoff's "lovely [and] quite unexpected discovery [which] opens the way to identifying the material composition of the sun and fixed stars with the same certainty as we determine that of [lithium chloride] etc. with our [chemical] reagents." Two days later, Helmholtz informed Thomson about Kirchhoff's work (James 1983). In the meantime, Kirchhoff was seeking to explain why the emission and absorption lines from a substance should, at any given temperature, coincide with one another (Siegel 1976). Once having developed a satisfactory theory, he returned to solar spectroscopy. His goal, he informed his brother, was "nothing less than chemically analyzing the sun" (Kirchhoff 1860).

The Formation of the Solar Physics Community, 1860–1870

By 1860, thanks both to the astronomers' establishment of solar observing programs and the physicists' creation of solar thermodynamics and solar spectroscopy, the conditions were propitious for a new specialty dedicated to research on the sun. During the next decade, a small band of enthusiasts capitalized on this opportunity. By 1870 they had given shape to the new specialty of "solar physics." In doing so they led the way in the establishment of astrophysics—the systematic study of the physicochemical constitution of all celestial bodies—as an integral part of astronomy (Meadows 1970, 1984a).

The Community's Founders

The pioneer solar physicists tended to be scientists who preferred exploration to demonstration. Most, in fact, did not have the skills, instruments, or temperament needed to compete in the more rigorous branches of physics and astronomy. Some were amateurs who, lacking scientific employment, depended on their own money and occasional grants to finance their investigations. Others were professionals who, though they earned their livings as scientists, were for one reason or another somewhat marginal within their disciplines (Hermann 1973; Lankford 1981; Hufbauer 1986).

Solar physics got off to a faster start in Great Britain than in any other nation for a variety of reasons (Williams 1987). First, the Royal Astronomical Society was an excellent forum for those eager to reach potential participants in and patrons of solar science. Second, the British were in a strong position for mounting eclipse expeditions because of their many colonies and investments around the globe. Last and probably most important, amateurs were numerous in Britain. Unhampered by institutional commitments to positional astronomy, they could easily enter solar research. About 1860, the amateurs Carrington and De la Rue were the leading British students of the sun. Although Carrington gave up his research a few years later (Forbes 1971), fresh talents were entering the field. Most successful among the new recruits were Balfour Stewart (1828–87), the superintendent of the Kew Observatory and as such one of the few professionals in early British solar physics, William Huggins (1824–1910), wealthy son of a London silk mercer, and J. Norman Lockyer (1836–1920), a civil servant in the War Office who founded the journal *Nature* in 1869 (Meadows 1972).

Although solar physics got off to a strong start in Britain, it was not neglected across the Channel. Marginal professionals rather than amateurs played the leading role in developments on the Continent. In France the most important solar physicists during the 1860s were Hervé Faye (1814–1903) and P. Jules Janssen (1824–1907). Faye had done prizewinning work on cometary orbits as a young man at the Paris Observatory. However, during the 1850s, particularly after 1854 when he went to remote Nancy as professor of astronomy, his interests expanded beyond positional astronomy. By the time he returned to Paris in 1862 as a member of the Bureau des Longitudes, he was ready to plunge into a new subject like solar physics. Janssen, who held degrees in mathematics and physics, depended on temporary scientific jobs and stipends for his living until 1865. Then, in recognition of his successes in spectroscopy, he was appointed to the chair of physics at the Ecole Spéciale d'Architecture in Paris.

To the south in Italy, the leading figure in solar physics throughout the 1860s was the Jesuit astronomer Angelo Secchi, who, as we have seen, led a successful eclipse expedition to Spain in 1860. Among the younger Italians joining solar research, the most active was Secchi's friend Pietro Tacchini (1838–1905), who in 1863 became adjunct astronomer at Palermo University's modest observatory. In German-speaking central Europe, the leading solar scientists about 1860 were Schwabe, the discoverer of the sunspot cycle, Wolf, now professor of astronomy in Zurich's Polytechnikum (Jaeggli 1968), and Kirchhoff, the Heidelberg physicist. In 1863 Kirchhoff, having wrung all he could from the spectral analysis of the sun, abandoned the field. Among the Germans beginning solar research during the decade, the most important were Gustav Spörer (1822–95) and J. C. Friedrich Zöllner (1834–82). Spörer, who had a Ph.D. in positional astronomy, taught mathematics and physics at the gymnasium in Anklam. Zöllner, a physicist by training, was an assistant in the observatory at Leipzig University and then from 1866 an associate professor of "physical astronomy" there (Hermann 1982; Hamel 1983). To the north in Scandinavia, the leading contributor to solar science was Anders Jonas Ångström (1814–74), whose retiring personality made him, despite his chair of physics at Uppsala University, a somewhat marginal figure in his own discipline.

Across the Atlantic in the United States—a nation that had gone further than Britain but less far than France, Italy, Germany, and Sweden in transforming science into an occupation—both amateurs and professionals played important roles in early solar physics. Most successful among the amateurs was Lewis M. Rutherfurd (1816–92), a New York attorney whose inheritance and wealthy wife enabled him to retire before reaching thirty-five (Warner 1971). The most important professional was Charles A. Young (1834–1908), who in 1866 succeeded his father as professor of physics and astronomy and director of the observatory at Dartmouth College in New Hampshire.

During the 1860s, therefore, scientists on both sides of the Atlantic joined in investigating the physics of the sun. The field's leaders—De la Rue, Stewart, Huggins, and Lockyer in Britain; Faye and Janssen in France; Secchi and Tacchini in Italy, Schwabe, Wolf, Spörer, and Zöllner in Germany and Switzerland; Ångström in Sweden; and Rutherfurd and Young in the United States—were mainly amateurs and marginal professionals. Comparatively free to set their own agendas, they applied the observational and theoretical tools developed during the 1840s and 1850s to the goal of advancing knowledge of the sun.

Research

The solar physicists obtained many striking observational results during the 1860s (Clerke 1902; Meadows 1970). At the beginning of the decade, for instance, De la Rue supported the idea that sunspots were depressions in the sun's luminous surface layer by using pictures taken twenty-six minutes apart with the Kew photoheliograph to produce a stereoscopic view of the photosphere. About 1864, Rutherfurd demonstrated the potential value of photography in spectroscopy by distributing a detailed photographic map of the solar spectrum (Warner 1971). Two years later, Huggins made a persuasive case that the photosphere had a granulated structure (Bartholomew 1976). In 1868, Ångström published the first comprehensive map of the solar spectrum to give the wavelengths of the Fraunhofer lines in metric units.[11]

That same year, Janssen and Lockyer devised a method for monitoring prominences outside eclipse (Meadows 1972). Independently of one another, they realized that the trick would be to eliminate all the light from the sun except that having the characteristic reddish hue of a prominence. To this end, each first used a powerful spectroscope to produce a high-dispersion spectrum and then looked at the sun's limb through a slit placed at the red part of the spectrum. Each was rewarded with a clear view of the prominences rising above the solar disk. Using this technique, Lockyer (fig. 2.8) soon went on to develop a strong case that prominences were ejected from the underlying reddish stratum or "sierra" discovered at the 1852 eclipse—a stratum that with characteristic verve he renamed the "chromosphere." In 1869, Young and William Harkness of the U.S. Naval Observatory capped off the decade's solar discoveries. At the eclipse that year, each detected a bright green line in the corona's spectrum. The origin of this line, the first of several coronal emission lines to be found, would remain a puzzle for seventy years.

Even as the solar physicists were making such observations, they were abandoning William Herschel's model of the sun for one based on the new insights from solar thermodynamics and Carrington's discovery of differential rotation. Secchi led the way in January 1864 by suggesting that the trend of research indicated that the sun was a gaseous heat engine (Clerke 1902). But it was Faye (1865) who most effectively developed this new model of the sun's physical constitution. After reviewing earlier conjectures and recent research, he insisted on the importance of keeping in mind that the sun was just one of a great multitude of stars. For all its grandeur, the sun was quite ordinary. A few universal laws ought, therefore, to explain most solar phenomena. Faye's starting point was the nebular cosmogony. Following Helmholtz,[12] he postulated that the gravitational collapse of matter scattered through space would necessarily give rise—through the conservation

11. His achievement was subsequently honored by spectroscopists who adopted the Ångström as the unit for measuring wavelengths—1 Å equals 10^{-10} m.

12. By the mid-1860s, Helmholtz's version of the contraction theory of the sun's energy generation had triumphed over the meteoric theory. Indeed, William Thomson had become one of its leading proponents (Burchfield 1975; Smith and Wise 1989).

Figure 2.8. Norman Lockyer observing prominences outside eclipse; his main telescope was equipped with a spectroscope—d indicates the collimator; c, the prism plate, which holds seven prisms; and e, the observing eyepiece. (Source: Lockyer 1874, 214.)

of energy and angular momentum—to extremely hot, slowly rotating stellar masses. The temperature in such a primeval sun would be so high that all its molecules would be dissociated into their constituent atoms. Progressive contraction and cooling,[13] however, would soon lower the temperature at the surface to the point that molecules and particles could form there, producing the photosphere. Thenceforth, over a period lasting millions of years, hot gaseous masses would rise to the surface from deep within the sun, cool there by radiating into space until solids began forming, and then—on account of the density of these precipitates—sink back into the body of the sun.

In making the case for his model, Faye maintained that it could easily account for the differential solar rotation Carrington discovered. One need only assume that the currents ascending to the equator emanated from a shallower average depth than those ascending to higher latitudes. This sug-

13. The American engineer J. Homer Lane soon developed a mathematical theory for a gas star that implied a contracting star's temperature would rise. His prescient paper was given little credence because solar physicists could not bring themselves to believe that the sun behaved like a pure gas (Powell 1988).

gestion, like many other details of Faye's theory, did not survive long. But his central ideas—that the sun was an ordinary star; that it originated in the gravitational collapse of matter; that the heat so engendered put it into the gaseous state; that its internal heat was carried to the surface by convection currents; and that its photosphere was a condensation surface—were all to remain central to the theorizing of solar physicists into the twentieth century.

Rhetoric and Ritual

Delighted with their observational and interpretive successes, the pioneers of solar physics began propounding the virtues of their approach to celestial phenomena in the mid-1860s. Secchi (1865, 146–47), for example, remarked that "however sublime . . . modern astronomy may be and whatever admiration its progress has justly excited," the discipline was seriously limited because it did not aspire to learning more about heavenly bodies than their orbits, masses, and distances. With the spectroscope, however, astronomers could now investigate the "physical structure" of these bodies, a subject "no less important for acquiring an understanding of the universe." Similarly, Huggins (1866a, 441, 1866b, 8) was happy that astronomers had the means to do more than study celestial motion. At long last, they could obtain "knowledge *from observation* of the structure and chemical constitution of the sun, the fixed stars, and the nebulae" by extending "the laws of terrestrial physics to . . . the heavenly bodies." Zöllner (1865, 316), the first to suggest that the emergent branch of astronomy be called "astrophysics," thought that the recent successes in uniting "physics and chemistry with astronomy [presaged] similar coalescences of distinct disciplines into a higher and more general unity."

Although these early spokesmen agreed on the importance of astronomy's new tools, they were not of a single mind regarding solar physics' place in astrophysics or, as it was also known, astronomical physics. Huggins and Zöllner, perhaps because they enjoyed closer relations with positional astronomers or perhaps because they were uncomfortable with the anthropocentrism involved in giving a privileged position to the sun, viewed it as but one among many important subjects in the new branch of astronomy. By contrast, Secchi, who represented the prevalent viewpoint among the founders of the solar physics community, assigned a particular significance to solar research. He maintained that the sun warranted special attention because its proximity made it both relatively knowable to scientists and profoundly important to humanity. This stance was manifest in the title Secchi (1870) gave his trailblazing treatise on solar physics—in translation, "The sun: Exposé of the principal modern discoveries about the structure of this star, its influence in the universe and its relations with the other celestial bodies."

Solar physicists were bound together by more than shared excitement over the rapid pace of discovery, pride in their fresh outlook on the sun's constitution, and enthusiasm about their prospects for enriching astronomy. They were also united by a significant competitive ritual—the eclipse expedition. Besides being a glorious spectacle, every eclipse provided solar physicists with an opportunity to discover something new about phenomena above the sun's dazzling photosphere. To reach a station along the path of totality required coordinated fund-raising and preparation. To witness an eclipse took some luck with the weather. To make the most of the moments of totality demanded uncommon adroitness. Those who went out to eclipses were legion. Those who came back with good results were few. Those who returned with discoveries were the elect. Whatever the outcome, participating in an eclipse expedition was becoming an important bonding ritual for solar physicists.

Fortuitously, three total eclipses occurred in the end of the decade (Ranyard 1879). In 1868 the path of totality swept through India; in 1869, through North America; and in 1870, through the Mediterranean region. The first two eclipses attracted a good number of observers, but the third, despite its occurrence during the Franco-Prussian War, had the largest draw. Well over a hundred observers took up stations along the path of the eclipse. Their number included Huggins, Lockyer, Janssen, Secchi, Tacchini, and Young as well as many men who would make names for themselves as solar physicists in the next three decades. Among those suffering disappointment, Janssen could claim first place. His journey to the eclipse began with a dramatic escape by balloon from besieged Paris. He made his way to Oran in North Africa. As luck would have it, however, he and the other observers there were beclouded (Meadows 1972). Among those rewarded with success, Young headed the list. At Jerez de la Frontera in Spain, he and his assistant got the first glimpse of the "flash" spectrum—an array of bright lines that, for a few moments before totality, replaced the dark Fraunhofer lines.

Consolidating Solar Physics' Position, 1870–1890

With the heady achievements of the 1860s behind them, solar physicists began thinking about how they could consolidate their field's position. They hoped that the uniqueness of solar physics' observational methods and the importance of its contributions would be recognized with substantial patronage. This, they presumed, would enable them and their recruits to sustain the brisk advance of solar research. Their quest for institutional support and interesting results did not go unrewarded. But they would learn that their expectations were unrealistically high.

Institutional Support

Solar physicists in several European nations and the United States opened campaigns for distinct astrophysical observatories and observing programs in the 1870s. At first their pursuit of such support received a sympathetic hearing. Soon, however, some positional astronomers, fearing that funding for their own research might be undercut, were arguing that astrophysical work should be conducted within existing observatories. The intensity and outcome of such debates varied from nation to nation. By 1890 seven observatories had been founded expressly for astrophysical research, including solar research. These ranged in size from the Royal Astrophysical Observatory in Potsdam near Berlin (Hermann 1975; Hassenstein 1941) and the Observatory for Physical Astronomy in Meudon near Paris (Janssen 1896) down to the Bellini Observatory, a station for solar observing atop Mount Etna. In addition, a dozen observatories that had originally been established for positional astronomy or meteorology were pursuing one or more solar observing programs. The most ambitious in this category were the Royal Observatory in Greenwich and the Zurich Polytechnikum's observatory. Besides the observatories, two higher schools—in London and Catania—had professors who were responsible for teaching astrophysics, including solar physics. Last, the American and French academies awarded special prizes for astrophysical research, and an Italian scientific society published an annual journal for celestial, especially solar, spectroscopy.

The British struggle over institutional support for solar physics was the sharpest and hence the most revealing (Meadows 1972, 1975; Williams 1987). Lockyer, as secretary to the special Royal Commission on Scientific Instruction and the Advancement of Science from 1870 to 1875, spearheaded the struggle for a "physical" observatory. He opened his campaign in May 1871 with a lecture at Cambridge University on the solar atmosphere, insisting that "all the knowledge we can ever hope to gain of the physical constitution of [the stars] must be got by the study of solar physics" (Lockyer 1874, 330). The following year he persuaded Thomson, De la Rue, Stewart, and others to endorse the idea of a national physical observatory. However, George B. Airy, the aging yet vigorous astronomer royal, refused to countenance such a rival to the Greenwich Observatory. He sought to improve Greenwich's strategic position by volunteering to take over the Kew Observatory's program of daily solar photography. Aided by Huggins and others who regarded Lockyer's conception of astronomical physics as narrow and self-serving, Airy secured a vote against a national physical observatory from the Royal Astronomical Society's Council.

Despite having the ear of the royal commission, Lockyer could not compete with Airy's coalition. Airy soon obtained funding from the Admiralty for a program of photographic monitoring of the sun. In the ensuing years, the Greenwich Observatory went on to establish cooperative programs with observatories in India and Mauritius. Lockyer's progress was slower. Not until 1878, three years after the royal commission recommended the establishment of a physical observatory, did he persuade the government to set up the Solar Physics Committee to disburse research grants. Not until 1879 or 1880 did he manage to get a modest annual grant of £500 for his own solar observing, including eclipse expeditions. And not until 1888, after seven years of lecturing, was he able to secure a chair for "astronomical physics" in the South Kensington Normal School for Science, the predecessor of the Royal College of Science. As of 1890 Lockyer finally had a professorship, facilities, and a budget; but neither his own solar observing program nor that of the Greenwich Observatory came up to his dreams of the early 1870s.

Research without Surprises

Turning from the institutional to the scientific arena, we find a similar pattern of solid yet unremarkable accomplishment. During the 1870s and 1880s, several mathematical physicists and astronomers polished the contraction theory of solar-energy generation (Burchfield 1975; DeVorkin 1984). Meanwhile, solar observers worked mainly at contriving and deploying more sensitive, reliable, and efficient instruments. They did not discover any major new phenomenon with these tools; but they did refine and extend knowledge of such familiar phenomena as the sun's differential rotation, its spots, prominences, and corona, its eleven-year cycle, its spectrum, and its radiative power (Meadows 1970, 1984b). The increasing role of Americans in astrophysics (Herrmann 1973; Brush 1979), as well as the rather prosaic character of solar research in this era, is illustrated by the careers and contributions of Samuel Pierpont Langley and Henry Augustus Rowland.

Shortly after the American Civil War, Langley (1834–1906) gave up an engineering career to pursue astronomy (Abbot 1906; Beardsley 1981; Osterbrock 1984; Eddy 1990). In 1867, having served a brief apprenticeship at Harvard College Observatory and the U.S. Naval Academy, he secured appointment as director of the Allegheny Observatory and professor of astronomy and physics at the Western University of Pennsylvania in Pittsburgh. He soon evinced an interest in solar physics by journeying to Oakland, Kentucky, and then to Jerez de la Frontera, Spain, to observe the 1869 and 1870 eclipses. He went on—despite the distractions of running a time service for the railroads as a means of raising funds—to win recognition with his vivid portrayals of sunspots.

In the mid-1870s, Langley turned from this descriptive work to more challenging inquiries. At first he divided his attention between thermal and spectroscopic studies, but he decided by 1880 to focus on the sun's heating power. He invented the "bolometer," which could detect minute changes in radiant intensity. It did so by comparing the electrical current through a platinum strip of known temperature with that through an irradiated black strip that, depending upon whether the incident radiation heated it to a lower or higher temperature than the reference strip, had a greater or smaller conductivity. The instrument turned out to be so sensitive that, when illuminated by a spectrograph, it could trace how the intensity of solar radiation varied with wavelength all along the solar spectrum from the ultraviolet into the deep infrared.

In 1881, eager to exploit the bolometer's capabilities, Langley staged an expedition to Mount Whitney in California. His goal was to measure atmospheric absorption across the spectrum, thereby removing a major element of uncertainty in Pouillet's determination of the solar constant. He and his assistants mapped the solar spectrum bolometrically from stations differing in altitude by over 3,000 m. After returning to Pittsburgh, they analyzed their data to ascertain how the intervening air had influenced the transmission of solar radiation. They found that atmospheric absorption was much greater at the ultraviolet end of the spectrum than at the infrared end (fig. 2.9). Taking account of this differential absorption, Langley estimated the solar constant at 3 cal/cm^2/min. This result would prove to be farther from the modern value than Pouillet's much cruder estimate. However, not foreseeing this turn of events, Langley's contemporaries applauded his sophisticated method. In 1886 the Royal Society of London and the National Academy of Sciences in Washington awarded him the Rumford Medal and the Draper Medal, and in 1893 the Paris Academy gave him its Janssen Prize.

Like Langley, Rowland (1848–1901) entered the world of work as an engineer. He knew from the outset, however, that he wanted a career in physics. At twenty-three, he got a job at the College of Wooster in Ohio. His ensuing rise was rapid. Five years later, thanks to European plaudits for his electromagnetic research, he landed the chair of physics at the brand-new Johns Hopkins University in Baltimore, Maryland (J. D. Miller 1970; Kargon 1986). Since this was the first American university to embrace the German commitment to graduate education and research, the appointment was a tremendous opportunity. Rowland made the most of it. During his quarter-

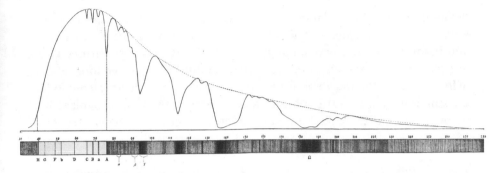

Figure 2.9. Langley's final bolometric map of the solar spectrum; the dashed line indicates the intensity without atmospheric absorption. (Langley 1883, plate 4 at end of issue.)

century at Johns Hopkins, he played a prominent part in the rise of American physicists to a respectable, if still secondary, place in the world of physics.

Although Rowland's research on electromagnetic phenomena was interesting, he was, and still is, best known for his contributions to precision spectroscopy (Warner 1986). His point of attack was diffraction gratings—mirrors with evenly spaced parallel rulings that could disperse incident light into a spectrum without any of the absorption that characterized spectra made with prisms. Since spectral resolution increased as the distance between ruled lines decreased, his initial goal was to devise a superlative ruling engine. He and his instrument maker produced a screw of singular uniformity to accomplish this end. By 1882, their engine was turning out gratings superior to those that Lewis Rutherfurd had been distributing to astronomers and physicists during the preceding decade. Rowland had met his first objective.

At this juncture, Rowland started to take an interest in the sun, or more precisely the solar spectrum. The best way to demonstrate the merits of his gratings was to map the solar spectrum with greater precision than had several predecessors. In February 1882, for instance, he wrote Charles Young that "my machine for taking gratings is now at work and surpasses Rutherfurd's as far as his surpasses others." By way of proof, he boasted of his success in resolving the solar E line into a doublet. A few weeks later, he went on to report that "I have invented *concave* gratings which need no telescopes. . . . [They] are a grand success for photographing the spectrum and I have now some photographs better than Rutherfurd's. . . . They are principally valuable in determination of the distribution of heat in the spectrum & for *photographing* the spectrum, seeing that the light need pass through no glass" (Rowland 1882). Following up on these expectations, Rowland soon furnished a concave grating for Langley's bolometric studies and began work on a new photographic map of the solar spectrum. This project, which he and his associates pursued until 1897, yielded long tables of line wavelengths and identifications of more than three dozen elements in the sun. Rowland's contributions to solar spectroscopy were recognized by many prizes, including the National Academy of Science's Draper Medal in 1890.

Langley, Rowland, and others succeeded in extending and refining the results of the field's founding era during the 1870s and 1880s, but they did not enrich solar physics with any major discoveries or theories. To have sustained the pace of the 1860s, they would have needed tools as novel as thermodynamics and spectroscopy were in that decade. In response to the field's unspectacular record, some recruits—most notably Hermann C. Vogel,[14] Edward C. Pickering,[15] and James E. Keeler[16]—transferred their allegiance to the nonsolar parts of astrophysics. Others gave up astrophysics altogether. But many, their optimism tempered by recent experience, continued trying to decipher the sun's mysteries.

Hale's Revitalization of Solar Physics, 1890–1910

"Some think," George Ellery Hale (1868–1938) wrote before his twenty-fifth birthday, "that solar work is pretty well played out—in reality it is only beginning" (Hale 1893). His achievements, both as a solar physicist and as an institution builder, gave substance to his prediction. Between 1890 and 1910, as the last of the solar physics community's founders were passing from the scene, he was at the center of the field's revitalization. By inventing monochromatic solar photography and the tower telescope, by demonstrating the presence of magnetic fields in sunspots, by establishing new obser-

14. Vogel (1841–1907), the director of the Potsdam Astrophysical Observatory from 1882, used his extensive work in stellar spectroscopy to propose a schema of stellar evolution that was widely used into the twentieth century (Herrmann 1976; DeVorkin 1984).

15. Pickering (1846–1919), the director of Harvard College Observatory from 1877, presided over a major program of photographing and classifying stellar spectra (Plotkin 1978; DeVorkin 1981).

16. Keeler (1857–1900), the first spectroscopist at Lick Observatory, then Langley's successor at the Allegheny Observatory, and finally the director of Lick Observatory, was best known for his spectroscopic studies of Saturn's rings and his exceptional celestial photographs that revealed the abundance of spiral nebulae (Osterbrock 1984).

vatories, journals, and organizations, he renewed the solar physicists' belief in their field's significance.

From Amateur to Professional

Hale's entry into solar physics was swift and sure (Wright 1966; Wright, Warnow, and Weiner 1972; Osterbrock 1984, 1986). The eldest son of one of Chicago's magnates, he took up astronomy as a boy of fourteen. His hobby soon developed into a passion. By 1886 he was mapping the solar spectrum with a flat diffraction grating purchased from John Brashear, an instrument maker in Pittsburgh who worked for Langley and Rowland. That spring he journeyed to Pittsburgh to see Brashear's offerings. While there, he met Langley and Keeler at the Allegheny Observatory. Back home, he wrote Rowland to inquire about buying a concave grating. Irrepressible despite Rowland's refusal, he took advantage of a family trip to Europe that summer to visit the leading London instrument shops and the Meudon Observatory for Physical Astronomy. After being shown around the observatory, he obtained a set of offprints and one of Janssen's classic photographs of photospheric granulation. That fall, already a well-equipped astronomer, young Hale enrolled at Massachusetts Institute of Technology.

During his formal education, Hale reached the frontier in solar research. As a sophomore, he persuaded Pickering to give him access to the Harvard College Observatory and his ever-generous father to provide him with a spectroscopic laboratory adjacent to the family's new mansion in the Chicago suburb of Kenwood. In the summer of 1888, he ventured to examine an issue that had not yet been settled—the validity of Lockyer's claims for carbon lines in the solar spectrum. His confirmation of the claims also confirmed his optimism about his prospects as a researcher. The following summer, he sought out Rowland at the Johns Hopkins University and Young at the College of New Jersey in Princeton to acquaint himself firsthand with their techniques and undertakings. Two months later, while riding a cable car in Chicago and mulling over Young's earlier failure to photograph solar prominences outside eclipse, Hale imagined how this problem might be solved. What he needed to do was isolate a single line in the prominence spectrum and let it build up a monochromatic image on a moving plate that kept pace with the sun's passage across the telescope's field of view. He struggled throughout his senior year to translate the idea into practice. In April 1890 he finally got a promising photograph. Writing up the project in his senior thesis, he graduated from MIT that June with a B.S. in physics.

At this point, most young Americans thinking seriously about research would have begun working toward a Ph.D. at, say, Heidelberg, Berlin, Johns Hopkins, or Harvard. Not Hale. With his father's blessing and support, he immediately married, took a honeymoon to California—so that he could see the largest refracting telescope in the world at the University of California's new Lick Observatory atop Mount Hamilton (Osterbrock, Gustafson, and Unruh 1988)—settled in with his parents, and started transforming his laboratory into the Kenwood Physical Observatory. He wanted a state-of-the-art facility for pursuing his interests in solar spectroscopy and prominence photography. He installed two Rowland concave gratings, a dynamo for producing comparison spectra, a photographic laboratory, various astronomical spectroscopes, and a twelve-inch refractor with an equatorial mounting. Soon Hale (1891) was happily writing a friend that "I got a prominence photo good enough to prove the success of the method, and the result is that I am just now feeling pretty *neat.*" That summer, after a dedication ceremony in which Young (1891, 321) praised him as a "true 'amateur,'" he visited the leading solar physicists in England, France, Germany, and Italy. Then he set about improving his "spectroheliograph." In 1892 he convincingly demonstrated the instrument's value by obtaining monochromatic photographs of prominences around the sun's entire circumference and of faculae—bright areas associated with sunspots that could be visually detected only near the sun's limbs—at all solar meridians. His achievement made a big impression. Over the next decade or so, it inspired spectroheliographic programs at several observatories and served as the chief justification for many academy memberships and prizes.

In 1891, even as he was perfecting prominence photography, Hale was launching a periodical for astrophysics and spectroscopy. He evidently got the idea for such a journal in conversation with Young at the dedication of his Kenwood observatory. That fall he reached an agreement with the editor of the *Sidereal Messenger* for joint publication of this popular periodical and his own astrophysical journal under the title *Astronomy and Astro-Physics.* This awkward arrangement lasted from 1892 through 1894. Then, with Keeler as coeditor and an editorial board that included Huggins, Pickering, Rowland, Tacchini, Vogel, and Young, Hale founded *The Astrophysical Journal: An International Review of Spectroscopy and Astronomical Physics.* Although the journal had financial problems for some years, it rapidly became the foremost periodical for observational astrophysics.

In the meantime, Hale was making the transition from brash amateur to established professional. His father's wealth combined with his own record, talent, and drive more than made up for his lack of Ph.D. In July 1892 he was appointed associate professor of astrophysics and director of the observatory at the new University of Chicago on two conditions. He was to serve without salary for three years. And providing that an additional $225,000 could be raised for astronomy in the next two years, his father was to donate the

Kenwood Physical Observatory to the university. Later that summer, upon hearing of the availability of a forty-inch blank for a lens, Hale offered the Chicago trolley tycoon Charles T. Yerkes the opportunity to finance a refracting telescope that would supplant the Lick Observatory's instrument as the largest in the world. Yerkes found the proposition irresistible (H. S. Miller 1970). As a consequence, Hale (fig. 2.10) spent much of his time over the next five years supervising the establishment of the University of Chicago's Yerkes Observatory at Williams Bay, Wisconsin. In 1897, as the observatory neared completion, he acquired the last trappings of the successful professional scientist. Thanks to Keeler, who had succeeded Langley as director of the Allegheny Observatory in Pittsburgh, he was awarded an honorary Sc.D. by the Western University of Pennsylvania. Shortly thereafter he was advanced to full professor of astrophysics at the University of Chicago. From our vantage point, it is difficult to imagine that the erstwhile amateur who presided that October over the dedication of the Yerkes Observatory had not yet reached thirty.

Following the Sun to California

Hale's next few years were, for him, comparatively quiet ones. He had the grief of losing his parents and the joy of acquiring a son. He helped organize the American Astronomical and Astrophysical Society, hosting its first meeting at Williams Bay in 1899. He continued editing the *Astrophysical Journal,* assuming full responsibility for the job when his friend Keeler suddenly died in 1900. He also resumed his solar research, giving particular attention to sunspot spectra and spectroheliography.

In 1902 Hale stepped up his pace. When he learned early that year of Andrew Carnegie's establishment of the Carnegie Institution of Washington with an endowment of $10,000,000 to promote science, he immediately began dreaming about a reflecting telescope much larger than the Yerkes refractor. His hopes were dashed by the Carnegie Institution's advisory committee for astronomy, which preferred a policy of funding several small projects. But they were revived at the end of the year when, in deference to Carnegie, a small committee for southern and solar observatories was established. Besides Hale, its members were the positional astronomer Lewis Boss, director of the Albany Observatory in New York, and the stellar spectroscopist William Wallace Campbell, Keeler's successor as director of the Lick Observatory in northern California. Hale set his sights on a solar observatory with a large reflecting telescope for stellar astrophysics. In pursuit of this objective, he first secured letters commenting on the need for such an observatory from eighteen solar and stellar physicists in America and Europe (Boss, Campbell, and Hale 1903). Then, lured to the West by the reports

HE STUDIES THE SUN.

Original and Interesting Researches of Prof. George E. Hale.

SOLVING A DIFFICULT PROBLEM.

By His Work the First Photograph of Solar Prominences Was Made— His Latest Scientific Device.

To be an able astronomer at twenty-four is some hing; to have acquired a special knowledge of a special subject in science that is rare even among scientific men is something more; but to be a discoverer, to be a man with origination, with singleness and fixity of purpose, to be facile princeps among experts in a widely covered field at that very youthful age, is to be one in thousands, a genius in a century. Such is Professor George E. Hale, of this city, the director of the Chicago University observatory. Professor Hale's remarkable and interesting discoveries in solar observation alone entitle him to a prominent place in the world of science, a world more aristocratic than any nobility on earth. But when his scientific success is coupled with

PROFESSOR GEORGE E. HALE.

his youth they surpass much that is recorded in that marvelous book, the history of modern astronomy. If the future of his work may be gathered from its past Professor Hale will certainly add more to the stock of human knowledge about the sun than any other one man.

Figure 2.10. George Hale as the subject of a Chicago newspaper article, about 1892; just as Hale captured this reporter's fancy, so did he capture the imagination, and support, of Charles Yerkes and many other philanthropists throughout his career. (Courtesy of Niels Bohr Library, American Institute of Physics.)

from a Lick staff member making a detailed site survey, he visited and was captivated by Mount Wilson near Pasadena.

In October 1903, when the committee met to prepare its report, Hale opened his case for a solar-stellar observatory with an argument that had its roots in the seventeenth-century insight that the sun is a star:

> In all reasoning on the physical constitution of the stars, especially in connection with the great problem of stellar evolution, we must start from the sun as a type object and elucidate stellar phenomena from an intimate acquaintance with solar phenomena. . . . Conversely the only means of studying the origin and development of the sun and of determining what it will become in the future is afforded by the phenomena of the stars and nebulae, for we find in the heavens stars in all stages of growth, illustrating every step in the process of evolution by which the sun has been developed from a nebula. Solar research should thus begin with the nebulae, proceed with a physical investigation of those celestial objects which represent the earliest stages of stellar growth, culminate in a study of solar structure and radiation, and conclude with an examination of the red stars, one of which the sun will some day become.

To understand the stars, the astrophysicist must decipher the sun; and to discern the sun's past and future, he must look to the nebulae and stars. Besides this cognitive argument for his proposed observatory, Hale pragmatically urged that powerful modern instruments presented "an exceptional opportunity for great advances" in solar research. His case was strong enough to win the endorsement of Boss and Campbell, but not so strong as to induce them to forsake their own hopes for an observatory in the Southern Hemisphere. Hence the committee urged support for both observatories (Boss, Campbell, and Hale 1903, 49–51).

That fall Hale, ever the optimist, sent his family to Pasadena and organized a small expedition to check out the winter observing from Mount Wilson. In mid-December he too made the journey to California, despite having just been informed that Carnegie was not yet prepared to finance the solar observatory. The setback was temporary. In early 1904 he received important indirect support when the Royal Astronomical Society and the National Academy of Sciences awarded him the Gold Medal and the Draper Medal. As the year passed, he obtained sustaining grants from the Carnegie Institution and individual philanthropists, made encouraging contacts with rich southern Californians, and took a ninety-nine-year lease on Mount Wilson. Finally, in December 1904 Hale was informed by the Carnegie Institution that it would fund construction. Thus was born the Mount Wilson Solar Observatory, which soon became and long remained the world's premier facility for observational research in solar and stellar physics.

International Cooperation and Solar Magnetism

Meanwhile, quite possibly as an offshoot of his campaign for the solar observatory, Hale had launched a campaign for international cooperation in solar research. His first step was to arrange for delegations from several national scientific societies to meet at the Saint Louis Exposition of September 1904 to explore the idea. Hale opened this meeting with a succinct account of cooperation in solar research since the formation of the Society of Italian Spectroscopists in 1871. Then, insisting that "great advances in our knowledge of the Sun merely await the application of instruments and methods already available," he argued that it should be possible to reap the benefits of cooperation without suppressing "individual initiative" (Hale 1904, 307, 310). In agreement with Hale, those present resolved to found the International Union for Co-operation in Solar Research. The Solar Union, as it was commonly known, held its first meeting at Oxford in 1905, its second at Meudon in 1907, and its third at Mount Wilson in 1910 (DeVorkin 1981). These general meetings provided the elite of the solar physics community with a face-to-face forum. Meanwhile, the Solar Union's committees on wavelength standards, solar radiation, spectroheliography, sunspot spectra, eclipses, and solar rotation were promoting cooperation and consensus among those working on specific topics.

Besides overseeing the development of the Mount Wilson Solar Observatory and the formation of the Solar Union, Hale somehow found time for research. In 1908, indeed, he and his colleagues opened a fresh line of inquiry by establishing the existence of solar magnetic fields. He did so while pursuing his long-standing interest in sunspot spectra. A natural place to begin the story is with the completion of Mount Wilson's second solar telescope in the fall of 1907. The first of its kind, this tower telescope (fig. 2.11) achieved high resolution at modest cost (Hale 1908a). Three features accounted for the telescope's power—its mirrors were thick enough to retain their focus when exposed to the sun; its mirrors were high enough to intercept the sun's rays before the solar image could be distorted by turbulent air from ground heating; and its spectrograph was large enough to obtain detailed spectra from whatever part of the sun's image might be selected for analysis. As Hale hoped, the tower telescope's spectrograph disclosed many new sunspot lines. It also produced the first photographs of closely spaced double lines, or doublets,[17] in sunspot spectra. Hale's interest was piqued.

The decisive clue for interpreting the doublets came from spectroheliography. About February 1908, Hale and his colleague Walter S. Adams

17. Young discovered spectral doublets by visual means in 1892.

Figure 2.11. Mount Wilson Solar Observatory's first sizable telescopes—the sixty-foot tower telescope (completed 1907) and, immediately behind it, the Snow horizontal telescope (completed 1905). (Source: Hale 1908a, facing p. 206. By permission of *The Astrophysical Journal.*)

were investigating the tower telescope's suitability for spectroheliographic work. Among other things, they tried using some new plates that were more sensitive in the red than existing plates. They found to their surprise that monochromatic photographs of the sun taken in the brightest known hydrogen line—a red line with a wavelength of 6563 Å—revealed many details that had not been visible in photographs taken in hydrogen lines of shorter wavelengths. Eager to get the best possible view of these details, Hale instructed the staff of the Snow Telescope, a horizontal solar telescope with larger mirrors than the tower telescope, to take daily spectroheliograms in hydrogen's red line. The observers soon obtained photographs of what appeared to be vortices surrounding sunspots (fig. 2.12). In describing the results, Hale (1908b, 100) pointed out that the vortical patterns resembled those presented by "iron filings . . . in a magnetic field." He promised an early investigation of the possibility that magnetic fields were responsible for the presence of doublets in sunspot spectra.

Hale's plan was inspired by his familiarity with the research of Pieter

Figure 2.12. A spectroheliogram of "vortices" around sunspots, taken by Ferdinand Ellerman with the Snow telescope; the first vortex photographs, which were obtained in April and May 1908, lacked the resolution of this one made on September 9. (Source: Wright, Warnow, and Weiner 1972, 56. By permission of the MIT Press © MIT Press.)

Zeeman, a Dutch physicist who had discovered in 1896 that spectral lines often split into doublets when originating in the presence of magnetic fields. Hale knew in particular that the components of doublets produced by the Zeeman effect should be polarized in opposite directions. In late June 1908, he tested many sunspot doublets for this property. He did so by passing the sunspot's image through crystals that, if correctly oriented, would almost extinguish one polarized component and then, if rotated by 90°, would almost extinguish the other. As Hale anticipated, the doublets tested had the polarization properties associated with the Zeeman effect. Also as he anticipated, doublets from sunspots with clockwise and counterclockwise vortices were oppositely polarized. The evidence, Hale (1908c, 341) suggested, "seems to indicate the probable existence of a magnetic field in sun-spots." Although he adopted a tentative stance, he was quite confident that sunspots were the seat of strong magnetic forces. In fact, he went to the trouble of comparing the separation of doublet components from sunspots with that of doublets produced under known laboratory conditions so that he could estimate the strength of spot fields. The result was about 3000 gauss, thousands of times greater than the terrestrial field at the earth's poles.

Hale's discovery of magnetic fields in sunspots culminated two decades of remarkable achievement within and on behalf of solar physics. During his twenties, Hale had developed spectroheliography, started the *Astrophysical Journal,* and established the Yerkes Observatory. Then, after a brief slowdown, he had freshened the rationale for research on the sun, founded the Mount Wilson Solar Observatory, organized the International Union for Cooperation in Solar Research, and installed the first tower telescope. Sometimes disparaged as a promoter, he must have been delighted to come up with convincing evidence for the role of magnetic forces in celestial phenomena. It would have been difficult to find a better way to demonstrate his scientific prowess. Serving as an exemplar as well as a spokesman, Hale had provided effective leadership in revitalizing research on the sun.

The State of Solar Physics about 1910

By 1910, solar science was no longer dominated by scientists with no deep commitment to understanding the sun. Rather, most solar research was done by solar physicists whose first scientific concern was to expand knowledge of the sun's constitution and behavior. In pursuing this objective, these men—for solar physicists were all men still—enjoyed considerable institutional support. Although some were amateurs with their own personal observatories, more were salaried professionals working in facilities and with apparatus funded by governments, universities, foundations, and philanthropists. Although most were still publishing in the proceedings of national academies and astronomical societies, they had come to regard the *Astrophysical Journal* as the field's central periodical. And though they were more likely to interact with their local and national peers, they were becoming—thanks to eclipse expeditions and Solar Union meetings—increasingly international in their dealings. In short, solar research had become the province of a specialty-oriented community that consisted mainly of professionals and possessed effective channels of communication.

Also by 1910, as the result of the solar research of the preceding century, knowledge of the sun went far beyond the positional astronomers' estimates of its distance, size, mass, rotation rate, and direction of motion through the heavens. To judge from contemporary surveys (e.g., Sampson 1911), solar physicists had good grounds for thinking that the sun consisted of terrestrial elements heated to the gaseous state, that its photosphere had a temperature of about 6000 K and radiated about 4×10^{24} cal/sec, that its angular velocity was greater at its equator than its poles, that the production of sunspots followed an eleven-year cycle, that chromospheric activity and coronal shape varied along with the sunspot cycle, and that sunspots were the seat of strong magnetic fields. And having long since rejected Herschel's conjectures that the sun was solid and habitable, they were now speculating about such physical issues as the causes of solar activity, the source of the sun's energy, and the sun's place in stellar evolution.

In looking to the future, solar physicists were once again optimistic about their field's prospects. Like Hale, they anticipated that improved instrumentation, additional observing programs, and allied physical experimentation would reveal the secrets of the sun. Also like Hale, they regarded the sun as the key to the stars. During the next three decades, however, they and their successors were in for some surprises both from the theoretical physicists and from a new breed, the theoretical astrophysicists.

CHAPTER THREE

The Maturing
of Solar Physics

1910–1940

Between the Solar Union's meeting at Mount Wilson in 1910 and the start of World War II in 1939–41, solar physicists improved their observational capabilities—both by joining forces in cooperative solar monitoring programs and by increasing the accuracy and versatility of existing solar instruments. They used their improved capabilities to refine and extend knowledge of the sun's radiative output, of photospheric, chromospheric, and coronal spectra, of sunspots and other transient phenomena, and of the solar cycle. For all their activity, however, solar physicists rarely had the lead roles in their field's most important advances during this period.

The initiative here was usually taken by outsiders—scientists whom no one, including themselves, regarded as solar physicists before their contributions to the field. Astronomers lacking a close prior familiarity with solar observations took the lead in introducing the coronagraph, solar cinematography, and monochromatic filters. Likewise, with few exceptions, astronomers and physicists who had done little or no solar theorizing led the way on such important problems as the sun's structure, composition, energy generation, and coronal spectrum. They did so by drawing upon contemporary physical research—especially Niels Bohr's establishment of the quantum theory of spectral lines, this theory's evolution into quantum mechanics, the development of nuclear-reaction theory, and the ongoing improvement of experimental knowledge of radiant, atomic, and nuclear phenomena. For instance, it was a theoretical physicist who made a compelling case that certain nuclear-reaction chains powered the sun and other stars. And it was a laboratory spectroscopist who set the stage for and played the key role in the identification of the hot ions giving rise to the coronal emission lines. Coming in the late 1930s, these achievements simultaneously added an essential element to the nineteenth-century model of the sun as a vast heat engine and initiated thoughts about the limits of this model.

The present chapter describes the maturing of solar physics during the three decades following the Solar Union's meeting at Mount Wilson. The focus is first on observational work, particularly solar monitoring and the invention of the coronagraph. Then the narrative turns to the concurrent development of solar theory, culminating with descriptions of the interpretive breakthroughs on the vexing questions of the source of the sun's energy and the production of the corona's emission spectrum. The conclusion takes stock of solar physics at the start of World War II.

Observational Solar Physics

Between 1910 and 1940, solar physicists devoted much thought and energy to monitoring the sun's behavior. They as well as scientists interested in the sun's influence on terrestrial phenomena supposed, and rightly so, that monitoring programs would expand knowledge of the sun and its effects. Even as the solar physicists were improving and exploiting their monitoring capabilities, they were acquiring more efficient, precise, and varied telescopes and auxiliary equipment. Their own efforts here were devoted primarily to upgrading proven instrumentation. It was generally outsiders who took the risks involved in developing significant new observing tools.

Cooperative Monitoring of the Sun

By 1910 solar physicists could regard the state of sunspot monitoring with considerable satisfaction. Spot statistics were being issued by Greenwich Observatory based on photographic monitoring by a network of observatories in the British Empire and also by the Swiss Federal Observatory in Zurich based on visual monitoring at several European observatories. Statistics for most other solar phenomena were in a much less satisfactory state. Typically a given phenomenon was monitored by no more than two or three observatories, and the resulting data were never coordinated. George Ellery Hale's International Union for Co-operation in Solar Research was trying to remedy this situation, but it had not yet made much headway.

With the outbreak of the Great War in August 1914, Hale's Solar Union and its commissions ceased work. There was no way scientists on opposite sides of the battlefront could or would cooperate. Before the war's end, in fact, Hale and other leading American scientists had joined the French and

British in thinking that postwar international organizations must exclude the Germans. The decisive step in implementing this belief was taken when scientific leaders from the allied nations met in Paris shortly after the armistice to organize the International Research Council. The delegates agreed that the council would foster the formation and cooperation of international disciplinary unions that excluded German and Austrian scientists (Kevles 1971; Schroeder-Gudehus 1978). The first such union was the International Astronomical Union, which had its constitutive meeting in July 1919. The astronomers present established thirty-two commissions to cover their discipline. The six solar commissions were responsible for the sun's radiant output, solar atmospheric motions, the structure and composition of the sun's atmosphere, eclipses, tables of solar spectra, and solar rotation. The distribution of chairmanships—three Americans, two Frenchmen, and one Briton—reflected current views about national standing within solar physics (International Research Council 1920, 165–68).

Although Hale nominally chaired only one of the six commissions, he still seems to have thought of himself as the spokesman for the whole field. He certainly assumed this role when preparing his commission's report for the International Astronomical Union's first congress, to be held in Rome during May 1922. He proposed that seven observatories—Cambridge, Greenwich, and Stonyhurst in England, Meudon in France, Arcetri-Florence in Italy, Mount Wilson in America, and Kodaikanal in British India—serve as "centres for the compilation, discussion and publication of various classes of solar statistics" (Hale 1922, 31). He also proposed that the Swiss Federal Observatory in Zurich be asked to resume its traditional oversight of sunspot statistics once Switzerland affiliated with the Union. However, as a supporter of the International Research Council's ostracism of Germany, he made no place in his plans for the Potsdam Astrophysical Observatory. Hale's health was too delicate for him to go to the Rome meeting; but his proposals, and the desire to unify solar physics that underlay them, struck a responsive chord. Indeed, the solar physicists present not only endorsed Hale's plan but also merged four of the commissions into one commission for solar physics, giving the chairmanship to Hale's trusted colleague C. Edward St. John.

Besides proposing a division of labor among the main solar centers, Hale (1922, 33) had suggested in his commission report that "arrangements be made, under the auspices of the International Research Council, for a systematic comparative study of solar and terrestrial phenomena, in cooperation with the International Geophysical Union, the International Union for Scientific Radiotelegraphy, and other bodies." This suggestion set in train developments that would lead, some six years later, to an ambitious scheme for cooperative solar monitoring. It was approved in Rome and then passed on to the Manchester physicist Sir Arthur Schuster, a friend of solar physics who was serving as the International Research Council's general secretary. In 1924 Schuster arranged for the establishment of a provisional Commission on Solar and Terrestrial Relationships under the chairmanship of Sydney Chapman (1888–1970), an applied mathematician who would play a prominent role in solar-terrestrial physics for the next forty-five years (Akasofu, Fogle, and Haurwitz 1968).

Chapman and his colleagues presented their report at the International Research Council's third general assembly in 1925. In their view, solar activity influenced the terrestrial magnetic field, aurorae, the weather, and probably atmospheric electricity, radio transmissions, and several other phenomena as well. They advanced a comprehensive plan for the study of solar-terrestrial relations, which included calling on the International Astronomical Union to issue timely statistical reports on solar activity. Their report was well received. Chapman's group was accorded status as a standing commission and granted funds for publishing its findings and recommendations (International Research Council 1925; Solar and terrestrial relationships 1926).

St. John, who was a member of the new solar-terrestrial commission as well as the chair of the International Astronomical Union's solar physics commission, had responsibility for arranging the publication of the desired solar statistics. A deliberate man whose own research in precision spectroscopy did not involve monitoring, he slowly gathered suggestions about how to proceed. Then at the union's third congress in 1928, he charged an ad hoc committee with formulating the final proposal. To ensure success, he gave the chairmanship to another member of Chapman's commission— Giorgio Abetti, the director of the Arcetri-Florence Astrophysical Observatory. Abetti and his colleagues recommended monitoring of sunspots and three other indicators of solar activity. They also recommended that the resulting data be sent to William O. Brunner, director of the Swiss Federal Observatory in Zurich, for publication. The committee's proposals were all accepted (St. John 1928). Before the year was out, Brunner published the first issue of the International Astronomical Union's *Bulletin for Character Figures of Solar Phenomena*. It reported data from thirteen observatories around the globe for January–March 1928.

Over the next decade, the *Bulletin's* statistical reporting on solar activity underwent numerous changes. One instigated by Hale turned out to be of particular importance. With the aid of the Mount Wilson staff, he had developed the spectrohelioscope, a modified spectroheliograph that permitted visual observation of chromospheric phenomena in a narrow wave-

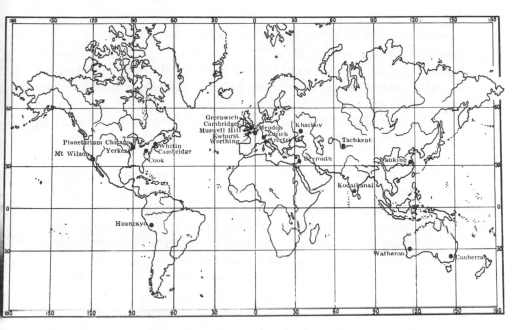

Figure 3.1 Observatories reporting to the International Astronomical Union's *Quarterly Bulletin on Solar Activity* in the late 1930s; two contributing observatories—Simeis in the Soviet Union and Pomona in the United States—were inadvertently left off the map. (Source: D'Azambuja 1939, 46.)

length band of choice (Hale 1929). In the early 1930s, after distributing his instrument to several observatories, Hale began urging the organization of a cooperative watch for "chromospheric eruptions"—sudden bursts in the chromosphere that would later be called flares (Hale 1931a, 1931b). Soon enough observatories around the world had spectrohelioscopes for flare coverage to be almost continuous. The job of editing a flare list for the *Bulletin* was taken up by Lucien d'Azambuja of the Meudon Astrophysical Observatory (D'Azambuja 1934).

Thanks to the recruiting efforts of Brunner and d'Azambuja, participation in the *Bulletin* continued to grow throughout the troubled 1930s. By the summer of 1939—shortly after it was renamed the *Quarterly Bulletin on Solar Activity*—thirty-one observatories around the globe were contributing to this cooperative enterprise (fig. 3.1).

Some Solar-Monitoring Findings

As solar monitoring capabilities improved between 1910 and 1940, solar physicists and other scientists interested in the sun's effect on terrestrial phenomena searched the data for interesting results. Numerous findings were announced. Some results never did inspire much confidence. Others, however, substantiated the value of solar monitoring programs.

Charles Greeley Abbot (1872–1973), the director of the Smithsonian Astrophysical Observatory from 1907 to 1944, has the dubious distinction of having drawn the most controversial result from his monitoring data. Throughout his long directorship, he sought to measure variations in the sun's radiative output and their influence on terrestrial weather (Jones 1965; Hoyt 1979; DeVorkin 1990b). His contacts with his predecessor Samuel Langley and the manifest variability in the weather had disposed him to believe that the solar constant was misnamed. He soon realized, however, that atmospheric conditions were affecting the measurements his group was obtaining. Starting in 1918, therefore, Abbot secured his data from monitoring stations on desert mountains in Chile, the American West, and Africa. His goal was to detect and eliminate atmospheric perturbations by comparing observations made several times each day at these dispersed stations.

From the 1920s into the 1950s, Abbot made numerous claims for solar variability (Hoyt 1979). Every time he and his staff subjected their accumulating data to a fresh analysis, however, they altered their estimates of the variability's amplitude. In his first report to the International Research Council's Solar-Terrestrial Commission, for instance, Abbot (1929b, 14) commented that "the range of variation formerly ascribed to solar change has been much reduced. But even in monthly mean values there still remains, as found in the 9-year period over which excellent results at all seasons are now available, a range of solar constant values from 1.91 to 1.97 calories. Daily observations exhibit a total range still wider" (fig. 3.2). A decade later, Abbot (1939, 11) was again obliged to report to the Solar-Terrestrial Commission that his staff was "re-reducing all solar constant observations made since 1923 [because] the first reduction . . . was done by an erroneous method." Given these and like reports, it is small wonder Abbot's announcements of solar variability were always controversial (e.g., Bernheimer 1929, 1936; Sterne 1942). In fact, as will become clear in chapter 7, reasonably convincing evidence of solar variability was not extracted from Abbot's data until the late 1970s.

By comparison, George Ellery Hale's program of monitoring the magnetic polarity of sunspots yielded results that survived contemporary and subsequent scrutiny. Following the discovery in 1908 that sunspots were the seat of strong magnetic fields, he and his colleagues at Mount Wilson regularly observed the magnetic properties of large spots. Their hope was to find some pattern in spot polarities. At first their search seemed hopeless because they found spots of both polarities on each side of the solar equator. But a pattern emerged once they stopped thinking of spots as individual entities. They

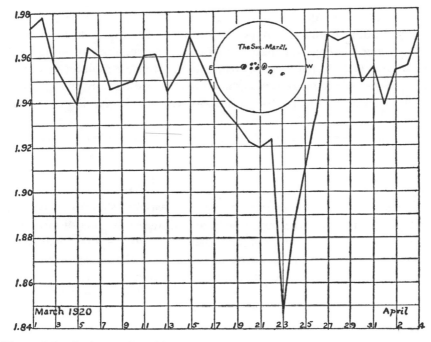

Figure 3.2. Charles Greeley Abbot's view of the influence of a large spot group on the solar constant; his data indicated that the solar constant decreased from 1.96 to 1.85 cal/cm²/min over an eight-day period in March 1920 as an exceptionally large spot group approached the solar meridian. (Source: Abbot 1929a, 148. By permission of the Smithsonian Institution Press.)

noticed that spots in pairs almost always had opposite polarities. Moreover, almost all the leading spots in pairs in the northern solar hemisphere had one polarity, and almost all the leading spots in pairs in the southern hemisphere had the other.

Hale was quite familiar with this pattern by 1912–13 when the sunspot cycle beginning in 1901 ended and a new one began. To his surprise, his staff found that the polarities of the first spots of the new cycle were reversed in both hemispheres. Distinguishing the spots of the old and new cycles was easy since, as Carrington had discovered back in the late 1850s, the old-cycle spots were near the equator and the new-cycle spots at higher latitudes. Hale's initial hypothesis (1915, 384) was that the reversal in polarities occurred at minimum because of "the change of latitude which a new cycle introduces." This notion evidently led him to expect that another reversal would occur once the spots of the new cycle got close enough to the solar equator—perhaps at the maximum that was expected about 1917.

Hale and his colleagues were ready to follow up on this possibility (Hale and Nicholson 1938). He had used his earlier discovery of sunspot fields to secure funding from the Carnegie Institution of Washington for a much larger tower telescope and more sensitive auxiliary equipment. From 1915, the Mount Wilson staff used this instrument for visual monitoring of sunspot polarities. The maximum came and went without any change in the polarity pattern. Anticipations shifted forward to the ensuing minimum when the next new cycle would begin (Hale et al. 1919). On June 24, 1922, Ferdinand Ellerman detected the new cycle's first spot at a high latitude in the sun's northern hemisphere. Its polarity was opposite that of the lead spots in the old cycle. Since single spots usually had the polarity of lead spots in pairs, Mount Wilson's solar physicists supposed that a polarity reversal was right around the corner (Summary of Mount Wilson observations 1922). Over the next year and a half, as a growing number of new-cycle spots appeared, the evidence became conclusive. In late 1923, Hale sent a full report to *Nature*. The discovery, he suggested (1924a, 110), warranted a redefinition of the solar cycle's period "as the interval between successive appearances of spots of the same magnetic polarity." In his view, this twenty-two year magnetic cycle (fig. 3.3) was more fundamental than the traditional eleven-year cycle.

A second important result to be extracted from solar monitoring data during the interwar period was the discovery of a relation between sudden fadeouts in shortwave radio reception and solar flares. During the 1920s, shortwave radio achieved importance as a means of long-distance communication. Radio engineers, who were responsible for maximizing reliability, were puzzled that from time to time toward the end of the decade reception

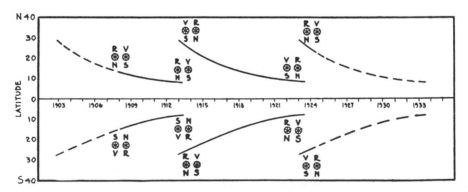

Figure 3.3. Hale's view of the law of sunspot polarity; the lines sweeping toward the solar equator traced the mean spot latitudes through each cycle; R and V indicate polarization toward red and violet; N and S, toward north and south; note that the polarity reversed with the beginning of each new cycle. (Source: Hale 1924b, 54.)

faded entirely away for fifteen to thirty minutes. Most presumed that equipment failures caused these sudden fadeouts. A few, however, knowing that the sunspot cycle was near maximum, believed some solar process was responsible. The most serious investigator of this possibility was E. Hans Mögel, a young engineer with the Transradio Corporation in Berlin (Appleton 1938; Dieminger 1974, 1986). He showed that the sudden fadeouts occurred only on the daylight side of the planet and suggested that solar disturbances caused them by producing intense bursts of ultraviolet radiation (e.g., Mögel 1931). So far as it went, Mögel's work was sound and insightful. Nonetheless it attracted virtually no contemporary attention. Few radio engineers were disposed to study possible solar effects, and few solar physicists were interested in what was going on in radio.

In the mid-1930s, when the number of sunspots was again on the rise, a similar story came to a very different conclusion. Late in the spring of 1935, J. Howard Dellinger (1886–1962), chief of the U.S. National Bureau of Standard's Radio Section, got a query from a correspondent in France about an abrupt fadeout at the main Paris receiving station on May 12. On checking into the matter, he learned first that fadeouts had occurred at the same time at the RCA and Bell receiving stations in New York and New Jersey and then that similar fadeouts had occurred on March 20 and July 6. After a fourth fadeout occurred on August 30, Dellinger (1935, 705) reported:

> A remarkable vagary of radio transmission has recently been found. . . . It is a world-wide phenomenon, or more accurately semi-world-wide, as it involves all high frequency radio transmission over the illuminated half of the globe and not the dark half. Depending apparently on some solar emanation lasting only a few minutes, it may be of interest to workers in sciences other than radio. In fact, its thorough elucidation appears to call for the study of such cosmic data as solar radiation intensities, terrestrial magnetism, atmospheric ionization, aurora, earth currents, etc.

In quest of the solar origin of fadeouts, Dellinger soon wrote Mount Wilson Observatory to inquire whether any unusual solar events had preceded their onset. As luck would have it, the sun had not been under observation at the times of the March and May events. But Dellinger was thrilled to learn that major flares had occurred just before the July and August fadeouts.

During the next year and a half, Dellinger set the pace in exploring the flare-fadeout relation. With the cooperation of numerous receiving stations, he identified 118 fadeouts between late 1934 and the end of 1936. Then, using d'Azambuja's flare lists in the International Astronomical Union's *Bulletin*, he examined their association with fadeouts. The correlation was far from perfect. He found that half of the fadeouts coincided with observed flares, but that only a quarter of the more intense flares were accompanied by fadeouts. These findings led him to conclude that some unknown solar phenomenon caused both the visible surge in the chromosphere and the disturbance in the terrestrial ionosphere that interrupted shortwave reception. Dellinger (1937, 137–38) postulated that this phenomenon generated "a very sudden burst of very penetrating . . . electromagnetic radiation of frequency far above visible light." The radiation could evidently pass right through the ionosphere's outer layers, which were responsible for reflecting radio signals over the horizon. It must penetrate down to and ionize "a level of the atmosphere where the air density is great enough to insure rapid recombination of the ions." His supposition was that this ionized layer absorbed rather than transmitted radio waves, thereby causing the fadeouts.

Dellinger (1937, 139) went on to point out that his theory implied that shortwave fadeouts provided "a means of studying a new class of invisible solar radiation, not hitherto accessible to detection or measurement." Seth B. Nicholson, a solar physicist at Mount Wilson who had been one of Dellinger's informants, was also impressed. He claimed that the "understanding of the unexpected phenomenon of sudden radio fade-outs was reached so quickly because [on] the solar side of the problem, the organization of observers was complete and ready to function" (Nicholson 1938, 114). Those participating in the cooperative monitoring of the sun had contributed to a noteworthy advance in the empirical knowledge of the sun and its influence on the earth.

New Instruments

Most of the new instruments used for observing the sun between 1910 and 1940 were variants of instruments already in the solar armamentarium. The lead role in their development was played by solar physicists who were, through constant use, intimately familiar with the limitations of the instruments already at their disposal and hence alert to whatever opportunities might arise for improving performance. Thanks to this work of refinement, the solar physics community's observational standards rose throughout the period.

While solar physicists were busy with improvements, a few comparative strangers to the specialty—at the time of their initial successes—were making some dramatic innovations in solar instrumentation. For instance, the American Robert R. McMath (1891–1962) was an amateur with a background in planetary and lunar cinematography when his group completed the first spectroheliokinematograph for taking motion pictures of solar phenomena in 1933 (Mohler and Dodson-Prince 1978; McMath and Petrie

1933). And the Swede K. Yngve Öhman (1903–88) was a stellar astro-physicist when he constructed the first narrow-band optical filter in 1937 (Öhman 1938a, 1938c; Wyller 1989). The endeavors of McMath and Öhman attest to the sun's continued allure for inventors of new instruments. The best interwar example of this allure, however, is provided by the career of the man who solved the stubborn problem of observing the sun's corona in full daylight.

Bernard Lyot (1897–1952), the son of a Parisian surgeon, displayed a youthful interest in astronomy (Chevalier 1952; D'Azambuja 1952; Dollfus 1983). He built a small observatory at the age of sixteen and won election to the amateur Société Astronomique de France two years later. Meanwhile, in deference to his family's wishes, he was studying engineering at the Ecole Supérieure d'Electricité. He graduated in the third year of the Great War. He soon found work in the physicist Alfred Pérot's program to develop naviga-tional aids for military aircraft. After the war, he worked as a teaching assistant in Pérot's laboratory at the Ecole Polytechnique and pursued a higher degree. In 1920, strongly recommended by Pérot on account of his dexterity and his insight into instruments, the young physicist secured a place on the staff of the Meudon Observatory for Physical Astronomy.

Lyot soon embarked on a project in planetary astrophysics that would eventually lead him to the problem of observing the corona outside eclipse. The project began with his surprise that, when observed with existing polar-imeters, the light reflected from the planets, unlike the light reflected from objects in the laboratory, gave no indication of being polarized, or more intense in one plane than in others. Provoked by this challenge to his skill as an instrument maker, he devised a polarimeter that had an accuracy of one part in a thousand. It revealed, as he had expected, that planetary light was indeed polarized. Moreover, the amount of polarization appeared to vary with the angle of the sunlight's incidence on the planet. Lyot proceeded to measure this variability as a means of investigating the nature of the visible surfaces of the moon and planets. He built a good case, for instance, that the surfaces of Mars and Mercury were similar to that of the moon by showing that the polarization of the light from all three bodies varied in the same way with the angle of the sunlight's incidence. By 1929, however, Lyot realized that to clinch the case for Mercury he would need to observe the planet in daylight, when its angular separation from the sun was small. He could not do so at Meudon because atmospheric dust was scattering sunlight into his polarimeter.

To get above the dust, he spent part of the summer at the Pic du Midi Observatory high in the Pyrenees. The trip was a success (Lyot 1929; Dauzère 1931). The skies were so clear that if he stopped the telescope's aperture down from 25 to 18 cm in order to prevent sunlight from scattering at the edge of the lens, he could observe Mercury to within 5° of the sun. Pleased with his initial results, he returned to the Pic du Midi the following summer to complete his study. To follow Mercury even closer to the sun, he refined his observing techniques in two ways. He diaphragmed the main lens down to 11 cm. He also shaded the lens with "a small circular disk supported by bamboo poles." The outcome was that he managed to observe Mercury to within 1.5° of the sun (Lyot 1930a, 705).

Not only did Lyot finish his prizewinning study of planetary polarization on this second trip to the Pic du Midi, but he also embarked on his solar investigations. Just four days after arriving, he detected a large solar prominence with his telescopic arrangement (Lyot 1931). He decided at once to follow up this tantalizing observation by investigating whether he could detect the sun's corona. Various acquaintances had suggested that his polarimeter might be sensitive enough to register the polarized fraction of the corona's light (Lyot 1939). He now had an excellent opportunity to pursue this possibility. Lyot's decision to seize it attested to his self-confidence. The corona was over a million times fainter than the sun. The year before, indeed, two German astrophysicists had maintained that atmospheric scattering of sunlight precluded detection of the corona except during eclipses (Kienle and Siedentopf 1929). However, suspecting that much of the scattered sunlight that had bedeviled prior attempts originated in the observing apparatus, Lyot had brought a fine 8-cm lens to the Pic that summer.

Working with this lens and other materials at hand, Lyot fashioned the first "coronagraph." It consisted of the lens, which was stopped down with a diaphragm to 3 cm, followed by a screen to block the image of the sun's disk, and at the end of the tube, attachments for an eyepiece, a polarimeter, a spectroscope, or a spectrograph. On July 25, 1930, he mounted his instrument on the observatory's equatorial telescope. For the next two days he used it to examine a large prominence that happened to be on the sun's limb. Once the prominence passed from view, he investigated whether the halo surrounding the occulting screen included coronal light. The work went quickly. First he established that some of the halo's light had the corona's characteristic polarization. Then he detected the corona's distinctive green emission line at 5303 Å. Finally, in early August he captured this and another line on the first photographs of the coronal spectrum outside eclipse (Lyot 1930b, 1931, 1939).

Upon returning to Meudon, Lyot set about building a new coronagraph (fig. 3.4) for the 1931 observing season at the Pic du Midi. His goal was to minimize the scattering of light within the instrument. He gave particular attention to the primary lens. To ensure that it did not "possess a single vein,

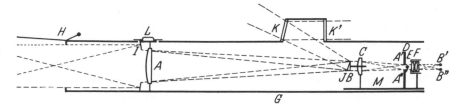

Figure 3.4. Bernard Lyot's design for his 1931 coronagraph; I indicates the first diaphragm; A, the primary lens; B, the occulting disk; C, the field lens; and D, the second diaphragm; it is a testament to Lyot's internationalism that he published his first full description of the coronagraph in the new German journal *Zeitschrift für Astrophysik*. (Source: Lyot 1932, 77.)

bubble, scratch, or surface hollow in its central part," he personally selected the glass blank, supervised its figuring at the Institute d'Optique, and did the final polishing (Lyot 1932, 76–77). He also gave considerable thought to the instrument's general design. For instance, he found that he could intercept the light diffracted at the edge of the diaphragm in front of the primary lens by inserting a second diaphragm behind the internal occulting disk and field lens.

Back at the Pic du Midi during the summer of 1931, Lyot used his new coronagraph to take the first photographs of the inner corona outside eclipse. He also tried to detect additional coronal emission lines. On account of the decline in the corona's intensity accompanying the sunspot cycle's approach to minimum, however, he could do no more than remeasure the wavelengths of the lines he had detected the year before. Disappointed, he decided to abandon his coronal studies until solar activity returned to moderately high levels.

Lyot's success in detecting the corona outside eclipse — a problem that had defeated many of the best observers since the 1870s—was enthusiastically received. In France his first paper on the corona in the Académie's *Comptes Rendus* had the unusual distinction of being followed by two notes of appreciation. Henri Deslandres, the former director of the Meudon Observatory, insisted "on the grand importance of his discovery," attributing it to Lyot's "experimental development [which] rested on continuous and prolonged contact with facts." Likewise, Deslandres's successor, Ernest Esclangon, remarked that Lyot had not only opened "a new stage in the study of the solar corona" but also shown the folly "of demonstrating the impossible" (Lyot 1930b, 837–38). Not long afterward, Deslandres commented at the Société Astronomique that Lyot's success did "very great honor to French science" (Hamon 1931, 73). Once astronomers in other countries saw Lyot's report of his 1931 observations, they added their accolades to those of his countrymen. In England, for instance, Harry H. Plaskett (1933, 278) acclaimed Lyot's "new and significant results" in the Royal Astronomical Society Council's annual report on the progress of astronomy. He praised Lyot's methods partly because they yielded coronal wavelengths of "an accuracy far greater than hitherto attained" and partly because they promised to reduce the need for costly eclipse expeditions. Soon afterward in Germany, Walter Grotrian (1933, 73) of the Potsdam Astrophysical Observatory waxed eloquent on "the great significance of [Lyot's] success for solar research." He hoped that the coronagraph would be widely adopted for coronal monitoring in the forthcoming solar cycle.

In 1935, the sunspot cycle having passed through its minimum, Lyot (fig. 3.5) installed an improved coronagraph at the Pic du Midi Observatory. During that summer and the next four, he sought to prove that his instrument could reveal properties of the corona that had eluded generations of eclipse observers (Lyot 1936b, 1936c, 1937, 1939). His success was greater toward the spectrum's red end because scattering decreases as wavelength increases. At wavelengths greater than 6000 Å, he located all three of the lines already found during eclipses (Mitchell 1936) and discovered four new lines. At shorter wavelengths, by contrast, he could only detect three of the sixteen known lines and found just one new line. He also established that the coronal emission lines were quite broad, astutely interpreting this fact as evidence that they were produced by atoms at very high temperatures.

Besides investigating the corona, Lyot (1936a) was seeking to establish the coronagraph's value for prominence cinematography. His rival here was McMath, who began getting films of prominence eruptions after the completion in mid-1936 of a tower telescope at the University of Michigan's McMath-Hulbert Observatory (McMath 1937; McMath and Pettit 1937). Lyot was soon making movies of comparable quality. In August 1938, indeed, his prominence films were a sensation at the Stockholm congress of the International Astronomical Union (Swings 1938). Impressed by the rapid advance of solar cinematography, the union established a commission including Lyot and McMath to lay plans for cooperative filming of prominence evolution (D'Azambuja and Newton 1938).

That same year, Lyot began work on a monochromatic filter for his coronagraph. His inspiration was probably Öhman's announcement (1938a) of a monochromator made of alternating layers of quartz plate and polaroid film. Having earlier published the basic principle for such a filter, he may well have been annoyed by the Swede's success. At any rate, he extracted a published acknowledgment of his conceptual priority (Öhman 1938b). Then, in preparation for the 1939 observing season, Lyot developed his own

Figure 3.5. Bernard Lyot about 1948, a decade after winning the Royal Astronomical Society's Gold Medal for his development of the coronagraph. (Courtesy of the Observatoire de Paris.)

quartz-polaroid filter. It could exclude all radiation outside a narrow bandwidth of 3 Å centered on any one of four chromospheric or two coronal lines. Using this filter in conjunction with his coronagraph, he obtained superlative movies of prominences and photographs of the corona. On account of the outbreak of war, however, this and his subsequent achievements with monochromatic filters were not widely known until near the end of World War II (Lyot 1944, 1945).

Among the many ways Lyot's contemporaries recognized the coronagraph's value to solar physics, two were especially striking. In the late 1930s, Max Waldmeier (b. 1912) of the Swiss Federal Observatory and Donald H. Menzel (1901–76) of the Harvard College Observatory established coronagraphs at mountain stations near Arosa, Switzerland, and Climax, Colorado (Waldmeier 1939; Menzel 1940; Kidwell 1990). Meanwhile in 1939 the Royal Astronomical Society awarded Lyot its Gold Medal, the highest award in astronomy, for proving himself "a worthy successor to men like Janssen and Deslandres, who have made the French school of solar physicists a part of the intellectual glory of their country" (Plummer 1939, 540). Thus, starting with an instrument (the polarimeter) and techniques (observing at a mountain site, severe diaphragming of lenses) that he had developed for the study of planetary polarization, Lyot had transformed himself within a decade into a leading solar observer.

Theoretical Solar Physics

While the solar physics community's observational capabilities were improving, its theoretical picture of the sun was being revolutionized. About 1910 most solar physicists regarded the sun as a turbulent gaseous sphere composed of familiar elements mixed in roughly terrestrial proportions, heated to a viscous incandescence by some unknown internal process and surrounded by a relatively cool chromosphere and cooler corona. During the next three decades, they came to entertain a substantially different picture of the sun. In the mid-1920s they were persuaded that the great bulk of solar matter, despite being much denser than water, behaved like a perfect gas. A few years later they were convinced that hydrogen, despite its comparative scarcity on the earth, was far and away the sun's most common element. Then in the late 1930s they were presented with a cogent case that the sun's energy originated in cyclical nuclear processes occurring near its center. Finally, as World War II was spreading around the globe, they were given strong evidence that the corona, despite its distance from the sun's energy-generating core, was much hotter than the photosphere and chromosphere.

The scientists who wrought this transformation in the solar physics community's picture of the sun did so by bringing recent physics to bear on problems of the sun's constitution. Like Lyot, most did not identify with the solar physics community when they began their pioneering work. Instead, these outsiders typically viewed the solar phenomena they worked on as revealing instances of important general problems or as interesting tests of their physical tools and insights. Their successes attest to the importance of maintaining lines of communication between neighboring specialties and across disciplinary boundaries.

The Internal Constitution of the Sun

In 1916 Arthur Stanley Eddington (1882–1944), the new director of Cambridge University's observatory, began research on the physical constitution of the stars (Hufbauer 1981; Kenat 1987). He brought a strong background in mathematical physics and stellar kinematics to this theoretical issue. As a boy, he had been sufficiently interested in the sun to give a talk about the contributions of eclipse observing to solar knowledge (Eddington 1898). Nonetheless, his initial research on stellar structure had nothing to do with our star. Rather, his starting point was the recent work by Henry Norris Russell (1877–1957) of Princeton University, who had established that some stars were both much larger and more rarefied than ordinary stars like the sun (DeVorkin 1984). Eddington reasoned that these "giant" stars were so ethereal that they must behave in accord with the laws for perfect gases. Proceeding on this assumption, he investigated how the outward flow of heat generated within a giant star would counterbalance the force of gravity. His analysis challenged the customary view that convective upwelling of heated matter was what prevented a star from collapsing. Instead, it appeared that the radiation traveling toward the star's surface exerted enough outward pressure to counterbalance gravity. Radiation pressure, in short, maintained the equilibrium of a perfect-gas star. While gunfire thundered across the Channel, Eddington (1916) presented his theory of "the radiative equilibrium of the stars" at the Royal Astronomical Society. His colleagues, despite their profound reservations about his pacifism, gave his paper an enthusiastic reception. For all their excitement, however, none of those present anticipated that a line of research had been opened that would be found in 1924 to bear directly on views of the sun's interior.

In developing his theory of giant stars during the next four years, Eddington incorporated various suggestions from his fellow astronomers as well as recent results from physics. His initial calculation of a giant's central temperature yielded a figure of about 7,000,000 K. Several colleagues pointed out that this result was at odds with his assumption that stellar gas had an average atomic weight of 54 (equivalent to iron). They were not objecting here to his supposition that the relative abundances of the elements in a giant star were the same as for the earth. At the time, this idea was almost axiomatic. Rather, they were taking him to task for neglecting the insights to be gained from Ernest Rutherford and Niels Bohr's new conception of the atom as consisting of electrons orbiting a central nucleus. This model indicated that at Eddington's high temperatures the nuclei of most atoms would have lost about half their electrons. The resulting gas of positive ions and electrons would, because an electron's mass was so low, have an average atomic weight much lower than 54. A gas consisting, for instance, of thirteen-times ionized iron atoms and the freed electrons would have an average atomic weight of about 4—that is, 55.8 ÷ (1 ion + 13 electrons). Eddington (1917) conceded the point, ending up with a value of about 5,000,000 K for a typical giant star's central temperature.

Although Eddington assented to the idea that ionization would be high in a giant star's interior, he tenaciously defended his initial conclusion that the star's total energy output, or luminosity, depended on its mass.[1] His chief critic was the mathematical physicist James Hopwood Jeans, who insisted that the claimed dependence was an implausible consequence of dubious premises about energy generation in stars (Jeans 1917). In rebutting Jeans, Eddington relied on Russell (1919), who suggested that a gaseous star could regulate the energy output from subatomic processes either by contracting, which would increase internal temperatures and hence reaction rates, or by expanding, which would decrease temperatures and reaction rates. Eddington (1919, 1920) proposed two processes that might behave satisfactorily. He first suggested that stellar energy originated in the mutual annihilation of electrons and protons. This hypothetical reaction implied, according to the relativistic relation $E = mc^2$, stellar lifetimes in the trillions of years. The alternative he proposed a year later was based on the latest research of two colleagues at Cambridge University's Cavendish Laboratory. Rutherford had knocked protons out of the nuclei of light elements by radioactive bombardment. And Francis Aston had shown that all nuclei possessed less mass than an equivalent number of protons, interpreting this result as supporting Rutherford's view that nuclei consisted of protons and electrons bound together by strong electromagnetic forces. Drawing on this research, Eddington suggested that stellar energy may be generated by atom building through the combination of protons into heavier nuclei. Such a process pointed to stellar lifetimes in the billions of years. Although neither Eddington nor his contemporaries had grounds at the time for choosing between the two suggested processes, his second proposal would later be recognized as prescient.

1. Ever since the discovery of radium's heating effect in 1903, Eddington had presumed that subatomic energies powered the stars (Hufbauer 1981). He believed, all the same, that his mass-luminosity relationship for giant stars was quite independent of the mechanism of energy generation, be it gravitational contraction (as was generally believed before the 1903 discovery) or atomic reactions (as was ever more widely believed thereafter).

During the early 1920s, Eddington redoubled his efforts to reconcile the astronomical and physical sides of his theory of stellar constitution. He pursued two objectives. One goal was to modify the theory of giant stars so it would apply to the sun and other ordinary stars. He did so by replacing the perfect-gas laws with Johannes van der Waals's equations for an imperfect gas. The most interesting consequence to emerge from this gambit was the possibility that the sun's central temperature was about 18,000,000 K and its central density about 13 g/cc (Eddington 1922). Eddington's second objective was to discover why the hot gases in stellar interiors were so much more effective in retarding the outward flow of radiation than theoretical and experimental physics predicted. He made little headway in this endeavor, but in making the attempt he stimulated his colleague E. Arthur Milne (1896–1950), a mathematical physicist with an assistantship at Cambridge University's Solar Physics Observatory, to carry out a fresh analysis of the degree of ionization in giant stars. Milne (1924) used the latest physics to investigate what would happen to the electrons of atoms under the conditions Eddington now said prevailed in a typical giant star—central temperature about 6,000,000 K and central density about 0.07 g/cc. His conclusion was that the atoms of most elements would possess no more than a few tightly bound electrons.

Milne's finding came as a welcome surprise to Eddington, for he had begun to doubt his supposition that stellar atoms retained about half their electrons. But early in 1924 he realized that the result was of far less interest than the equation it was deduced from. This equation indicated that the dependence of ionization on density was so weak that highly ionized atoms should be abundant at the very high temperatures he had found for the center of the sun. If so, the bare nuclei, ions, and electrons composing the gases of ordinary stars must comply with the perfect-gas laws after all. His theory of giant stars, including his mass-luminosity relation, should apply to ordinary stars.

Testing this claim would not have been possible a few years earlier, when there were only a few stars for which astronomers had values of both the mass and the luminosity. In fact, all Eddington's early theorizing about giant stars was done without the benefit of first-class measurements for a single giant. This situation was remedied in 1922 when Paul W. Merrill of Mount Wilson Observatory published data for the two giant stars in the binary system of Capella (DeVorkin 1975; Merrill 1922). In carrying through his test in 1924, Eddington (fig. 3.6) used the Capellan data to determine the constant in the mass-luminosity relation and graphed the logarithmic version of the equation. Then he plotted the position of the sun and other stars with well-determined masses and luminosities. The results were stunning. All the points, including that for the sun, fell near the theoretical curve. As

Figure 3.6. Arthur Stanley Eddington about 1927, a few years after he extended the mass-luminosity relation to ordinary stars. (Source: Abbot 1929a, facing p. 301. By permission of the Smithsonian Institution Press.)

Eddington had hoped (Crelinsten 1982, 364–67), his theory applied not only to giant stars but also, despite their greater density, to the sun and other ordinary stars.

In March, Eddington (1924) presented his results at the Royal Astro-

nomical Society. Shortly thereafter, sensing that his research on stellar structure had reached its apogee, he started a monograph expounding his theory.
The Internal Constitution of the Stars (1926) was an instant classic. Although
its purview was the stars in general, it soon became the authority on the
sun's interior. The sun, according to Eddington, consisted of a terrestrial mix
of elements in a state of high enough ionization to behave in accord with the
perfect-gas laws. Energy in the form of intense radiation originated near its
center in a process sensitive to temperature and density—either mutual
annihilation of electrons and protons or element building by the transmuta-
tion of hydrogen into helium and heavier elements. The outward flow of
radiation created a pressure that counterbalanced the superincumbent matter's immense weight. Only near the surface, where the temperature and
density were far below the central values of 39,500,000 K and 76.5 g/cc,
would convection supplement radiation as an important means by which
the sun's heat continued its outward journey.

As one indication of the validity of this novel view of the sun's internal
constitution, Eddington (1926, 149) pointed to his theory's success in predicting the output of solar energy. His equations enabled him to calculate the
solar luminosity from three astronomical measurements—the mass and
luminosity of Capella's larger component and the mass of the sun. The result
was that the sun should generate energy at the rate of 5.62×10^{33} ergs/sec.
This theoretical value, which was about 50 percent larger than the observed
value, was not at all bad, considering all the approximations that had gone
into the estimate. The theory was far from perfect. Still, Eddington, despite
being an outsider to solar physics, had forged a strong link between the sun's
mass and luminosity.

Hydrogen in the Sun

As Eddington was extending his theory of perfect-gas stars to the sun,
inquiries were going forward that would undermine the view that the stars
had approximately the same composition as the earth (DeVorkin and Kenat
1983b). This common presumption was abandoned in two stages. Between
1923 and 1929, concurrent efforts to understand atomic spectra and interpret astronomical spectra culminated in Henry Norris Russell's persuasive
case that hydrogen was superabundant in the *atmospheres* of the sun and
stars. Then in 1932, Eddington and Bengt Strömgren independently developed a strong argument that hydrogen was also abundant in stellar *interiors*.

In the six decades following Kirchhoff and Bunsen's initiation of spectral
analysis, solar physicists measured more than 20,000 Fraunhofer lines and,
by comparing the results with laboratory data, managed to attribute almost
6,000 of these lines to some fifty elements. As of 1920, however, they still

lacked theoretical tools that would enable them to work out the relative abundances of the elements constituting the sun's atmosphere. Many suspected that the darkness of an element's Fraunhofer lines was somehow connected with its abundance. But none could see how to translate this intuition into a convincing quantitative analysis.

During the 1920s, the rapid advance of atomic physics provided tools for deducing approximate elemental abundances from solar and stellar spectra (DeVorkin and Kenat 1983a, 1983b). Meghnad Saha, an Indian physicist doing postdoctoral research in London, initiated such work. He combined statistical thermodynamics with Bohr's theory of the atom to derive an equation that indicated how the degree of ionization of an element increased with temperature and decreased with pressure. In 1923 Milne and Ralph Howard Fowler, another mathematical physicist at Cambridge University, went considerably beyond Saha. They examined the production of line spectra by gases consisting of populations of several elements, each of which in turn consisted of subpopulations of atoms having different degrees of ionization. To be specific, they developed equations from which the fraction of an element's atoms capable of absorbing radiation of a given wavelength could be calculated from the ambient temperature and pressure.

Although Fowler and Milne pointed out the relevance of their theory to the abundance problem, Cecilia H. Payne (1900–79) was the most successful in using it for this purpose (DeVorkin and Kenat 1983a, 1983b). She learned of their work during her undergraduate studies at Cambridge University. Feeling that a woman had no prospects in British astronomy, however, she followed up on it in a dissertation on stellar atmospheres completed at Radcliffe College in 1925. There she not only had a somewhat more supportive environment but also had access at Harvard College Observatory to the world's richest collection of stellar spectra (Kidwell 1987). In her dissertation, Payne first used the Fowler-Milne theory to determine how the atmospheric pressure and temperature varied with the spectral type of ordinary and giant stars. She then made two assumptions: that all stars had roughly the same composition, and that, when observing equipment and conditions were constant, all marginally visible lines—lines so faint they could barely be detected—were produced by the same number of atoms. Proceeding from these assumptions, she applied the Fowler-Milne theory to stars with different temperatures and hence different marginal lines to derive the relative abundances of eighteen elements. The outcome, to her surprise, was that hydrogen and helium were much more abundant than all other elements in stellar atmospheres. It is clear in retrospect that she had extracted an extremely significant finding about the sun from stellar data. Although she included her startling result in her dissertation, however, Payne (1925, 186) branded it "spurious."

Payne's repudiation of what would come to be regarded as her most significant result was motivated by the strongly expressed doubts of Princeton's Russell (DeVorkin and Kenat 1983b). Her deference was only prudent. Russell stood alongside Eddington and Jeans as one of the pioneers of theoretical astrophysics. Moreover, he had worked on and around the abundance problem for over a decade. Russell's conviction that the composition of the sun and stars matched that of the earth was rooted in the rough agreement of early estimates of solar-stellar and terrestrial elemental abundances. It was strengthened by Payne's finding that the relative abundances of the metals in the stars agreed fairly well with those in the earth's crust. It also appeared to receive independent corroboration from an intriguing consequence of Eddington's stellar theory. In 1921 Eddington had found that radiation pressure replaced convection as the main factor in stellar equilibrium precisely within the rather narrow range of masses that corresponded to the observed stellar masses. He thought this result, which rested on the assumption that stars were composed of a terrestrial mix of elements, must somehow explain the existence of stars. Like Eddington, Russell thought this success in "explaining" the stars could not be fortuitous.

Until late 1928, Russell's belief that the sun's composition resembled the earth's was so strong that he steadfastly discounted the growing evidence for Payne's high hydrogen abundance (DeVorkin and Kenat 1983b). Some of this evidence came from his own research. In 1925 Russell helped formulate the quantum rules for calculating the transition probabilities that atoms in a given state would produce each of two or more closely spaced lines constituting a spectral multiplet. Over the next two years, aided by his assistant Charlotte Moore and Mount Wilson's Walter S. Adams, he used these rules to ascertain the relative numbers of atoms producing lines of different intensities in multiplets appearing in the solar spectrum. This method of "calibrating" Rowland's estimates of line intensities enabled Russell and his colleagues to calculate the relative abundances of several elements in solar and stellar spectra. However, they presumed they had an anomalous result when their calibration indicated the presence of large amounts of hydrogen.

Russell finally shifted his grounds in December 1928, just after arriving in Pasadena for two months of research at Mount Wilson Observatory. Among the postdoctoral fellows there was Albrecht Unsöld, who was preparing to migrate into theoretical astrophysics. Thanks to his training in theoretical physics with Arnold Sommerfeld, the young German was conversant with the new quantum mechanics. He had already used this theory to derive elemental abundances from the intensity profiles of the darkest solar lines (DeVorkin and Kenat 1983b). His results for six elements, including hydrogen, agreed well with Payne's findings from stellar data. Upon becoming acquainted with Unsöld and his research, Russell concluded that the evidence for hydrogen's great abundance in the solar atmosphere had become overwhelming. He saw at once that he had a grand opportunity to revolutionize thinking about the solar abundance problem. Perhaps hoping to avoid competition, he stopped attending a set of lectures that Unsöld was giving and began work on a full-scale review of the question (Russell 1929b).

In February 1929, after finishing the first draft of his paper (Russell 1929a; Menzel 1929), Russell visited Donald Menzel, who was then at Lick Observatory in northern California. While there, he learned that his former student had recently obtained an important result from the Lick eclipse plates of flash spectra—that the gases composing the chromosphere had an average atomic weight of about 2. He appreciated that this was yet another indication of hydrogen's abundance. Once back at Princeton, Russell (1929c) incorporated a discussion of Menzel's result in the final version of his magisterial review of the solar abundance problem. His chief objective was to establish hydrogen's abundance in the solar atmosphere. Although neither his methods nor his results were strikingly novel, his powerful advocacy brought the idea into the mainstream of astrophysics.

Russell (fig. 3.7) had four main arguments for hydrogen's preponderance in the solar atmosphere. First he showed that, at the temperature of the sun's atmosphere, sizable fractions of the atoms of almost all the fifty-seven known solar elements were in the excited states required for producing their most intense lines. By contrast, minuscule fractions of the hydrogen atoms were in such states. Yet the principal hydrogen lines were among the darkest in the solar spectrum. Hydrogen must therefore be extraordinarily abundant. Second, he calculated approximate figures for elemental abundances in the solar atmosphere by using Unsöld's precise results to anchor his prior calibration of Rowland's line intensities. Hydrogen atoms turned out to be orders of magnitude more abundant than those of any other element. Third, he drew attention to the close agreement between his and Payne's abundances for sixteen elements. This concordance struck Russell (1929c, 65) as "very gratifying . . . , especially when it is considered that Miss Payne's results were determined by a different theoretical method, with instruments of a quite different type . . . , and even on different bodies—a long list of stars, almost all of which are giants." And fourth, he pointed out that Menzel's result of an average atomic weight of 2 for the chromosphere was consistent with a high hydrogen abundance. For instance, a gas consisting by weight of 45 percent hydrogen and 55 percent heavier elements would have the indicated atomic weight. Thus, whatever angle he approached the issue from, Russell found indications of hydrogen's abundance in the atmospheres of the sun and other stars. Although he seems to have suspected that

Figure 3.7. Henry Norris Russell about 1929, when he made his case for hydrogen's preponderance in the sun's atmosphere. (By permission of the Art Museum, Princeton University.)

hydrogen was equally abundant *inside* the stars, he left the possibility of homogeneous composition undiscussed. By default, therefore, his case for hydrogen's predominance was initially interpreted as demonstrating that most of a star's hydrogen is somehow pushed to its surface (DeVorkin and Kenat 1983b).

Not long after Russell's landmark article appeared in the July 1929 issue of the *Astrophysical Journal*, the opening shots were fired in a battle over stellar theory that would lead to recognition of hydrogen's abundance in stellar interiors. Milne, who had obtained a chair at Oxford University, was concerned that allegiance to Eddington's model was perniciously foreclosing serious consideration of alternatives that might possibly resolve some outstanding stellar problems. In November 1929 he began an all-out campaign against his former colleague's theory (Milne 1929). A seasoned controversialist, Eddington (1930a) was not slow to launch a counterattack. At first bystanders tended to root for Milne. After a year or so of furious battle, however, most could see that he was losing his way in mathematical thickets. By 1932 it was clear that Eddington's theory would emerge from Milne's critique essentially unscathed (Chandrasekhar 1980).

One benefit of Milne's battle with Eddington was that it helped lure four talented young scientists into theoretical astrophysics—the Englishman Thomas George Cowling (b. 1906), the German Ludwig F. Biermann (1907–86), the Dane Bengt Strömgren (1908–87), and the Indian Subrahmanyan Chandrasekhar (b. 1910). Another was that it induced several theorists to try to explain the magnitude of stellar opacity—that is, the capacity of stellar matter to retard the outward flow of radiation. During 1930 and 1931, nothing came of their endeavors. But in the winter of 1932, Eddington decided to examine the consequences of assuming a high hydrogen abundance in stellar interiors. His decision indicated his exasperation with the opacity problem, for he had dismissed this possibility a decade earlier on the grounds that it was at odds with his explanation of the narrow range of stellar masses. He soon found that the discrepancy between the laboratory and astrophysical values for the opacity vanished when he assumed that stellar gases consisted by weight of about one-third hydrogen and two-thirds heavier elements (Eddington 1932). Midway through the analysis, he received a manuscript from Strömgren, who had carried through similar calculations and reached the same conclusion (Strömgren 1932). This agreement settled the matter so far as Eddington, Strömgren, and most other theoretical astrophysicists were concerned. Thenceforth hydrogen was regarded as the predominant constituent not only *on* but also *in* the sun and other stars obeying the mass-luminosity relation.

The Sun's Energy Source

During the four years preceding the establishment of hydrogen's abundance in the stars, quite separate developments increased the likelihood that element building, not annihilation, would turn out to be the source of stellar energy. New astronomical evidence and theorizing favored a cosmological time scale of billions of years over that, inspired by the annihilation hypothesis, of trillions of years. In early 1929, Edwin P. Hubble of Mount Wilson Observatory reported evidence indicating that most galaxies were receding from our own with velocities proportional to their distances (Smith 1982). A year later Eddington proposed that a dynamic solution to Einstein's equations of general relativity might account for Hubble's distance-velocity relation. He soon learned that his former student Georges Lemaître had published just such a solution three years earlier in an obscure Belgian journal. Championing Lemaître's theory as if it were his own, Eddington pointed out that Hubble's data indicated the universe began expanding billions of years ago. If so, the universe was much younger than advocates of the annihilation theory were wont to believe. "Perhaps," suggested Eddington (1930b, 30), "the lesson of the galaxies is to wake us from our dream of leisured evolution through [trillions] of years."

Physicists, meanwhile, were buttressing the idea that protons can penetrate nuclei at the temperatures prevailing in stellar interiors. Starting in 1928, the Russian George Gamow (1904–68) and other theorists were increasingly able to account for nuclear phenomena. They devised explanations of alpha decay—that is, the emission of a high-speed helium ion from a radioactive nucleus—and of alpha penetration—the entrance of an alpha particle into a nucleus. They argued that the emitted or incident alpha particle must, despite its comparatively low energy, leak through the nucleus's barrier field in accord with the equations of the new quantum mechanics. Almost immediately, Robert d'E. Atkinson (1898–1981), an English experimental physicist working in Berlin, decided to apply these ideas to the stars. In discussion with Gamow's friend Fritz Houtermans, he realized that if a low-energy alpha particle could leak into a nucleus, so could a low-energy proton. At the stellar central temperatures given by Eddington, some protons should be energetic enough to penetrate and transmute nuclei. Atkinson and Houtermans (1929) soon laid out the basic idea. Two years later Atkinson (1931), having taken a position at Rutgers University, where he enjoyed Russell's counsel and encouragement, refined the case for element building as the source of stellar energy. A year later yet, the Cambridge experimentalists John Douglas Cockcroft and Ernest Thomas Sinton Walton added fresh impetus to Atkinson's approach by confirming the possibility of proton penetration with their new particle accelerator.

As of 1932, Atkinson's hypothesis of energy generation through element building—despite being favored by trends in both cosmological and nuclear research—was still more akin to a good hunch than a robust theory. During the next few years, astrophysicists furthered the idea in three chief ways. Observers strengthened the case for a time scale in the billions of years. The most important work here was done by the Dutch emigré Bart Bok of Harvard College Observatory. He showed that open clusters of stars in our galaxy were flying apart so rapidly that these clusters would no longer be in existence if their constituent stars had ages of trillions, instead of billions of years (Bok 1934). Meanwhile, theorists were adjusting their estimates of conditions near the center of stars where, they supposed, most energy generation occurred. For instance, in response to the new view of hydrogen's abundance, they reduced the figure for the sun's central temperature from 40,000,000 K to 20,000,000 K (e.g., Eddington 1935, 171). A few younger theorists were also laying to rest Eddington and Jeans's notion that stars drawing energy from subatomic processes having a high temperature dependence would be subject to runaway oscillations. In particular, Cowling (1935) demonstrated that a star powered by element building, a process having a temperature sensitivity of about T^{20}, would be stable against vibrational breakup.

Although astrophysicists worked around the edges of the stellar-energy problem, they knew too little about nuclear physics to tackle it directly. Nuclear physicists, despite possibly having the requisite physical knowledge, were too busy with their experimentation to get involved with the question. Unlike the astrophysicists and the nuclear physicists, some of the most able young theoretical physicists had not only tools but also incentives for giving the problem a try (Hufbauer 1985). Their training in problem solving and their subsequent endeavors to keep abreast of nuclear experimentation gave them the needed agility and expertise. Moreover, their hopes of matching the achievements of the creators of quantum mechanics inspired them to regard important physical enigmas as exciting challenges. It is small wonder, therefore, that during 1936–38 such theoretical physicists as Hans A. Bethe (b. 1906), George Gamow, J. Robert Oppenheimer, and Edward Teller in America, Carl Friedrich von Weizsäcker in Germany, and Lev Landau in the Soviet Union all tried solving the stellar-energy problem.

The most successful of these competitors was Bethe. Indeed, his solution was swiftly recognized as a significant breakthrough and subsequently served as the justification for his Nobel Prize (Bethe 1968). A student of Sommerfeld's, Bethe was making his way up the German academic ladder

Figure 3.8. The Fourth Washington Conference on Theoretical Physics: held March 21-23, 1938, this interdisciplinary conference on the sources of stellar energy brought together specialists from theoretical astrophysics—for example, Donald Menzel (front row, third from left), Subrahmanyan Chandrasekhar (front row, ninth from left), and Bengt Strömgren (second row, seventh from left)—and nuclear theory—for example, Hans Bethe (second row, third from left), Edward Teller (second row, fifth from left), George Gamow (back row, far left), and Charles Critchfield (back row, far right). (Courtesy of the Department of Terrestrial Magnetism, Carnegie Institution of Washington.)

until the Nazi party took power (Bernstein 1980). Then, half-Jewish by background, he was forced to look elsewhere for work. In 1934 he got a position at Cornell University. He spent the next few years moving to the forefront of nuclear theory. He demonstrated his mastery of this fast-moving area in three major review articles that came to be known as "Bethe's Bible." Although aware of attempts by Weizsäcker and Gamow to identify the nuclear reactions powering the stars, he was loath to give any time to such a speculative venture. In fact, had it not been for the special urging of his friend Teller, he would have passed up Gamow's Fourth Washington Conference on Theoretical Physics (fig. 3.8), which was devoted particularly to the stellar-energy problem (Chandrasekhar, Gamow, and Tuve 1938).

It was while attending this conference in March 1938 that Bethe realized his command of nuclear data and theory might enable him to resolve the long-standing conundrum. He acquired an up-to-date view of stellar conditions from the theoretical astrophysicists Strömgren and Chandrasekhar. He also became acquainted with the shortcomings of recent solutions. Most important, he learned that Teller's student Charles Critchfield was independently urging investigation of the process that he himself regarded as most promising—that is, a chain of reactions beginning with proton-proton collisions and culminating in the synthesis of helium nuclei. Before leaving Washington, Bethe arranged to collaborate with Critchfield in exploring this possibility.

On the train journey home, according to Gamow's colorful telling (1940, 112–13), Bethe concluded,

> "But it should not be so difficult after all to find the reaction which would just fit our old Sun . . . ; I must surely be able to figure it out before dinner!" And taking out a piece of paper, he began to cover it with rows of formulas and numerals, no doubt to the great surprise of his fellow-passengers. One nuclear reaction after another he discarded from the list of possible candidates for the solar life supply; and as the Sun, all unaware of the trouble it was causing, began to sink slowly under the horizon, the problem was still unsolved. But Hans Bethe is not the man to miss a good meal simply because of some difficulties with the Sun and, redoubling his efforts, he had the correct answer at the very moment when the passing dining-car steward announced the first call for dinner.

Bethe actually took somewhat longer (Bernstein 1980). Still, within six weeks of the conference, he was giving colloquia on the stellar-energy problem and corresponding about his solution (Bethe 1938a, 1938b).

In searching for the energy-generating reactions, Bethe (1939) systematically examined how protons would interact with nuclei of increasing mass up to the heaviest isotope of chlorine. He found two reaction chains that would generate energy at the temperatures prevailing in ordinary stars. The first of these thermonuclear chains was the one he was investigating with Critchfield. It began with a proton-proton reaction and culminated, after two more protons tunneled successively into the resultant nucleus, with the formation of a helium nucleus. Energy was liberated all along the chain by the conversion of mass into kinetic and radiant energy. Bethe's calculations indicated that Critchfield's chain would be the predominant source of energy in stars that were less massive than the sun and hence cooler. His second chain was actually a cycle in which carbon nuclei acted, so to speak, as pots

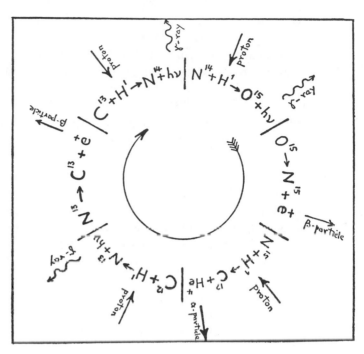

Figure 3.9. Gamow's illustration of the carbon cycle, which, following Bethe, he identified as the "cyclic nuclear reaction chain responsible for the energy generation in the Sun." About 1950, theoretical astrophysicists concluded that the solar central temperature was closer to 15,000,000 K than 20,000,000 K and hence that the sun was mainly powered by the Critchfield-Bethe nuclear chain beginning with proton-proton reactions. (Source: Gamow 1940, 114.)

for cooking protons up into helium nuclei. It began with a proton tunneling into a carbon nucleus and culminated, after three more protons tunneled one by one into the nucleus, with the division of the resultant nucleus into a carbon nucleus and a helium nucleus (fig. 3.9). As with the proton-proton chain, the energy liberated in the carbon cycle came from the conversion of mass into kinetic and radiant energy. Bethe's calculations indicated that his cycle, which required higher temperatures than Critchfield's chain, would predominate in the sun and more massive stars. In making the case for his theory, Bethe compared the central temperatures of several stars yielded by Eddington's theory and his own. The agreement was impressive. In the sun, for instance, the central temperatures, assuming a composition of 35 percent by weight of hydrogen, were 19,000,000 K and 18,500,000 K, respectively!

Among the theory's many consequences, two were of particular interest. First, since neither energy-generating chain did more than transform hydrogen into helium, the elements heavier than helium must have been created before the sun and stars. Bethe was too cautious to go beyond drawing attention to this point. But Weizsäcker (1938), who independently discovered the carbon cycle, daringly speculated about the origin of the heavier elements in a primeval explosion that also caused the recession of the galaxies. Second, the theory indicated that the sun would shine for another twelve billion years. It was Gamow who capitalized on this idea with his

Birth and Death of the Sun (1940)—the most widely read book on the sun during the next two decades. Thanks, therefore, to the interlopers from theoretical physics, particularly to Bethe, Weizsäcker, and Gamow, the solar physicists had new connections to develop and fresh vistas to explore.

Coronal Emission Lines

While nuclear theorists were investigating the energy-generating reactions in the sun, stellar spectroscopists were grappling with another of solar physics' most stubborn problems—the origin of the corona's emission spectrum. Charles Young and William Harkness had discovered the corona's green emission line almost seven decades earlier at the 1869 eclipse. Since then, eclipse observers had detected eighteen additional lines, and Lyot with his coronagraph had found another five (Mitchell 1936; Lyot 1939). In all this time, however, not a single one of the growing list of coronal lines had been convincingly explained.

For a good while, solar physicists had believed that the mysterious lines might be produced by an unknown element called, in analogy to helium, "coronium." But confidence in coronium's existence declined after 1910 as physicists and chemists filled out the few remaining gaps in the periodic table of elements. It vanished in 1927 when Ira S. Bowen, a spectroscopist at the California Institute of Technology, succeeded in explaining the emission lines from gaseous nebulae (Hirsh 1979). He showed that these lines, which had commonly been ascribed to the unknown element "nebulium," were produced when the electrons in oxygen and nitrogen ions made transitions that had to be inferred on theoretical grounds, since they could not be observed under laboratory conditions. After Bowen's success, efforts to identify coronal lines focused on analogous transitions in the ions of abundant solar elements (Milne 1945). Invariably, however, the resulting claims that particular transitions gave rise to particular lines were soon discredited. Frustration mounted. The Belgian astrophysicist Pol Swings (1939, 69, 78) captured the prevalent mood when he concluded that it would take "a brilliant inspiration" to resolve this "grand enigma of astronomical spectroscopy."

As it happened, Walter Grotrian, the Potsdam spectroscopist who had acclaimed Lyot's coronagraph, had already had the key inspiration. Most investigators, including Swings, made the commonsense assumption that the solar atmosphere's temperature declined with distance from the photosphere and hence that the corona was comparatively cool. Grotrian, by contrast, had concluded as early as 1934 that the corona might be a good deal hotter than the underlying chromosphere and photosphere. He had come to this counterintuitive position in the course of trying to explain the corona's continuous spectrum, or light radiated at wavelengths other than those of the bright emission lines. Grotrian (1934) found he could account for the coronal intensity's observed variation with wavelength by assuming that the continuous spectrum was produced by the scattering of photospheric radiation from electrons having an average velocity of 4000 km/sec. The trouble was that such speeds corresponded to a temperature of some 350,000 K, over fifty times higher than the temperature of the photosphere.

During the next few years, Grotrian remained on the lookout for further indications of high coronal temperatures. His vigilance did not go unrewarded. In 1937 he read two papers by the Swedish laboratory spectroscopist Bengt Edlén (b. 1906) reporting emission spectra from several elements whose atoms had been stripped of many of their electrons by high-voltage sparks. In mulling over Edlén's articles, he noticed that iron atoms bereft of nine or ten of their electrons might undergo rare transitions that would produce the coronal emission lines with wavelengths of 6375 Å and 7892 Å.[2] Grotrian (1937) communicated his tentative identifications to Edlén. He went on to confess, however, that he had failed in his attempts to pursue the idea. "I am," he went on, "consequently quite skeptical whether the above-mentioned numerical agreements are more than coincidental. But perhaps it would be worthwhile to look into the question, and you are certainly the only one who can do so." In reply Edlén seems to have stressed the difficulty of securing more pertinent data (Edlén 1986). Taking the course of prudence, Grotrian decided against immediate publication. In fact, he did nothing with his idea for two full years. Then Bowen and Edlén (1939) announced that rare transitions of six-times ionized iron atoms had apparently given rise to nine lines in the emission spectrum from a stellar nova in 1925. To ensure his priority, Grotrian (1939a) rushed his proposed coronal-line identifications into print.

Although Grotrian's 1939 article did not go much beyond his 1937 letter, it did stimulate Edlén to take another look at the coronal-line problem (Edlén 1986). He quickly found that some of his unpublished results from a prior study of calcium's spark spectrum indicated that eleven- and twelve-times ionized calcium atoms produced two additional coronal lines. He began to think that Grotrian had indeed been on the right track, but he lacked the experimental data needed to follow up this possibility directly. To get around this difficulty, he determined the energy levels of ions that might be responsible for coronal lines by extrapolating from spectral data for ions possessing the same electronic configurations as these candidate ions. His

---

2. Grotrian was relying here on Lyot's recent measurements.

Figure 3.10. A spectrogram showing the coronal line at 5303 Å—the bright white arc— taken by Lyot with his coronagraph; the dark absorption lines were from prominences and the chromosphere. (Source: Lyot 1939, facing p. 590. By permission of the Royal Astronomical Society.)

extrapolations soon yielded promising results. His lingering doubts dissolved when he found that thirteen-times ionized iron atoms could produce the corona's most intense line—the green line at 5303 Å (see fig. 3.10).

By June 1939, Edlén was sharing a preliminary list of line identifications with friends (Swings 1943). At a French conference on novae and white dwarfs the following month, however, he limited himself to endorsing Grotrian's results and hinting at his own methods (Edlén 1941a). Slowed down by another project and by insurmountable difficulties in measuring the energy levels of highly ionized iron atoms (Edlén 1986), he did not write up his coronal results until 1941. By this time Edlén (1941b) had identified thirteen lines in addition to Grotrian's two lines. He found that these lines could be traced back to rare transitions in nine ions of iron, nickel, and calcium. Edlén went on to use the identifications in conjunction with eclipse and coronagraph data on line intensities to discuss the temperature and composition of coronal matter. He pointed out that not only the energies needed to produce the most abundant ions but also the breadth of the iron lines indicated that the corona's temperature exceeded 2,000,000 K.

Despite the mounting fury of World War II, Edlén's results—the line identifications and high temperature estimate—were rapidly disseminated (e.g., Menzel 1941; Russell 1941). Astrophysicists were quick to see that his definitive treatment of the emission spectrum demolished the commonsense idea of a cool corona. However, as has often been the case in solar physics, his success in resolving this enigma opened up a more profound, and recalcitrant, issue—the origin of coronal heating.

The State of Solar Physics about 1940

By the beginning of World War II, solar physics had acquired a certain maturity. Solar physicists in over thirty observatories around the globe were helping to keep the sun under constant surveillance. They were using the results, which appeared regularly in the International Astronomical Union's *Quarterly Bulletin on Solar Activity,* to refine knowledge of solar activity. Together with the radio physicists and geophysicists, they were also using data on solar activity to deepen and extend knowledge of the sun's influence on radio transmissions and geomagnetism. Meanwhile, a few venturesome migrants into the field, as well as solar physicists inspired by the migrants' successes, were busily developing new means of monitoring the sun: Lyot, Waldmeier, and Menzel, the coronagraph; Lyot and McMath, solar cinematography; Lyot and Öhman, the monochromatic filter.

By 1940 solar physics also possessed a fairly sturdy interpretive framework. During the preceding three decades, various scientists—mostly outsiders familiar with recent physical research—had significantly augmented the field's store of theoretical tools and results. Eddington's theory of radiative equilibrium was the starting point for more subtle inquiries into the sun's internal constitution. Russell's work on the sun's composition was serving as a benchmark for more sophisticated investigations of elemental abundances. Bethe's identification of thermonuclear reactions capable of powering the sun for billions of years was being accepted as a fundamental breakthrough. And Edlén's work on the origins of coronal lines was on the verge of sweeping the field.

PART TWO

Solar Physics' Evolution into a Subdiscipline

In looking at the thirty-three decades from Galileo's first telescopic studies down to World War II, part 1 emphasized major innovations in solar science, highlighting significant advances in solar observational capabilities, empirical knowledge of solar phenomena, and theoretical understanding of the sun's structure and operation. Part 1 also gave considerable attention to contextualizing solar science's development. It examined the emergence of the solar physics community in the mid-nineteenth century and illuminated this specialty-oriented community's subsequent organizational endeavors and continuing dependence on outsiders for novel instruments and interpretations.

Since the beginning of World War II, solar research has evolved along so many different paths at the same time that there is a significant advantage to focusing on only a few lines of development. Doing so ensures that space is available for treating chronology, biography, patronage, and communication patterns with the sort of attention that distinguishes historical narratives from scientific review essays. Accordingly, I have been quite selective in parts 2 and 3, giving little or no attention to research on such important questions as the sun's origin and developmental path, the solar dynamo's production of the activity cycle, the temperature and density structure of the solar atmosphere, the genesis and evolution of flares, and the solar-stellar connection. Limiting coverage in this way frees up the space needed for historical treatment of the subjects that are discussed.

Part 2 addresses two related subjects—the diversification, refinement, and expansion of solar observational capabilities since 1939 and the consequent transformation of solar physics from a small specialty into a broad subdiscipline. The focus on instruments and techniques has ample warrant because solar physics is above all an observational science. Large shares of the field's funding and its practitioners' time have gone into preparing for and carrying through observations of the sun's rich and variable output.

Furthermore, the resulting profusion of findings has acted—more than in any other branch of astronomy—as both an awesome challenge to theory and an ever more stringent set of constraints.

By the 1970s, solar physics possessed not only a richly varied observational armamentarium but also its own journal and sections in disciplinary associations. In addressing the field's evolution into a subdiscipline, part 2 follows up on the attention given in part 1 to this community's cooperative endeavors and collective aspirations. Such issues are of interest for two main reasons. Studying them illuminates how trends in patronage have influenced the way solar physicists have interacted with one another and with outsiders. More important, it illuminates the ethos underlying the solar physicists' ongoing labors to enlarge knowledge of the sun. We shall see that observational virtuosity has not, for all its importance, been the only force contributing to the advance of solar science during the past half-century.

CHAPTER FOUR

Solar Physics as Beneficiary of War

1939–1957

Beginning in September 1939 with Germany's blitzkrieg in Poland and ending in August 1945 with the United States' nuclear bombardment of Hiroshima and Nagasaki and the Soviet Union's defeat of Japanese forces in Manchuria, World War II had a profound impact on all the sciences (Roland 1985), including astronomy (DeVorkin 1980; Kidwell 1990). The war cut into customary sources of support, undermined commitments to personal and disciplinary research agendas, obstructed channels of communication, and interfered with recruitment. At the same time, however, it engendered unprecedented government patronage, directed attention to fresh problems, and gave rise to new instruments and techniques. World War II's effect on solar physics was consistent with this general pattern of disruption and invigoration. On the one hand, the war hampered solar physicists in countless ways. On the other hand, it prepared the way for a postwar flourishing of solar research.

This chapter follows solar physics through World War II and its postwar expansion to the beginning of the International Geophysical Year in mid-1957. The account of wartime developments emphasizes the participation of German and Allied solar physicists in ionospheric forecasting projects, the frustrated efforts of a few German scientists to observe the sun from above the atmosphere with the aid of V-2 rockets, and the discovery by British and American scientists that the sun emitted radio waves. The description of the postwar burgeoning of solar physics first spotlights the rise of new solar centers in Colorado and New Mexico. It then examines how various scientists enriched solar observations by using rockets as instrument platforms, by creating radio analogues to the spectrograph and spectroheliograph, and by devising a means of mapping weak magnetic fields on the sun. Last, it recounts preparations for solar monitoring during the International Geophysical Year. The conclusion assesses solar observational capabilities at the start of this major cooperative venture.

Solar Physics during World War II, 1939–1945

When the fighting began in September 1939, most solar physicists in Europe rallied to their flags. As the war spread, their colleagues around the globe came to presume that military imperatives took precedence over scientific interests. The last vestige of the cooperative order of the interwar years disappeared in 1942. Unable to get observational reports, William Brunner of the Swiss Federal Observatory in Zurich stopped publishing the *Quarterly Bulletin on Solar Activity.* By this time, indeed, authorities in the warring nations regarded solar observational capabilities and data as military assets. At least one dissertation in solar physics was "born classified" (Roberts 1943, 1983; Kidwell 1990). The prevailing attitude, as expressed by a writer for the American Astronomical Society's wartime newsletter, was that much solar research was "of such a nature that some important results may not become known publicly until after the war" (The Sun 1944, 1).

Studies of radio transmission during the 1930s had revealed two ways solar activity influenced the ionosphere's ability to propagate radio waves (e.g., Grotrian 1939b). The higher the number of sunspots, the lower was the maximum frequency that could be used in shortwave transmissions. And the greater the number and intensity of flares, the larger was the possibility that shortwave fadeouts would interrupt reception. Then, as the war was getting under way, Zurich's Max Waldmeier (1939) argued that coronal observations could be used to predict terrestrial magnetic storms. Radio scientists kept these influences in mind when, in the early years of the war, they persuaded their respective governments and military forces to establish agencies to forecast conditions for radio communications. In particular, they arranged for solar physicists to provide the forecasting agencies with daily observations of solar activity.

Solar and radio scientists also laid the groundwork during the war for two new solar observing techniques. Scientists in Germany made extensive preparations to use a V-2 rocket as a platform for observing solar ultraviolet radiation that could not penetrate the atmosphere, and scientists in Britain and the United States discovered that the sun emitted radio waves.

The Third Reich

Germany led the way in transforming solar observing into a military asset after 1939 (Kuiper 1946; Kiepenheuer 1948; Dieminger 1948). The key figures in the German effort were the radio physicist Johannes (Hans) N. Plendl and the solar physicist Karl-Otto Kiepenheuer (1910–75). When the war began, Plendl was with the German Aeronautic Research Institute as director of radio research and development at the Luftwaffe's test station in Rechlin, about 100 km north of Berlin (Plendl 1985; Kröber 1984). His group, which had developed aerial guidance beams in the mid-1930s, was working on the problem of forecasting radio conditions. Plendl presumed— based on his own earlier work on the influence of solar activity on short-wave communications (e.g., Plendl 1931)—that accurate forecasting would depend on timely access to solar observations. Germany, however, had few solar observatories and none with up-to-date equipment.

Wartime mobilization gave Plendl, who had won the confidence of Luftwaffe General Wolfgang Martini, the opportunity to develop the requisite observational capabilities. In the fall of 1939 he arranged for Kiepenheuer— a recent Göttingen Ph.D. who had a strong interest in solar physics (Bruzek 1975; de Jager 1978)—to be seconded from the Luftwaffe to his group at Rechlin (Kiepenheuer 1945; Kröber 1984). Plendl and Kiepenheuer immediately visited the radio division's ionospheric observatory at Wendelstein in the Bavarian Alps to see whether it would be a good site for solar monitoring (fig. 4.1). Their conclusion was positive. Within four months, Plendl obtained funding for a solar station there.

As construction proceeded at Wendelstein, Kiepenheuer was increasingly busy (Kiepenheuer 1945; Kuiper 1946). He decided on observing routines with the advice of the Potsdam astrophysicist Walter Grotrian who, as a Luftwaffe officer, had been assigned to the project (Kienle 1955). He supervised the completion of Germany's first spectroheliograph at Göttingen. And he took charge of the solar observatory at Meudon in August 1940 to ensure that the German occupying forces did not interfere with its operations. Feeling too isolated at Rechlin to be effective, Kiepenheuer and several assistants relocated to Göttingen in the spring of 1941. Soon afterward, he obtained Plendl's backing for additional solar stations at Kanzelhöhe in the Austrian Alps, Zugspitze in the Bavarian Alps, and Syracuse in Sicily.

Meanwhile, Plendl was enlarging his ambit (Plendl 1985; Kröber 1984). Besides organizing the shortwave forecasting program with the aid of Kiepenheuer and the radio physicist Walter Dieminger, he was hard at work on any number of radio-related problems that were arising in the Luftwaffe's far-flung campaigns. His achievements, and his forthrightness in pointing out problems with policies affecting military research and development, led Reichsmarschall Hermann Göring to appoint Plendl plenipotentiary for high-frequency research in late 1942. Notwithstanding his many respon-

Figure 4.1. Karl-Otto Kiepenheuer (left) and Hans Plendl at the Wendelstein ionospheric observatory in December 1939; they soon arranged for this station to be equipped for wartime solar monitoring. (Courtesy of Rudolf Kröber.)

sibilities, Plendl maintained an active interest in the radio-forecasting problem. He put Dieminger and Kiepenheuer in charge of institutes to run this project. In the case of the solar institute, he also approved construction of a headquarters observatory at Schauinsland near Freiburg im Breisgau.

Kiepenheuer and his colleagues at the Fraunhofer Institute, as it was named in April 1943, managed to provide Dieminger's forecasting group with a stream of solar observational reports (Kiepenheuer 1945, 1948; Kuiper 1946; Dieminger 1948). They obtained some solar data from observatories in Switzerland and the occupied nations, sheltering astronomers at Meudon, the Pic du Midi, Belgrade, and Prague-Ondrejov from the German military in exchange for their cooperation (De Jager 1978; Roberts 1983).

They also obtained solar data from observatories in northern Germany—Göttingen, Hamburg, and especially Potsdam. But unable to depend on the loyalty of the foreign observers or the weather at the northern observatories, Kiepenheuer's group relied most heavily on the Fraunhofer Institute's own stations at Wendelstein, Kanzelhöhe, Zugspitze, and Schauinsland.[1] Despite wartime shortages, this network of stations was equipped by war's end with spectroheliographs, photoelectric photometers, and coronagraphs.

As Plendl, Kiepenheuer, and their assistants were establishing the solar-monitoring network, plans were under way to use the A-IV ballistic rocket—now generally known as the V-2—to investigate conditions in the upper atmosphere (DeVorkin 1990c). In the spring of 1942, Wernher von Braun, the technical director for the V-2 at Peenemünde in northern Germany, decided to give the program a scientific dimension. His main objective was to maximize the rocket's performance by improving knowledge of the medium through which it would be traveling. But he was also intrigued by the challenge of making scientific observations at much higher altitudes than ever reached before. Von Braun asked the distinguished physicist Erich Regener, a leader in scientific ballooning throughout the 1930s (DeVorkin 1989a), whether he might be interested in such an undertaking. Regener, despite his open opposition to Nazi anti-Semitism, could not resist the opportunity to get instruments above the atmosphere. In July 1942, his institute in Friedrichshafen entered into an agreement with Peenemünde to develop a scientific payload and special nose cone for a V-2 flight.

Regener's group began work at once on instruments for determining the change with altitude in the upper atmosphere's pressure, temperature, and composition (DeVorkin 1990c). The most sophisticated and, from the point of view of solar physics, interesting instrument was an ultraviolet spectrograph. Its purpose was to ascertain the altitude dependence of ozone's abundance. Since the early 1920s, solar physicists had believed that ozone in the upper atmosphere was responsible for preventing most solar ultraviolet radiation from reaching the earth's surface (Bernheimer 1929, 1936; DeVorkin 1989a).[2] If so, the altitude dependence of ozone could be calculated from the increase in the solar ultraviolet's intensity with height. The plan was for the spectrograph to record this increase by taking spectrograms in rapid succession during the V-2's ascent. Since no one in Regener's group had the expertise to construct such an instrument, Peenemünde arranged

---

1. The Fraunhofer Institute's station in Sicily had to be abandoned in June 1943 because of the impending Allied invasion.

2. Since the early 1970s, there has been increasing concern about the threat of artificial chemicals to the ozone in the upper atmosphere.

for Hans-Karl Paetzold, a specialist in physical optics, to be released from combat duty for the task.

Once Plendl learned of von Braun's plans for the scientific mission— either on a visit to Peenemünde to witness the first V-2 firing in October 1942 or, more likely, on a subsequent inspection visit in May 1943 (Dornberger 1958, xxvii)—he was eager to get his own people involved. He had made inquiries before the war about the possibility of using rockets for investigating the origins of and conditions in the ionosphere (Plendl 1986). His agency soon arranged for Kiepenheuer, who had participated in one of Regener's prewar balloon flights, to help design a second ultraviolet spectrograph (Kiepenheuer 1945, 1948; Kuiper 1946; DeVorkin 1990c). Kiepenheuer set himself the goal of ascertaining whether, as was widely thought, some component of the solar ultraviolet was intense enough to engender the terrestrial ionosphere. For instance, emission at 1216 Å, which was radiated by hydrogen when making the "alpha" transition from its first excited state to its ground state,[3] should be very strong because of hydrogen's abundance in the sun. Kiepenheuer could not use quartz or glass in the spectrograph, since these materials were opaque to the extreme ultraviolet containing this and other spectral regions of interest. He decided to use a lithium-fluoride optical system to disperse the incident radiation and sensitized crystals that would darken upon ultraviolet exposure to record the spectrum. As with Regener's spectrograph, Paetzold had prime responsibility for implementing Kiepenheuer's plans.

All the while, hundreds of V-2s were being readied for use. Germany sent the first V-2s against London in September 1944. Two months later, paying no heed to the rapid deterioration in Germany's strategic position, von Braun drew up a firing schedule for Regener's mission (DeVorkin 1990c). His plan was to send a prototype aloft in the winter of 1945 and the scientific payload in the spring. In anticipation of the test, Regener's group got the instruments and nose cone to Peenemünde by mid-January 1945. After one final test of the integration, the instruments were packed away for the long-awaited launch. It never occurred. The last V-2 was fired from Peenemünde in late March 1945. Two months later, the Allied armies entered Berlin.

The Allied Nations

Like the Germans, radio scientists in several of the Allied nations established agencies during the war for forecasting radio conditions. Unlike the Germans, however, they were able to rely on existing observatories for most of

3. This line was commonly known as the "Lyman-alpha" line after Harvard's Theodore Lyman, who first studied it in the laboratory (Hevly 1987).

their solar data. Thus Britain's Inter-Services Ionosphere Bureau, founded in 1941, relied on the Royal Greenwich Observatory; Australia's Radio Propagation Committee, founded in 1942, relied on the Commonwealth Solar Observatory at Mount Stromlo (Mellor 1958; Gascoigne 1984); the United States' Interservice Radio Propagation Laboratory, founded by Howard Dellinger in 1942, relied on the McMath-Hulbert Observatory in Michigan, Harvard's new coronagraph station in Colorado, and the Mount Wilson Observatory in California[4] (Dellinger and Smith 1948; Cochrane 1966; Roberts 1983); and the Soviet Union's Sun Service relied on the Moscow, Abastumani, and Tashkent observatories (Astronomy in Moscow 1944).

The Allied radio scientists used the solar data at their disposal in two chief ways (Shapley 1945; Dellinger and Smith 1948; Roberts 1983; Kidwell 1990). They developed means of predicting the ionosphere's transmission characteristics at any place on the globe from relative sunspot numbers. And depending primarily on reports from Walter O. Roberts (fig. 4.2) on chromospheric and coronal activity from the Harvard station at Climax, Colorado, they succeeded in making fairly reliable predictions of magnetic storms in the ionosphere. These uses of solar data, though they did not play a decisive role in any major battles, were important enough to establish solar physics as a military asset that deserved long-range support.

While the monitoring programs were laying the basis for postwar patronage of solar research, three British and American investigators independently demonstrated that the sun emitted radio waves. This achievement was not entirely beyond the technical capabilities of prewar radio (Sullivan 1990). In 1933, after all, Karl G. Jansky of Bell Laboratories had announced the detection of radio waves originating somewhere in the reaches of the galaxy. More particularly, near the sunspot maximum of 1937, a few radio buffs had detected what they thought were solar radio signals. No one in the scientific community, however, paid any heed to the claims of these amateurs. The general presumption seems to have been that the photosphere's low temperature and the terrestrial atmosphere's opacity precluded radio detection of the sun.

The first of the three scientists to establish the existence of solar radio emission was James S. Hey, an erstwhile solid-state physicist who was serving in the British Army's Operational Research Group as a troubleshooter for the Anti-Aircraft Command's radar net (Hey 1973; Sullivan 1990). On two successive days in late February 1942, he received reports from all along the

4. Toward the end of the war, however, the American agency also obtained sunspot numbers from a network of amateurs organized by the American Association of Variable Star Observers (Heines 1944).

Figure 4.2. Walter Roberts about 1945 with the coronagraph at the Harvard station near Climax, Colorado; at an alitude of 3500 m, this was the highest permanent astronomical observatory in the world. (Courtesy of the National Center of Atmospheric Research/National Science Foundation.)

net that severe noise was blocking reception at every wavelength within the operating range of 3.5 to 5.5 m. At first he and others supposed the Germans had invented a powerful technique for jamming radar. Hey noticed, however, that the most intense noise was coming from the direction of the sun. Upon checking with the Royal Greenwich Observatory, he learned that the interference in the radar net coincided with the passage of an exceptionally large sunspot group across the solar meridian. He concluded that the radar must be picking up radio waves emitted by the region. Although Hey soon wrote up his discovery, he was obliged by security considerations to delay publication until after the war.

Some four months after Hey's discovery, Bell Laboratories' radio physicist

George C. Southworth detected a distinct type of solar radio emission (Sullivan 1990). Southworth headed a group engaged in hardware development and propagation studies for microwave radar. In June 1942, having earlier been intrigued by a colleague's suggestion that stars must emit radio waves, he had an assistant test the sensitivity of a new directional antenna by seeing whether its signal increased when it was pointed at the sun. The antenna, which was configured for waves of 3.2 cm, registered a small yet definite increase. Southworth calculated how much power a perfect radiator at the temperature of the sun would emit at this wavelength and, thanks to compensating errors, obtained a figure in rough accord with the observed result. From then until the fall of 1943, he and assistants investigated the phenomenon whenever time could be spared from the press of military work. They found the sun's thermal emission to be greater at 1.25 cm and less at 9.8 cm in about the ratios the model predicted. No matter how hard he tried, however, Southworth could not get an article about his discovery through wartime censorship until April 1945.

Unlike Hey and Southworth, the third discoverer of solar radio emission had no connection with radar. Indeed, Grote Reber (b. 1911) was a real rarity—an American radio engineer who somehow had not been swept up by the national war effort (Hey 1973; Reber 1983). Reber's interest in what would come to be called radio astronomy had been stimulated by Jansky's 1933 announcement that some radio noise was of extraterrestrial origin. Starting in the mid-1930s, he had followed up on Jansky's discovery by building a succession of radio telescopes behind the family home in Wheaton, Illinois. In 1943, despite wartime scarcities, he completed an instrument for 1.85-m waves that had sufficient sensitivity and directionality for mapping the distribution of radio sources within the galaxy (fig. 4.3).

Because the radio waves generated by spark plugs interfered with his reception, Reber (1944) did his observing late at night after most automobiles were off the streets. Once his project was well along, however, he looked for radio emission from the sun. His goal was to test the hypothesis—probably suggested by someone at nearby Yerkes Observatory who was familiar with Bengt Edlén's work on the solar corona—that sunlike stars in the Milky Way produced the galactic signal. Radio emission from the sun turned out to be well above his instrument's threshold of detection,[5] but it was not nearly high enough for all the ordinary stars together to account for the intensity of galactic radio emission. Reber sent a description of his new radio telescope and its various results to the *Astrophysical Journal* in May

5. He had found, in fact, the same phenomenon that Southworth had detected the year before.

Figure 4.3. Grote Reber's radio telescope in Wheaton, Illinois; mounted on an east-west axis, the 10-m sheet-metal dish could be directed to any desired declination between 90° N and 32.5° S. (Source: Reber 1944, 280. By permission of Grote Reber and *The Astrophysical Journal.*)

1944. The article's appearance half a year later gave him priority in announcing the radio sun.

Postwar Institution Building, 1945–1957

By August 1945, statesmen, officers, bureaucrats, and managers all around the world were keenly aware of the potential of science for contributing to national might, prosperity, and health (Kevles 1979; Forman 1987; De Maria, Grilli, and Sebastiani 1989). In the ensuing decade, confronted by the problems of economic recovery and superpower rivalry, they assigned science a higher priority than ever before in their budgeting. Besides the new funds, two other circumstances figured prominently in the postwar burgeoning of science. First, many scientists came away from the war with an opportunistic mind-set. In particular, they were eager to exploit wartime skills and contacts, willing to forsake prewar endeavors for more promising lines of inquiry, and readier than before to enter into team projects. And second, thanks to the war's stimulation of electronics, scientists were able to assemble or purchase better instruments for pursuing their research objectives (DeVorkin 1985). Like other scientists, solar physicists made the most of their new situation.

Immediately after the war, most solar physicists gave high priority to reviving the cooperation of the interwar decades (Stratton 1946; Struve 1946). Often assisted by radio-propagation agencies, they restored many monitoring programs that had been discontinued for want of observers and put new wartime programs on a regular footing. By 1947, solar observatories around the globe were once again sending periodic reports to Zurich. That July, Brunner's successor at the Swiss Federal Observatory, Max Waldmeier, resumed publication of the *Quarterly Bulletin on Solar Activity*. In the meantime, motivated not only by their concern with monitoring but also by their interest in investigating specific phenomena, solar physicists were improving their observational capabilities. At first they could do little more than refurbish or upgrade instruments that had survived the hostilities. After 1948, however, they found it easier than ever before to obtain funds for new telescopes and auxiliary apparatus. By 1953, for instance, fourteen of the world's fifty observatories and stations conducting visual studies of the sun were equipped with coronagraphs (Coutrez 1953).

Most of the solar physicists who led the postwar drive to improve solar observing capabilities were in their thirties. Partly as a result of their wartime experience and partly because of their need for facilities, these comparatively young men tended to be more alert than the field's elder statesmen to emerging opportunities for government patronage (DeVorkin 1990b). Moreover, they often had a greater awareness of new technical opportunities for refining auxiliary instrumentation. By 1957 several of these entrepreneurs—Leo Goldberg (1913–87), chair of the University of Michigan's Department of Astronomy; Walter O. Roberts (1915–90), director of the High Altitude Observatory in Boulder, Colorado; John W. Evans (b. 1909), superintendent of the Air Force's Sacramento Peak Observatory in New Mexico; Andrei Borisovich Severny (1913–86), director of the Soviet Academy's Crimean Astrophysical Observatory in Simeis; and Kiepenheuer,[6] director of the Fraunhofer Institute in Freiburg im Breisgau—had given their observatories important places in the gazetteer of solar physics.

Although each entrepreneur worked in a particular context and hence followed a unique path, the general phenomenon of postwar institution building is nicely illustrated by the interrelated histories of the High Altitude and Sacramento Peak observatories. The key figure in the initial stages of both these observatories was Harvard University's Donald H. Menzel (Kidwell 1990). Between the wars, it will be recalled, Menzel had buttressed Henry Norris Russell's case for hydrogen's abundance in the sun and established Harvard's new coronagraph station in Colorado (see chap. 3). During the war, it will also be recalled, Menzel's protégé Walter Roberts had provided the Interservice Radio Propagation Laboratory with regular reports from the Climax station on coronal conditions. This was not Menzel's only contact with America's wartime ionospheric program. He had also headed a naval communications group that used the ionospheric forecasts for scheduling the radio frequencies used in antisubmarine warfare (Shapley 1945; Goldberg 1977). With the war's end, Menzel (fig. 4.4) returned to Harvard enthusiastic about expanding the Climax station. His objectives were, it seems, both institutional and personal—to strengthen Harvard's solar observational capabilities and to position himself to succeed Harlow Shapley as director of the Harvard College Observatory.

Menzel, who was soon appointed chair of Harvard's Department of Astronomy and associate director for solar research at the Harvard College Observatory, made rapid headway during the first year of his campaign to expand the Climax station (Shapley 1946, 1947; Hallgren 1974; DeVorkin 1990c). He and Roberts arranged for Harvard and the University of Colorado to share responsibility for setting up a nonprofit corporation, with a headquarters office on the Boulder campus, to run the High Altitude Observatory at Climax. They persuaded the National Bureau of Standards' Central Radio

6. Kiepenheuer conducted himself with sufficient probity during the war that he suffered virtually no postwar difficulties.

Figure 4.4. Lieutenant Commander Donald Menzel in Canada for the total eclipse of July 9, 1945. (Source: Goldberg 1977, 249.)

Propagation Laboratory, the successor of Dellinger's wartime program, to cover the High Altitude Observatory's operating costs. They convinced John Evans, an exceptional instrument builder who had taken his doctorate at Harvard in 1938 and provided the Climax coronagraph with a monochromatic filter in 1941, to move out from Rochester to help Roberts develop new instruments. Finally, they prevailed upon the Climax Molybdenum Company to provide the station with a new site farther from the company's dusty open-pit mines.

During 1947, however, Menzel and Roberts ran up against a major obstacle (Shapley 1947, 1948; Hallgren 1974; Bushnell 1962; DeVorkin 1990c). The National Bureau of Standards and the Navy were ready to pay for operations and new instruments, respectively, but neither had the authority to fund a permanent observatory at the new site. In seeking to get around this problem, Menzel approached the physicist Marcus D. O'Day, who headed the Navigation Laboratory at the Air Force's Cambridge Field Station (Liebowitz 1985). O'Day saw no clear advantage to helping out the High Altitude Observatory, but he liked Menzel's suggestion that a new solar station near the White Sands Proving Ground in New Mexico would aid the Air Force's entry into rocket research on the sun. That summer Menzel, Roberts, and Evans followed up on O'Day's interest by joining him in an aerial reconnaissance of the region. They singled out Sacramento Peak, which was adjacent to the Air Force's Holloman Field and not far from White Sands, as the most promising site for the station. After field tests, Menzel and Roberts negotiated a contract with the Air Force for Harvard and the High Altitude Observatory to establish a solar station near the peak's summit.

Between 1948 and 1952, Menzel remained an important figure in the development of the "western solar stations," as he and his colleagues (e.g., Shapley 1950, 1) thought of the observatories at Climax and Sacramento Peak. But his role evolved into that of a high-level strategist, negotiator, and arbiter. In 1949, for instance, he played the key role in persuading the National Bureau of Standards to relocate the Central Radio Propagation Laboratory to Boulder (Roberts 1950; Goldberg 1977). By doing so, he did much to ensure the continuation of the bureau's support for the High Altitude Observatory's coronal monitoring program.

In the meantime, Roberts and Evans were busy at the western stations (Hallgren 1974; Bushnell 1962). They had major contracts from the Office of Naval Research and the Air Force. Their greatest challenge was developing two 40-cm coronagraphs, one each for the Climax and "Sac Peak" observatories. These instruments would be twice the size of Lyot's coronagraph at the Pic du Midi Observatory, which was still the largest in the world. In

1950, after intensive design work, they established shops in Boulder for producing the optical components and subcontracted with Westinghouse to construct the large spars that would carry the instruments. Two years later, the Sac Peak station was ready for its big coronagraph—the protective turret was up and the spar in place. With full-scale operation about to begin, the Air Force informed Menzel that Harvard University would have to provide tenured faculty positions for the senior scientists if it wished to manage Sac Peak (Evans 1986). Harvard balked. So the Air Force took over, appointing Evans (fig. 4.5)—upon Menzel's recommendation—as superintendent of its new Upper Air Research Observatory.

Assisted by the steady flow of Air Force funds, Evans developed Sac Peak into a premier solar observatory (Evans 1956; Bushnell 1962). He and his staff soon mounted the 40-cm coronagraph and a 38-cm tubeless chromosphere telescope. Of equal importance, they equipped the station's telescopes with a growing array of sensitive auxiliary instruments. Evans's performance earned him increasing independence. By 1956 he had sufficient leverage to convince the Air Force to redesignate the Upper Air Research Observatory as the Sacramento Peak Observatory. He conceded that "the primary purpose of the observatory is a practical one: to study solar phenomena so disturbances in the ionosphere can be predicted" (Evans 1956, 437). But he wanted no misunderstanding; studying the sun was Sac Peak's sole responsibility in this enterprise. As he put it, "The permanent community of Sunspot, N.M., has come into being over the past few years, with one purpose in life—solar research" (441).

Unlike Evans, Roberts in Boulder did not have an openhanded patron. His life was an unrelenting campaign for the wherewithal needed for expansion (Hallgren 1974; Roberts 1952–57). In 1954, realizing that the association with Harvard was no longer of much advantage in fund-raising, he convinced the University of Colorado to assume full responsibility for the observatory. Thereafter he did everything possible to promote Boulder as a center for solar research. For a brief while he had the inside track for obtaining the Smithsonian Astrophysical Observatory, which the Smithsonian's leadership had decided, after more than a half-century of solar-constant studies under the leadership of Langley, Abbot, and Loyal B. Aldrich, to move out of Washington, D.C. (Doel 1990). However, he lost out to Harvard University with its strong record in graduate training in astrophysics. Roberts was disappointed but not daunted. He launched an Institute on Solar-Terrestrial Relations, convinced the organizing committee for the International Geophysical Year to designate the High Altitude Observatory and Central Radio Propagation Laboratory as World Data Centers for Solar Activity, and persuaded the University of Colorado's regents to establish a

Figure 4.5. John Evans in 1952 supervising work on Sacramento Peak's coronagraph. (Courtesy of the National Solar Observatory/Sacramento Peak.)

graduate department of astro-geophysics. He built up a strong staff that, by 1957, included R. Grant Athay (b. 1923), Gordon A. Newkirk, Jr. (1928–85), Harold Zirin (b. 1929), and as a part-timer, the indefatigable Sydney Chapman, who had assumed the presidency of the International Geophysical Year. And he instituted regular reviews by important solar physicists— Michigan's Leo Goldberg in 1956 and Utrecht's Marcel Minnaert in 1957— as a means of obtaining not only their expert counsel but also their good opinion.

Thus, by embracing and transforming Menzel's dreams for solar stations in Colorado and New Mexico, Roberts and Evans established significant new centers for solar research in Boulder and atop Sacramento Peak. In part their successes were the consequences of their personal qualities—their determination, their ability to motivate colleagues, their insight as recruiters (especially Roberts) and instrumentalists (especially Evans), and their skill at inspiring the confidence of patrons. In part as well, their successes were the consequence of the significant postwar expansion of military patronage for solar observing.

Extending the Observational Range, 1945–1957

As solar physicists resumed monitoring of the sun and developed more sophisticated versions of prewar instruments, a medley of outsiders were extending the observational range. Some fulfilled Regener and Kiepenheuer's hope of using rockets as platforms for observing energetic solar radiations that could not penetrate the atmosphere. Others followed up on the wartime work of Hey, Southworth, and Reber by studying the sun's radio emissions. And a father-son pair developed an instrument capable of measuring solar magnetic fields much weaker than those Hale had found in sunspots.

Rockets as Solar Observing Platforms

Between 1945 and 1957, teams of physicists and engineers made the first observations of the sun's extreme-ultraviolet and X-ray emissions by sending instruments above the atmosphere with the aid of rockets. Their endeavors—as well as those of other rocket scientists who investigated the properties of the earth's upper atmosphere—did much to prepare the way for the rapid blossoming of space science after *Sputnik* (DeVorkin 1987, 1990c; Needell 1987; Hevly 1987). Three themes in this story must be emphasized at the outset. First, the very existence of a rocket research program in the

United States depended on the military's decision to acquire the capability to conduct missile warfare. In other words, the military gave scientists the opportunity to do basic research from rockets in order to have ready access to the techniques they developed and the results they obtained. Second, the chances of getting results were so slight that the only scientists to stay with rocket research for any length of time were physicists and engineers working in military laboratories or with long-range military contracts. And third, the technical difficulties were so severe that almost all those conducting research from rockets were obliged to devote far more attention to the development of their instrumentation and observing techniques than to the analysis of their findings.

As Germany collapsed during the spring of 1945, a special U.S. Army intelligence unit led by Colonel Holger N. Toftoy rounded up Wernher von Braun's team and all available V-2 components—enough for about a hundred rockets (Lasby 1971; DeVorkin 1990c). Toftoy was quick to see that these spoils might be used to help maintain postwar ties between the military and scientific communities. That fall, consequently, he charged Major James G. Bain with finding scientists who would be interested in producing payloads for the V-2s that were to be fired at the Army's new White Sands Proving Ground.

In some quarters interest was keen. For instance, Goldberg (1945) wrote Menzel in September that

> if anyone asked you what technological development could, at one stroke, make obsolete almost all of the textbooks written in astronomy, I am sure your answer and mine would be the same, namely, the spectroscopy of the sun outside of the earth's atmosphere. Now I am sure that this very unoriginal thought is shared by hundreds of other people, but it seems to me that we have reached the stage where at least serious developmental work along these lines is decidedly practical. [The] V-2 Rocket has attained a height of 60 miles, and with the host of control mechanisms that have come out of the war it should be possible to point the rocket at the sun within the required limits. . . . I would like nothing more than to be involved in such a project, even if it meant shaving my head and working in a cell for the next ten or fifteen years.

In reply Menzel assured Goldberg that, although the details were still obscure, such a project was under way.

By early 1946, Bain had located a sizable constituency for V-2 research in the universities and especially the military laboratories (DeVorkin 1990c; Hevly 1987). That January he briefed a meeting of more than three dozen scientists from a dozen institutions about plans for using the V-2s. The Army and its contractor, General Electric, would be firing the V-2s on a rapid schedule so as to maximize their value for the Army's own missile program. If those present could meet the schedule's deadlines, they would be welcome to send instruments up with the rockets. Six weeks later, representatives from eight institutions met in Princeton to organize themselves as a users' committee (Megerian 1946). As their chair they selected Ernest H. Krause, the physicist who headed the new Rocket Sonde Research Subdivision of the Naval Research Laboratory. They also recommended how payload berths on the first twenty-five rockets should be distributed among their institutions for experiments on such subjects as variation in atmospheric pressure, temperature, and composition with altitude, radio propagation, cosmic rays, and solar radiation. Thus began the V-2 Panel, an informal committee that would guide American sounding rocket research for more than a decade (DeVorkin 1987).

Almost all the scientists who ended up trying to make solar observations from V-2s during 1946 and 1947 worked for either the Naval Research Laboratory or the Johns Hopkins University's Applied Physics Laboratory, which was also funded by the Navy (Newell 1953). Their dependence on the Navy was not fortuitous (Hevly 1987; DeVorkin 1990c). The Navy regarded participation in the V-2 science program as essential for catching up with the Army in acquiring a missile capability. In addition, important scientific administrators at both laboratories—Edward O. Hulburt at the Naval Research Laboratory and Merle Tuve at the Applied Physics Laboratory—had long been interested in the ionosphere. Hulburt in particular had been arguing since the late 1920s that the ionosphere was produced by energetic radiation from the sun. The long-standing interest of these senior physicists in the upper atmosphere nurtured related interests in their younger colleagues.

Besides having generous and encouraging sponsors, most of the physicists and engineers who initiated the use of rockets as solar observing platforms had two specific qualifications for the task (DeVorkin 1990c; Hevly 1987). A prior acquaintance with solar physics was not one of them. Instead, their special qualifications were a skill in devising instruments capable of delivering results under extreme conditions and a capacity to continue working in the face of frequent disappointment. Virtuosity and determination were crucial for success. The pioneers of rocket solar physics managed to get instruments for photographing the sun's ultraviolet spectrum onto eleven of the twenty-eight V-2s launched during 1946–47 (Newell 1953). They were fortunate in that all of their V-2s were among the 70 percent attaining altitudes over 100 km. But frustrated by failures during flight and the destruction or loss of instruments upon their free-fall return, they obtained ultraviolet spectrograms on only five occasions.

The physicist Richard Tousey (b. 1908) was the pacesetter in rocket solar physics (DeVorkin 1986a, 1990c; Hevly 1987). He had earned a Harvard Ph.D. in 1933 with a dissertation written under Theodore Lyman's supervision on extreme ultraviolet radiation—to be precise, on the response of the mineral fluorite to Lyman-alpha emission from hydrogen at 1216 Å. In 1940, after teaching at Harvard University and Tufts College, he had landed his first solid job at the Naval Research Laboratory, which was building up its staff in anticipation of the United States' involvement in the war against Germany. During World War II, he had worked in Hulburt's division developing optical instruments and techniques. In late January 1946, two weeks after Bain described the opportunities for V-2 research at the laboratory, Krause met with Hulburt and Tousey to explore the possibility of developing an ultraviolet spectrograph for the rocket's payload. Hulburt liked the idea. He was eager to obtain data that would clarify the role of the sun's ultraviolet radiation in forming and maintaining the ionosphere. In discussion, Tousey convinced Krause and Hulburt that the desirability of reaching down to and beyond Lyman-alpha at 1216 Å warranted development of an instrument with a lithium-fluoride optical system.

Tousey's group came up with an ingenious design to compensate for the fact that the payload's orientation to the sun would not be constant (DeVorkin 1986a, 1990c; Newell 1953). Two tiny lithium-fluoride spheres would alternately direct the incident solar radiation to a reflection grating, where it would be dispersed into a spectrum that would be captured on film (fig. 4.6). The design, as Menzel soon pointed out, did have one serious flaw—the resultant spectra would have poor resolution. Faced by launch deadlines, however, Tousey and his colleagues thought it was better to secure ill-defined spectra than no spectra at all. Their pragmatism kept them in the game. Early in June 1946, they put their first spectrograph through its paces in the laboratory. It had no trouble detecting Lyman-alpha in the extreme ultraviolet. Some three weeks later, they flew the instrument on V-2 number 6 (figs. 4.7, 4.8). The rocket reached 108 km, but not a single identifiable piece of the spectrograph could be retrieved from the impact crater.

During the summer, Tousey's team produced a second spectrograph of similar design (DeVorkin 1986a, 1990c). Meanwhile, the launch personnel at White Sands found that the chances of payload recovery were much improved when the instruments were mounted in the tail section and the streamlined nose cone was blown off the rocket at the top of the trajectory. Tousey got his second spectrograph onto V-2 number 12, launched on October 10, 1946. It took six days to find the payload in the desert and two more to get the film cannister back to the Naval Research Laboratory. The results were exhilarating. The best spectrograms extended well below the atmospheric cutoff at 2950 Å, providing the first glimpse of the extreme ultraviolet radiation of the sun (fig. 4.9). Tousey soon reported his success to Krause, giving a boost to the Naval Research Laboratory's campaign to get the V-2 program extended beyond twenty-five launchings.

Although Tousey wanted to make the most of his data, he was too busy keeping up with the Army's launch schedule to acquire the requisite expertise in solar spectroscopy (DeVorkin 1990c). Hence he set the spectrograms aside. A month later, Menzel, Goldberg, and Lyman Spitzer of Yale University descended on Washington, both to take stock of the entire V-2 program and to evaluate Tousey's results. They found, as Menzel soon informed Roberts in Boulder, that Tousey's achievement had increased the Navy's interest in funding solar work. They also found that Tousey had strong proprietary feelings about his data. Despite their annoyance, Menzel and Goldberg were so eager to encourage Navy patronage of solar physics and to participate in the analysis of the V-2 spectra that they assiduously courted Tousey and his superiors. In February 1947 their campaign began to bear fruit—Goldberg got a contract from the Office of Naval Research to lay the theoretical groundwork for the analysis of solar ultraviolet spectra. Over the next year and a half, Goldberg and his associates positioned themselves to analyze any high-resolution spectrograms that Tousey might secure.

Tousey succeeded in obtaining ultraviolet spectrograms on V-2 flights in March and October 1947 and April 1948 (Newell 1953; Tousey 1953; Johnson et al. 1954). None, however, were of the quality desired. When it came time to renew his contract, Goldberg (1948) informed Spitzer that

> I have been somewhat disappointed in the development of the rocket project since its inception. It had been my hope and I think also yours that the rockets would open a new field of solar research which would justify the existence of an analysis group on a long range basis. At least thus far, however, the rockets have made an interesting contribution to the field of solar spectroscopy, [but] they have hardly opened up a new field of investigation. That, of course, may come with time but at the moment I should not want to ask for a renewal of the contract solely on the basis of the V-2 investigation.

The problem was that the early rocket spectra had such modest resolution— 1.5 Å at best—and such limited range—none penetrated below 2100 Å— that no really significant physical information about the solar atmosphere could be extracted from them.

Meanwhile, the American sounding rocket program had come under new leadership. In late 1947, Krause and several of his closest colleagues transferred into nuclear-weapons diagnostics (DeVorkin 1990c; Hevly

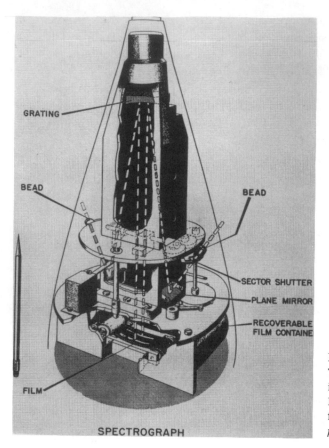

GRATING

BEAD

BEAD

SECTOR SHUTTER

PLANE MIRROR

RECOVERABLE
FILM CONTAINE

FILM

SPECTROGRAPH

Figure 4.6. Schematic of Richard Tousey's V-2 spectrograph; the scale is indicated by the pencil. (Source: Durand, Oberly, and Tousey 1949, fig. 2. By permission of *The Astrophysical Journal.*)

1987). Upon their departure, James A. Van Allen of the Applied Physics Laboratory assumed the chairmanship of the V-2 Panel and Homer E. Newell, Jr., became head of the Naval Research Laboratory's Rocket-Sonde Research Section. These able scientist-administrators presided over the American program of research from rockets for the next decade (DeVorkin 1987; Newell 1980; Van Allen 1983). During their tenure, German V-2s were succeeded by American Aerobees, Vikings, Rockoons, and Nikes (Newell 1953, 1959a). Also during their tenure, the preeminence of Tousey's group was challenged by two new groups.

Tousey's first serious competitor may well have been prompted to enter into rocket solar research by Tousey himself. In 1948, frustrated by repeated failures to get below 2100 Å with rocket spectrographs, Tousey and his associates had begun experimenting with an alternative means of detecting solar radiation at short wavelengths (Hevly 1987). Their plan was to make detectors from the thermoluminescent phosphors—substances that, if first

Figure 4.7. V-2 number 6 being readied for flight at White Sands Proving Ground in June 1946. (Source: William A. Baum photographic album, "White Sands, NM, 1946-1947," Space History Department, National Air and Space Museum.)

Figure 4.8. Launch of V-2 number 6 on June 28, 1946; the first V-2 to carry a spectrograph, it attained an altitude of 108 km, but the payload in the nose cone was destroyed upon the return to earth. (Source: William A. Baum photographic album, "White Sands, NM, 1946-1947," Space History Department, National Air and Space Museum.)

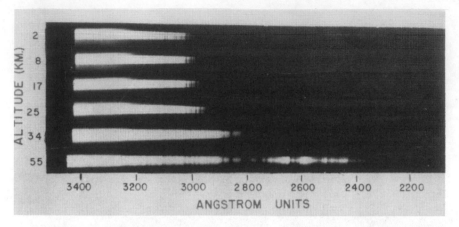

Figure 4.9. Solar ultraviolet spectrograms secured on the V-2 flight of October 10, 1946; note that the spectrum extended farther into the ultraviolet (to the right) as the rocket climbed out of the absorbing terrestrial atmosphere. (Source: Tousey 1986, 180. By permission of Pergamon Press.)

exposed to extreme ultraviolet or X rays and then heated, glowed with an intensity that indicated the level of exposure. In November 1948 and again in February 1949, Tousey's group sent phosphor-coated metal strips aloft on V-2s. The phosphor strips sitting behind filters that admitted only extreme ultraviolet radiation were exposed on both flights. But the phosphor strips behind the beryllium filters, which blocked all radiation except X rays with wavelengths less than 10 Å, were exposed only on the second flight. This flight, by good fortune, coincided with high solar activity and a radio fadeout. It seemed clear, as the Naval Research Laboratory announced in March 1949, that bursts of solar X rays caused fadeouts by disrupting the ionosphere. Here at last was both a confirmation and a refinement of the hypothesis—first advanced by Mögel and Dellinger in the 1930s (see chap. 3)—that energetic solar radiation caused shortwave fadeouts.

Herbert Friedman (b. 1916), one of Tousey's colleagues in Hulburt's division at the Naval Research Laboratory, was in an excellent position to follow up on Tousey's announcement (Hevly 1987). During and after the war, he had specialized in electronic radiation detectors, applying his instruments to an ever wider range of problems—for example, finding crystal cleavage planes, analyzing antifouling paints, measuring fuel levels in self-sealing tanks, and monitoring the radiative output of nuclear weapons. Upon learning of Tousey's achievement, Friedman immediately thought of the advantages that electronic methods of detection would have over phosphors. Electronic methods would be much more sensitive and rapid. Moreover, the resultant measurements could be relayed to the ground during flight via the rocket's telemetry system. Encouraged by Hulburt, Friedman and his group set out to determine the altitude dependence of ultraviolet and X-ray penetration into the terrestrial atmosphere. Their idea was to use photon counters in association with various filters to measure the changing intensity of solar radiation within four narrow wavelength bands. This would enable them to trace how intensity increased in each band as the rocket ascended. Friedman's team opened up this new approach to rocket spectroscopy with the flight of V-2 number 49 on September 19, 1949. Their counters confirmed that the sun's emission was intense in both X rays of wavelengths around 8 Å and ultraviolet in the Lyman-alpha region. The counters also measured altitude dependence, indicating that the solar X rays and Lyman-alpha were absorbed in the course of forming the ionosphere's E and D layers, respectively.

Friedman was slow to pursue these promising results (Hevly 1987). His expertise in radiation detection was much in demand after the first Soviet nuclear test and the outbreak of war in Korea. Moreover, as the V-2s were being phased out and American rockets phased in, newcomers to rocket science evidently had some difficulty securing payload berths. It was in 1952 that Friedman's group managed to resume work in rocket spectroscopy. Using improved counters, they refined their initial results on the sun's role in the formation of the ionosphere. They also provided the first measurements of Lyman-alpha's absolute intensity. Three years later, having increased the sensitivity of their counters, Friedman and his associates tried observing the nighttime sky (Friedman 1958, 1981; Hirsh 1983). They detected discrete ultraviolet sources and, in so doing, inaugurated ultraviolet astronomy.

In the meantime, a University of Colorado group had joined Tousey's and Friedman's groups as a pacemaker in the use of rockets for solar observing. This new group's rise was made possible by the Air Force physicist Marcus O'Day (O'Day 1954; DeVorkin 1990c). In 1947, shortly after getting work started on the solar station at Sacramento Peak, he had the idea that a good way for the Air Force to leapfrog to the forefront of rocket science would be to fly a coronagraph. Many obstacles stood in the way of this ambitious— one might say foolhardy—plan. The most obvious was finding a way to keep the coronagraph aimed at the sun while the rocket was in the upper part of its trajectory. O'Day apparently asked Menzel and Roberts to suggest persons capable of directing the coronagraph project. He ended up choosing William B. Pietenpol, the chair of the University of Colorado's Department of Physics.

The choice turned out, despite Pietenpol's somewhat lackluster career, to be an excellent one (DeVorkin 1990c). Pietenpol wisely dropped the idea of flying a coronagraph. This enabled the Colorado group to focus on developing a pointing-control system. Success was essential for transforming rockets into stable platforms for observing the sun. Pietenpol and his associates developed a biaxial pointing control, an electromechanical system that relied on photocells to maintain a constant orientation to the sun. The first trial—on Air Force Aerobee number 11 launched from Holloman Field below Sacramento Peak in April 1951—ended at 12 km, giving the pointing control no chance to prove itself. But the trial on December 12, 1952, with Air Force Aerobee number 33 was an unqualified success. Pietenpol's pointing control maintained the Aerobee's orientation vis-à-vis the sun throughout the upper part of flight (Tousey 1953; DeVorkin 1990c). In doing so, it kept the grazing-incidence spectrograph designed by his associate William A. Rense at the proper angle to the incident sunlight. The reward was in Rense's film cassette. It contained the first spectrogram that reached down to the solar Lyman-alpha line. Four years later, the Colorado group scored another triumph (Tousey 1958; Swings 1961). Thanks to Rense's instrumental ingenuity as well as the pointing control, they got the first Lyman-alpha photographs of the sun.

As Friedman's and Pietenpol's groups were establishing their presence, Tousey's group was steadfastly pursuing the goal of obtaining high-resolution spectrograms that extended deep into the extreme ultraviolet. They came closest to realizing their goal in early 1955 during the flight of Navy Aerobee number 29 (Tousey 1955, 1958). Using a Colorado pointing control, they secured a spectrogram that included some forty-five emission lines between 2006 and 977 Å. Tousey must have been pleased.

The record of solar observing from rockets down to 1957 was mixed. On one hand, the knowledge acquired about solar ultraviolet and X-ray emission with the aid of rockets turned out, contrary to the expectations of Goldberg and others, not to be of great import. It did, to be sure, reinforce belief in a hot corona and enhance the credibility and specificity of current theories about the sun's influence on the terrestrial ionosphere. But partly because solar theory was already fairly sophisticated and partly because the new data lacked the spectral, spatial, and temporal resolution of ground-based solar data, the use of rockets as observing platforms did not have a transforming effect on solar physics. On the other hand, we know in retrospect that Tousey, Friedman, Rense, their collaborators, and the many administrators, colleagues, and technicians who made their work possible were significant pioneers. They, more than any others, laid the technical basis for solar physics' rapid expansion in the space age.

Radio Telescopes as Solar Instruments

Those who extended solar observing to radio wavelengths proceeded quite independently of the rocket scientists. For one thing, solar radio waves were readily accessible, since they penetrated the earth's atmosphere.[7] Again, the antennas these waves were detected with came out of a technical tradition that had only the remotest connection with optics.

Starting in 1945, a growing number of physicists and engineers—most of them with backgrounds in wartime radio and radar projects—took up the study of extraterrestrial radio waves (Hey 1973; Edge and Mulkay 1976). They soon found that the sun was not the only discrete radio source in the heavens. Meteors entering the atmosphere and various unidentified celestial objects also emitted radio waves. Still, the comparative strength of solar radio emissions made the sun an attractive subject of investigation for most of the early groups using antennas as radio telescopes. Between 1945 and 1951, about 70 percent of all publications in the emerging speciality of radio astronomy were devoted to the sun (Sullivan 1990). Interest in nonsolar sources grew rapidly thereafter, driving down the fraction of solar papers. Yet through the mid-1950s, about ten groups around the globe were sending regular reports on solar radio emissions to Zurich for inclusion in the *Quarterly Bulletin on Solar Activity.* And when the International Geophysical Year commenced in July 1957, scientists at thirty-four stations were ready to monitor the sun with radio telescopes (Catalog of [solar] data 1963; Smerd 1969).

A group in Sydney, Australia, led the way in solar radio astronomy between 1945 and 1957 (Sullivan 1990; Wild 1987). That the Australians should achieve such prominence was surprising in light of the nation's modest standing in international science. However, British and American radio scientists were not as competitive as might have been predicted. The British who took up radio astronomy wanted to prove their field's value to the Royal Astronomical Society, whose leaders were much more interested in stellar and extragalactic problems than in solar physics. American scientists made a weak showing across the board in radio astronomy, evidently because the military—the chief patron of radio research in the United States—had little interest in sponsoring work at the longer wavelengths where celestial sources were best observed.

The scientists in Sydney who went into solar radio astronomy skillfully exploited the opportunity to lead in the field (Bowen 1984; Sullivan 1990;

7. Their ability to penetrate cloud cover meant that not even weather had to be taken into consideration in the choice of observing sites.

Wild 1987). The founding members of the group had spent the war at the Australian government's top-secret Radiophysics Laboratory at Chippendale near Sydney. On account of the laboratory's wartime contributions to radio and especially radar, the government was ready in 1945 to do all that was necessary to ensure its vitality in the postwar era. The consensus among the laboratory's administrators was that the scientific staff would maintain their edge if they turned to nonmilitary projects that excited their interest. One of the laboratory's radar groups—led by the physicist Joseph L. Pawsey (1908–62)—decided to see whether reports of solar radio emission could be confirmed with the war-surplus equipment at its disposal. In October 1945, Pawsey, Ruby V. Payne-Scott, and Lindsay L. McCready succeeded in their very first attempt. Using radar equipment at a hill overlooking the sea near Sydney, they detected 1.5-m radio emissions of varying intensity from the direction of the sun. The average intensity correlated nicely with total sunspot area (fig. 4.10) as reported by Clabon W. Allen, a solar physicist at Mount Stromlo Observatory near Canberra. Confident that their emissions had a solar origin, Pawsey, Payne-Scott, and McCready (1946a) rushed a letter off to *Nature*.

Pawsey's team, capitalizing on the Radiophysics Laboratory's policy of nurturing success, continued observing "solar noise." In 1946 they followed up their initial work in two significant ways (Sullivan 1990). They strengthened the connection between sunspots and solar radio emission by showing that a particular spot group was associated with enhanced emission. More important, they addressed the issue of the corona's temperature. Many months of observing the sun's emission at 1.5 m indicated that its intensity, though highly variable, almost never fell below a threshold of 0.5×10^{-15} watts/m^2. Pawsey asked David F. Martyn, an ionospheric physicist at Mount Stromlo, what to make of this threshold. Martyn, who had taken a temporary interest in the solar radio work, suggested that the antennae were picking up two signals—a steady component from the undisturbed solar corona that gave rise to the threshold intensity and a variable component that was somehow associated with sunspots. If so, the intensity of the steady component pointed to a coronal temperature of 600,000 to 1,200,000 K, in line with the coronal temperature implied by Edlén's prewar line identifications (see chap. 3). Pawsey (1946, 633) soon announced the "observation of million degree radiation from the sun."

Building on these early successes, Pawsey's expanding group maintained its lead in the exploration of the radio sun through the next decade (Pawsey 1953, 1961; Lovell 1964; Wild 1980, 1987; Christiansen 1984; Sullivan 1990). In 1947 three members of the group found that radio bursts from solar flares could be detected first at short wavelengths, then at longer ones.

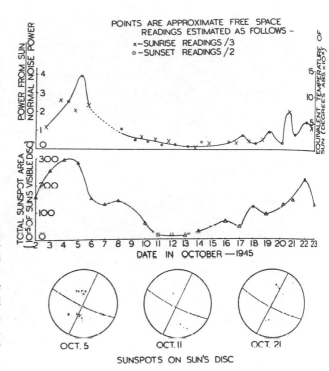

Figure 4.10. Graphs of the intensity of the solar radio signal at 1.5 m and the total area of visible sunspots in October 1945; the general similarity of the two curves clinched the case that solar activity engendered solar radio waves. (Source: Pawsey, Payne-Scott, and McCready 1946, 158. By permission of *Nature*, © 1946 Macmillan Magazines Ltd.)

Two years later, J. Paul Wild (b. 1923) and another group member substantiated this result with an instrument that could swiftly sweep from 2.8 to 4.3 m—the first radiospectrograph. In 1950 the theorist Stefan F. Smerd predicted that, in contrast to the solar limb darkening at visual wavelengths, there would be substantial limb brightening at radio wavelengths between 10 cm and 1 m. During the next seven years, W. N. "Chris" Christiansen led the group's development of antenna arrays for mapping the brightness distribution on the solar disk. As early as 1951, they were able to buttress Smerd's prediction with data on 21-cm emission obtained with a thirty-two antenna array. And in June 1957, with the completion of a crossed array at Fleurs (fig. 4.11)—the first radioheliograph—they began making 21-cm radio "pictures" of the lower corona (fig. 4.12).

Thus, like the rocket scientists, the radio scientists—to judge from this brief examination of Pawsey's group—displayed great technical virtuosity in inaugurating research on the radio sun. But in the final analysis their early work, like that of the rocket scientists, did not have a major impact on solar physics (Sullivan 1990). The radio astronomers, to be sure, did provide additional evidence of high coronal temperatures. They also added to the

Figure 4.11. One arm of the radioheliograph completed at Fleurs, Australia, in 1957; making radio pictures of the sun was soon a routine matter with this instrument (see fig. 4.12). (Source: Christiansen, Mathewson, and Pawsey 1957, 945. By permission of *Nature*, © 1957 Macmillan Magazines Ltd.)

store of knowledge about flares and many other solar phenomena. By and large, however, their contributions confirmed and augmented rather than transformed the solar physics community's view of the sun.

The Solar Magnetograph

While teams of rocket and radio scientists employed by institutions new to solar research took the lead in extending the spectral range, a father-son duo at Mount Wilson Observatory played the central role in increasing the magnetic range. In 1908, as we have seen (chap. 2), George Ellery Hale and his colleagues succeeded in detecting intense magnetic fields in sunspots. This success enabled them in the ensuing years to establish that the eleven-year sunspot cycle was only half of a more fundamental twenty-two-year magnetic cycle (chap. 3). However, their claims to have detected a general solar field that, like the terrestrial field, had a dipolar configuration were viewed with skepticism (Hetherington 1975a). Hale's estimate that the strength of

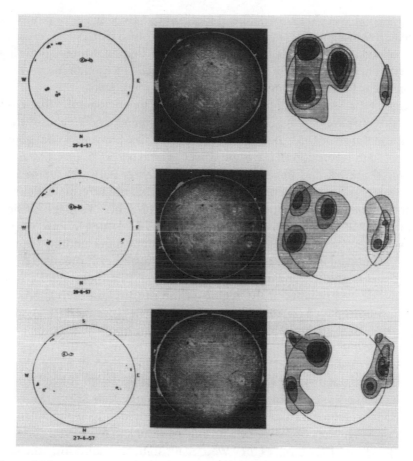

Figure 4.12. Comparison of sunspot diagrams, spectoheliograms, and radioheliograph pictures of the lower corona obtained in June 1957. (Source: Christiansen, Mathewson, and Pawsey 1957, 946. By permission of *Nature,* © 1957 Macmillan Magazines Ltd.)

the sun's polar field was about 50 gauss seemed too near the threshold of measuring sensitivity. An instrument with greater sensitivity was needed to resolve the issue.

In 1951 the retired solar physicist Harold D. Babcock (1882–1968) asked his son, the stellar astrophysicist Horace W. Babcock (b. 1912), to help him devise equipment with the requisite sensitivity (Babcock 1977). The elder Babcock—like Hale in his later years—thought that the conflicting values reported for the general field's intensity were signs of the field's variability (Hetherington 1975a). Some Mount Wilson colleagues, however, had joined the ranks of the skeptics. Babcock turned to his son for two reasons. The younger Babcock combined expertise in making spectroscopic gratings with a good working knowledge of electronics.[8] He was also a pioneer in research on magnetic stars. Although deeply involved in his stellar research, he decided to assist his father. He wanted to be helpful. Moreover, despite the comparative weakness of the sun's general field, he thought he might get some ideas for interpreting the behavior of magnetic stars.

Horace Babcock designed and built a "solar magnetograph" that could map the sun's weak magnetic fields—those between 1 and 20 gauss[9] (Babcock and Babcock 1952; Babcock 1953). His starting point was a flat grating of very high resolution that was produced on the observatory's new ruling engine (Babcock and Babcock 1951; Bowen 1952; Babcock 1986). Even with this grating, however, the Zeeman effect could not be visually detected in those solar lines that, if they originated in the presence of a magnetic field, should be split. Babcock reasoned that the sun's field outside sunspots was so weak that the split components were overlapping and blending. He drew upon his optical and electronics know-how to distinguish the overlapping components and thereby obtain an output signal that indicated the field's magnitude and polarity at the region of the sun under observation. This signal was displayed on an oscilloscope that, when monitored for about an hour by a time-exposure camera, built up a picture of the solar magnetic field as the telescope scanned the sun along some twenty parallel tracks. The resulting "magnetograms" provided a permanent record of the field's configuration (fig. 4.13).

The Babcocks set up the first magnetograph at the Hale Solar Laboratory in Pasadena, which Harold Babcock, who had retired in 1948, was using for his observing. They began getting promising results in May 1952 (Babcock 1977). Within three months they were ready to begin monitoring the sun's field. In their first report Babcock and Babcock (1952) stressed the complexity of the field patterns. They were impressed that an active region could be detected on the daily magnetograms both well before the appearance of an associated sunspot group and long after the group's disappearance. They also addressed the issue of the sun's general field. At the higher solar latitudes, the field intensity along the line of sight was between 1 and 5 gauss and the predominant polarities were opposite at the sun's north and south poles. This evidence suggested "the existence of a weak and distorted dipolar field" during August and September 1952 (286). The Babcocks predicted that it would take years of magnetic mapping to answer all the questions raised by their preliminary observations.

8. He began acquiring his knowledge of electronics while with the American radar project during the first two years of the war (Babcock 1977).

9. For reference, the earth has a polar field of about 0.5 gauss; sunspots, 1000–2000 gauss; and magnetic stars, 1000–35,000 gauss.

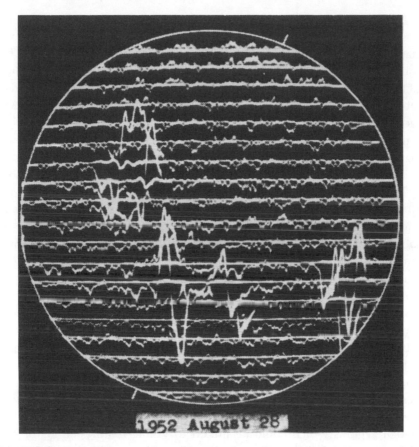

Figure 4.13. One of Horace and Harold Babcock's earliest magnetograms; upward deflections from the fiduciary lines indicate where the polarity was predominantly outward, and downward deflections, vice versa. (Source: Babcock and Babcock 1952, facing p. 286.)

After this initial report, Harold Babcock pressed on with the monitoring program while Horace worked at enhancing the magnetograph's sensitivity. In late 1954 they sent off the results of their first two years of mapping (Babcock and Babcock 1955a, 1955b). They reported that the sun's general field had a dipolar configuration with magnetic poles at the north and south rotational poles. The field's mean intensity at the poles was about 1 gauss, and its polarity was opposite the earth's polarity. These results contradicted Hale's earlier claims about the sun's magnetic characteristics at a similar stage in the twenty-two-year cycle. Although they passed silently over this contradiction, the Babcocks implied throughout that their observations were the first reliable measurements of the sun's general field.

Despite this attention to the general field, the Babcocks' primary interest was in the complex, ever-changing field pattern that the magnetograph was mapping at lower solar latitudes. Their magnetograms provided the first "objective evidence [for the] hypothesis that magnetic fields are fundamental to sunspots, *plages*,[10] prominences, chromospheric fine structure, bright coronal emissions, and, probably, regions of strong radio emission" (Babcock and Babcock 1955a, 349). In supporting this claim, they made detailed comparisons of their magnetic maps with concurrent Mount Wilson photographs of the photosphere and chromosphere, High Altitude Observatory measurements of coronal intensity, and Australian data on radio emissions.

The Babcocks' discussion of magnetic phenomena between 50° N and 50° S solar latitude gave special attention to bipolar and unipolar regions. The bipolar regions were areas where the magnetic flux leaving the sun about equaled that entering it. When such regions were young, they sometimes produced sunspots. As they aged over periods lasting as long as nine months, they expanded and absorbed any spots they might have produced. Not surprisingly, given their relation to spots, these regions obeyed Hale's familiar laws of sunspot polarity (see chap. 3). All this led the Babcocks to conceive of bipolar regions as "areas of opposite polarity [where] loops of a submerged toroidal field [have been] brought to the surface by rising material" (Babcock and Babcock 1955a, 349). Unipolar regions were areas of a single polarity where magnetic flux was either leaving or entering the sun. Despite being similar in size and longevity to bipolar regions, they were comparatively rare. Moreover, their fields were rather weak. Finally, unlike bipolar regions which could be linked to visible manifestations of solar activity, unipolar regions appeared to "constitute a kind of solar phenomenon not directly observable by other means." The word "directly" was chosen advisedly. The Babcocks believed these "newly observed" regions were "the sources of neutral but ionized corpuscular streams [that caused] terrestrial magnetic storms of the recurrent type" (1955a, 359).

The Babcocks were referring here to a recalcitrant puzzle in solar-terrestrial physics—the source of terrestrial magnetic storms having a twenty-seven-day recurrence tendency. These disturbances were thought—because their recurrence period matched the solar rotation rate—to originate at specific "M" (magnetic) regions on the sun (see chap. 6). The Babcocks' proposal was that unipolar regions could "be identified with the hypothetical 'M' regions" (1955a, 359). They supported this idea with observational and theoretical arguments. On the observational side, they reported that for eight months in 1953 a prominent unipolar region had regularly

10. Plages are bright areas in the chromosphere.

crossed the sun's central meridian about three days before the onset of a recurring magnetic storm. Since then, an absence of prominent unipolar regions had been matched by an absence of significant magnetic-storm sequences. On the theoretical side, they suggested that a hydromagnetic process propelled ions and electrons away from the sun. Although the ambient field lines above bipolar regions would cause these particles to collide "generating radio noise [and] forming visible prominences," those above unipolar regions would guide corpuscular streams "more or less radially from the sun, without pronounced collisions." If such streams should meet the earth, they would "cause M-region phenomena" (1955b, 296).

Turning to the sun's poles, the Babcocks speculated that these regions were unipolar regions "of a special class." If so, their extent, persistence, and average field strength would make the polar regions "even more copious sources" of corpuscular streams than the low-latitude unipolar regions. Knowing the field lines from the solar poles must connect far out in the solar system, the Babcocks suggested that these lines of force would guide the "diffuse corpuscular streams ejected from the polar regions to the general region of the equatorial plane" (1955a, 366). These speculations about the role of the sun's polar regions in solar corpuscular emission did not take hold at the time, probably because no avenues were available for giving them an empirical underpinning. Nonetheless, as we will see below (chap. 6), they anticiated later models of the solar wind's topology.

In 1955 the Babcocks' collaboration came to an end (Babcock 1977). Harold Babcock (1959) continued observing with the solar magnetograph in Pasadena, discovering the reversal of the sun's general field during the solar maximum of 1957–58. Horace, who wanted to finish his catalog of magnetic stars, limited his involvement with the magnetograph to supervising the installation of a second, much improved instrument at the large tower telescope atop Mount Wilson (Bowen 1956). Thereafter he concentrated on stellar astrophysics and observatory administration. Except for a significant interpretive piece on the evolution of the sun's magnetic field (Babcock 1961), his stint as a solar physicist was done.

By extending the observational range to weak magnetic fields, the Babcocks' magnetograph opened the way for detailed studies of the relation between magnetic field patterns and other solar phenomena. This was an important achievement. It would be fair to say, indeed, that since the mid-1950s magnetographic observations have contributed as much to solar physics as have radio, ultraviolet, and X-ray observations. The Babcocks, as we have seen, seized this opportunity. Since their day, numerous scientists have followed suit (e.g., see chap. 6).

Planning for Solar Observations during the International Geophysical Year, 1950–1957

In April 1950, during a gathering at James A. Van Allen's home near Washington, D.C., the ionospheric physicist Lloyd V. Berkner suggested a Third International Polar Year to coincide with the seventy-fifth anniversary of the First International Polar Year and, more important, the solar maximum expected in 1957–58 (Spencer-Jones 1959; Newell 1980; Needell 1989). The proposal appealed to everyone present, especially Sydney Chapman, who had served as the founding chair of the interwar solar-terrestrial commission (see chap. 3). Such an undertaking would not only enlarge knowledge of polar phenomena but also foster tension-reducing cooperation among scientists on both sides of the Iron Curtain. During the next two and a half years, Berkner's idea inspired much discussion and many resolutions. The outcome was that in October 1952 the International Council of Scientific Unions[11] established the Comité Spécial de l'Année Géophysique Internationale. In mid-1953, meeting for the first time in Brussels, the Comité elected Chapman president, Berkner vice president, and the Belgian solar-terrestrial physicist Marcel Nicolet general secretary (Nicolet 1958). Thereafter, these three men (fig. 4.14) presided over an increasingly complex planning process that included national, disciplinary, and data-handling committees. By July 1957, a network of over 2,000 stations staffed by some 20,000 scientists stood ready to begin synoptic observations of dozens of solar and terrestrial phenomena (Odishaw 1958).

Most of the Comité's planning for solar monitoring was handled by two subcommittees (Nicolet 1958). The working group for solar activity, which was established at the 1953 meeting, coordinated the preparations for ground-based observations of the sun during the International Geophysical Year. This working group's initial focus was on promoting the use of the latest instruments. It urged optical astronomers to utilize monochromatic filters with automatic cameras for observing chromospheric flares, polarizing spectrophotometers for recording coronal intensities, and Babcock magnetographs for mapping solar fields. It also encouraged radio astronomers to employ radiospectrographs and what would later be called radioheliographs for monitoring the sun's radio emissions. At the 1955 meeting of the Comité in Brussels, however, the working group's attention shifted from the issues of instrumentation toward the problems of coordinating observations, reporting data, and disseminating results. It proposed that specific

11. The International Council of Scientific Unions replaced the anti-German International Research Council in 1931 (Schroeder-Gudehus 1978).

Figure 4.14. Sidney Chapman (far right), president of the Comité Spécial de l'Année Géophysique Internationale, meeting with CSAGI directorate —including Marcel Nicolet (at the table's head) and Lloyd Berkner (to Nicolet's right)—just before the mid-1957 opening of the International Geophysical Year. (Source: Akasofu, Fogle, and Haurwitz 1968, 152.)

centers be put in charge of data management on particular subjects. The Comité soon generalized this proposal, directing all participating disciplines to name three world data centers—one in the United States, one in the Soviet Union, and one in a third country—for each specific subject of investigation and to give one such center the status of chief world center with responsibility for publishing the summary findings. In the case of solar physics, complex negotiations involving national committees, an advisory group from the International Astronomical Union, and the Comité's working group on solar activity led to the naming of data centers for ten subjects (see table 4.1). The prominence of the High Altitude Observatory and Central Radio Propagation Laboratory on the list of data centers attested to Walter Roberts's success in promoting Boulder as a center for solar observing.

The Comité's planning for solar observations from above the atmosphere was the province of the working group on rockets (Nicolet 1958). When this group first convened at the 1954 meeting of the Comité in Rome, representatives from the United States and France announced plans for using rockets as solar observing platforms during the International Geophysical Year. Their announcements were upstaged, however, by discussions at the same

Table 4.1
World Data Centers
for IGY Solar Phenomena, July 1957

Solar Phenomenon	Center A	Center B	Center C
Wolf sunspot numbers	CRPL Boulder	Moscow	Zurich[a]
Special sunspot numbers	CRPL Boulder[a]	Moscow	Zurich
Solar plage activity	HAO Boulder	Moscow	Arcetri[a]
Solar magnetic fields	Mount Wilson	Moscow[b]	Cambridge
Solar flare activity	HAO Boulder	Moscow	Meudon[a]
Flare and plage intensities	HAO Boulder	Crimea[a]	Meudon
Surges and prominence activity	HAO Boulder[a]	Moscow	Meudon
Chromospheric structures	HAO Boulder	Moscow	Freiburg[a]
Coronal intensities	HAO Boulder[a]	Moscow	Pic du Midi
Radio emissions	CRPL Boulder	Moscow	Sydney[a]

Source: Öhman 1958, 253.

[a]Chief world data center.

[b]Moscow was replaced by Crimea, which eventually became the world data center for solar magnetism.

meeting within the main Comité. In response to a campaign launched by S. Fred Singer (Newell 1980), an American who had been using rockets as platforms for investigating the earth's magnetic field, the Comité resolved that "in view of the great importance of observations during extended periods of time of extraterrestrial radiations [solar emissions] and geophysical phenomena in the upper atmosphere, and in view of the advanced state of present rocket techniques, the Comité recommends that thought be given to the launching of small satellite vehicles, to their scientific instrumentation, and to the new problems associated with satellite experiments, such as power supply, telemetering, and orientation of the vehicle" (Nicolet 1958, 171). With this resolution in mind, the Comité's directorate soon renamed the working group on rockets as the working group on rockets and satellites.

At the working group's meeting with the Comité in Brussels in September 1955 (Nicolet 1958), Homer Newell described American plans[12] for research with satellites during the International Geophysical Year. He emphasized at the outset that the United States had decided to proceed only because it seemed that satellites would "lead to worthwhile scientific results and not reduce to a mere stunt" (Newell 1958, 267). The satellite program, he insisted, was intended to complement, not replace, the rocket program.

12. It was in July 1955 that the White House announced that the United States would be launching satellites during the International Geophysical Year.

Figure 4.15. The International Geophysical Year's logotype; adopted before IGY's beginning in July 1957, it highlighted the role that American and Soviet satellites were expected to play in the research. (Source: Roberts 1957, frontispiece.)

Still, presuming that the United States would launch six to ten satellites weighing about 50 kg each, Newell speculated on their possible uses for investigating, in ascending order of difficulty, seven subjects. Solar observing was third on the list, right after the two subjects that even a passive satellite could clarify—the earth's shape and the outer atmosphere's density. In giving solar studies this priority, Newell tacitly recognized Friedman's successes in retrieving solar data from rockets by telemetry. "It will be possible," he remarked, "to look for fluctuations in the solar intensities as a function of time [and] when a solar flare occurs" (270). Once he had run through the remainder of the list, Newell revealed that he was not really indifferent to stunts by asserting that "above and beyond the scientific work that may be done for the IGY, the launching of the first satellite will be intrinsically important in and of itself, for it will signal the opening of a new frontier for exploration" (271). Sharing this enthusiasm, the Comité's directorate later chose a satellite circling the globe as the logo for the International Geophysical Year (fig. 4.15).

At the Comité's meeting in Barcelona a year later, a special symposium featured plans for observations from rockets and satellites during 1957–58 (Nicolet 1958). Berkner gave a detailed account of the American program. English, French, and Japanese representatives also described preparations to launch several rockets. And surprising nearly everyone, the Soviet representative announced that the U.S.S.R. would be launching not only rockets but also satellites to observe "cosmic rays, micro-meteorites, the geomagnetic field and solar radiation" (310–11). At the final plenary session, Chapman characterized the Soviet announcement as the meeting's chief highlight. He also expressed pleasure that the American and Soviet delegations had reached agreements that would permit the standardization of equipment for tracking satellites. Relieved by the apparent easing of Cold War tensions, he remarked that "this first announcement of a new satellite program and of this international co-operation is of the highest importance for the IGY program" (343).

The Comité's plans for monitoring the sun during the International Geophysical Year provoked mixed reactions from solar observers. Some of those who already had good facilities or who had little interest in time-dependent phenomena regarded the ambitious plans as a bothersome chore (Goldberg 1986a). Most solar observers, however, saw the International Geophysical Year as a marvelous opportunity both to improve their observing capabilities and to increase their knowledge of solar activity. When work commenced on July 1, 1957, ninety-five observatories and stations around the globe were set to participate in the program of monitoring the sun. Sixty-nine were equipped to observe only one or two of the ten solar phenomena designated for coverage. But twenty-six—including the High Altitude Observatory in Boulder, its Climax station, and the Sacramento Peak Observatory—had the instruments and staff for monitoring three or more such phenomena (Catalog of [solar] data 1963). Besides the optical and radio observers, rocket scientists interested in the sun were also ready to undertake work (Berkner 1958). Friedman's group, in fact, marked the start of the International Geophysical Year by firing the first of fifteen Nike-Deacons dedicated to the study of solar flares.

Hence, motivated by desires to enlarge knowledge of the earth and to promote scientific internationalism, an immense planning effort culminated in the International Geophysical Year, which ran from mid-1957 through 1958. The planners assumed from the outset that a comprehensive program of solar monitoring should be included in the overall program of observations. Research during the preceding century had established that the sun influenced a wide range of terrestrial phenomena. Planning for solar monitoring during the International Geophysical Year benefited solar research in several ways. It provided a fresh rationale for patronage and recruitment. It promoted the diffusion of new instruments and techniques. It bolstered cooperation among solar physicists. And it nurtured interdisciplinary interactions with geophysicists.

Conclusion

During World War II most solar research came to a halt. Even as the war disrupted observational and interpretive programs, however, it was setting the stage for postwar expansion. At a general level, governments around the world came to put a much higher value on scientists and their work. Scientists, conversely, became accustomed to increased government patronage and the larger undertakings it made possible. More specifically, three developments promised to redound to the advantage of solar research. Both German and Allied scientists established an important role for solar monitoring in programs for forecasting the ionosphere's shortwave transmission characteristics. In addition, German production of the V-2 rocket provided a means of sending instruments above the atmosphere to observe solar extreme ultraviolet and X-ray radiations. Meanwhile, British and American radio scientists prepared the way for studies of solar radio emissions.

In the decade following World War II, both solar physicists and outside specialists drew upon the war's legacy of utilitarian appreciation for science to improve solar observational capabilities. Solar physicists concentrated on optical telescopes and instrumentation, revitalizing many old observatories and starting several new ones. In the meantime, outsiders developed instruments and techniques for observing hitherto inaccessible solar phenomena. Physicists and engineers used sounding rockets and radio telescopes for exploring the sun at new wavelengths, and a stellar astrophysicist devised the magnetograph for studying solar magnetism outside sunspots. The postwar growth of solar physics culminated in the 1950s with hectic preparations for the International Geophysical Year. By mid-1957, solar physicists were obtaining data not only from a global network of nearly a hundred optical and radio observatories, but also from frequent sounding rocket flights.

As the International Geophysical Year commenced, most solar physicists had high hopes. They presumed that the acquisition of more comprehensive and precise data would advance knowledge of solar activity and its influence on terrestrial phenomena. Although they were too optimistic, the solar observations during the International Geophysical Year did prove of value. In particular, Friedman's study of flares by means of rockets conclusively demonstrated that solar X-ray bursts caused shortwave fadeouts (Hevly 1987). As it happened, however, the most important contribution of the International Geophysical Year to solar physics was an indirect one— *Sputnik.*

CHAPTER FIVE

The Solar Physics Community since Sputnik

1957–1990

Launched on October 4, 1957, *Sputnik* was billed by the Soviet Union as a contribution to the International Geophysical Year. It was that and much more (McDougall 1985). The satellite was a tribute to modern dreams of space exploration. It was a hitchhiker on the Soviet Union's new intercontinental ballistic missile. It was an avowal of that nation's desire to be second to none. Together with subsequent Soviet space firsts, *Sputnik* triggered a rapid buildup of American missile forces, motivated the creation of the National Aeronautics and Space Administration, and stimulated a major expansion in funding for scientific and engineering research and education (Killian 1977; Clowse 1981). Nothing less, most Americans believed, would maintain the nation's security and prestige (fig. 5.1).

By catalyzing so many developments, *Sputnik* hastened the growth of solar observational capabilities. The scientists who took advantage of the fresh opportunities for solar research pursued three main strategies. Some placed instruments for studying the sun on stratospheric balloons, high-altitude aircraft, rockets, and spacecraft. Others obtained new telescopes or substantially improved existing ones at ground-based observatories. And a very few sought to establish solar-neutrino detectors in subterranean facilities. As solar observing evolved into a sizable enterprise between 1957 and 1975, the solar physics community grew apace. Before long, solar physicists wanted special organizations and forums to represent their interests in the larger world and to expedite their communications with one another. Once in place, however, these subdisciplinary institutions fostered a certain parochialism.

As national priorities shifted away from science in the 1970s, solar physics moved from an era of exuberant growth into one of frustrating limits. The solar physics community's adjustment to this new context was not without

Figure 5.1. This cartoon (from the London *Daily Mail*), which appeared in *Newsweek* along with the accompanying editorial shortly after the launch of *Sputnik 1*, gives some indication of the Anglo-American response to the Soviet success in opening the space age. (Source: Bringing the space-spin down to earth 1957, 35. By permission of the London *Daily Mail*.)

pain. Indeed, lean budgets forced some solar physicists out of the field. Nonetheless, the community seems to have emerged from this difficult transition stronger than ever. Solar physicists have become more alert to their field's symbiotic relationship with the rest of astronomy. They have also managed, on balance, to continue improving their observational capabilities.

This chapter focuses on the relation between patronage for solar observing and the development of the solar physics community since *Sputnik*. It first considers the rise between 1957 and 1975 of what may be called the international space-science movement. Then it focuses in on solar physics, describing the rapid growth of the field's observational capabilities and the consequent emergence of subdisciplinary institutions. Next it explores the solar physics community's successful adaptation to the lean times that began in the 1970s. Finally, the chapter illustrates how all this affected solar physicists by tracing the careers of two of the community's leaders.

The Rise of Space Science, 1957–1975

After *Sputnik*, politicians and military men set the basic agenda for competition in space. This is not to say they proceeded without the benefit of scientific counsel. In fact, many scientific instruments were placed in orbit during 1957 and 1958 by the Soviet Union on *Sputnik 1, 2*, and *3* (Wukelic 1968) and by the United States on *Explorer 1, 3*, and *4, Vanguard 1*, and *Pioneer 1* (Corliss 1967).

Although scientists played some part in the early Soviet space program, they seem to have had a greater influence on the American program. It might have been otherwise. Eisenhower and many of his closest advisers thought space should remain the province of the Department of Defense. Significant opposition to this viewpoint emerged, however, from the President's Science Advisory Committee, which was formed in November 1957 after the Soviet launch of *Sputnik 2* (Killian 1977). The committee, which was chaired by James R. Killian, Jr., wanted a space agency that, like the Atomic Energy Commission, would have considerable autonomy from the military. Eisenhower resisted this idea on military and fiscal grounds. For instance, according to an official minute-taker, in early 1958 the President told Killian and others that "space objectives relating to Defense are those to which the highest priority attaches because they bear on our immediate safety. . . . He [Eisenhower] did not think that large operating activities should be put in another organization, because of duplication, and did not feel that we should put talent etc. into crash programs outside the Defense establishment" (Stares 1985, 42). But a month later, perhaps wanting to buttress the case for the freedom of space, Eisenhower endorsed a civilian agency. The upshot of intense congressional scrutiny and revision was a bill reconstituting the National Advisory Committee for Aeronautics as the National Aeronautics and Space Administration, or simply NASA. On October 1, 1958, just before the first anniversary of *Sputnik 1*, the new agency began operations (Newell 1980; McDougall 1985).

As Congress was shaping NASA, the National Academy of Sciences was doing everything possible to ensure that scientific participation in setting priorities for space continued after the International Geophysical Year (Berkner and Odishaw 1961; Hetherington 1975b). Throughout the winter and spring of 1958, not only the academy's committees for the International Geophysical Year but also the National Advisory Committee for Aeronautics and the National Science Foundation had been expressing concern about this issue. In June the academy's president, Detlev W. Bronk, responded by establishing the Space Science Board. He appointed Lloyd V. Berkner, the first proponent of the International Geophysical Year (see chap. 4), as chair and asked the board "to survey in concert the problems, the opportunities and the implications of man's advance into space, and to find ways to further a wise and vigorous national scientific program in the field" (Berkner and Odishaw 1961, 430). Over the summer, the board canvassed scientists across the nation for research proposals; and during NASA's start-up in the fall and winter of 1958–59, the board actively represented the scientific community's interests in discussions of the agency's priorities.[1]

In the meantime, Berkner and others had been campaigning for the creation of an international equivalent to the Space Science Board (Porter 1971; Massey and Robins 1986). The International Council of Scientific Unions responded by establishing a provisional Committee on Space Research in order "to provide the world scientific community with the means whereby it may exploit the possibilities of satellites and space probes of all kinds for scientific purposes and exchange the resulting data on a cooperative basis" (Porter 1971, 533). The council's executive committee asked Homer E. Newell, Jr., who was in the midst of transferring from the Naval Research Laboratory to NASA, to convene the new committee's organizing meeting in London. All went smoothly at the London meeting in November 1958. Those present agreed to seek permanent status, drafted a charter, and elected the Dutch theoretical astrophysicist Hendrik C. van de Hulst president. The Committee on Space Research's second meeting—at The Hague in

1. For a description of the Space Science Board's attempt to shape NASA's research agenda in the case of the sun's corpuscular emissions, see chapter 6.

March 1959—was less harmonious. The Soviet delegate started off by complaining that all the other delegates were from the advanced capitalist nations. Heated debate led to a decision to redraft the charter in order to ensure Soviet participation. Later that year, the revised charter opened membership to all interested nations and reserved certain positions on the executive committee for representatives from countries in the Eastern bloc.

Besides forcing a revision in the charter, the Soviet complaint prompted the United States to reveal its plans for cooperation in the exploration of space (Frutkin 1965; Newell 1980). Richard W. Porter, the leader of the American delegation, was piqued by Soviet insinuations that the United States wanted to exclude the smaller nations from space. He knew that NASA and the National Academy had been working out the details for an international program. Urgent telephone conferences with Washington officials enabled him to announce the next day that NASA stood ready to help other nations put instruments, or even complete scientific satellites, into orbit. His announcement was welcome news indeed. European scientists no longer needed to worry about being excluded from space research until their own countries had developed launch capabilities.

During the next few years, motivated both by a desire to take advantage of NASA's cooperative program and by concern about being left behind, scientists around the globe won support for their own national space science programs. By 1965 the number of Western nations with programs to use spacecraft as platforms for scientific investigations had climbed to eleven (Frutkin 1965). Thereafter, thanks in good measure to the Soviet Union's decision to establish a cooperative program of its own, space research spread into several Eastern-bloc countries and one developing nation—India (U.S. Congress 1977, 1982). By 1975 scientists in more than twenty nations had joined those in the United States and Soviet Union in studying natural phenomena with the aid of spacecraft.

As space research was spreading around the globe, scientists in the United States and the Soviet Union were using the ample resources made available to them to increase the sophistication of the American and Soviet programs. With the aid of rocket manufacturers and engineers, they were sending larger spacecraft on longer journeys. With the aid of radio and computer specialists, they were communicating with their spacecraft more rapidly and reliably. With the aid of specialized shops and firms, they were including more capable and diverse instruments in their scientific payloads. This was not all. As the programs matured, American and Soviet space scientists came to see that continuing success depended not only on the progress of space technology but also on the progress of ground-based observing programs and theoretical research (Tatarewicz 1990). To make the most of the resources at their disposal, they both promoted greater support for scientists using traditional approaches and cultivated closer ties with them. By the 1970s, consequently, the space science programs of the United States and, to a lesser degree, the Soviet Union had evolved into proficient enterprises that were tackling ever more varied, difficult, and interesting problems.

Bountiful Times, 1957–1975

For some fifteen years after *Sputnik,* scientists seeking to expand solar observational capabilities fared well in their pursuit of funds and resources. They got increasingly capable instruments into space. Less dramatic but every bit as important, they constructed a new generation of ground-based telescopes and steadily enhanced the efficacy of their auxiliary instrumentation. A very few scientists even developed subterranean facilities in an attempt to observe the energy-generating processes in the sun's interior.

Spacecraft as Solar Observing Platforms

The case for using spacecraft as platforms for observing the radiations and particles emitted by the sun was strong from the outset of the space age. The glimpses of the sun obtained with rocket-borne instruments before the International Geophysical Year had hinted at the potential of prolonged observations from above the atmosphere (see chap. 4). In addition, the sun's relative brightness and size made it the easiest of all celestial objects to observe with the early, rather simple satellites. Finally, knowledge of solar emissions would be essential both for designing adequate shielding for astronauts (Glasstone 1965) and for distinguishing surreptitious nuclear tests from violent solar events (Singer 1965). In short, solar observations from space were seen as having scientific merit, technical feasibility, and practical utility.

For a time, Soviet scientists led the way in using spacecraft as solar observing platforms. As early as November 1957, indeed, S. L. Mandel'shtam and his colleagues at the Lebedev Physical Institute in Moscow were able to orbit a set of electronic counters (fig. 5.2) for measuring the intensities of solar X rays and Lyman-alpha radiation aboard *Sputnik 2* (Guendel 1968). They succeeded in getting a rich set of readings from their instrument. Subsequently, however, it was recognized that their measurements were compromised by radiation from the Van Allen radiation belts (Mandel'shtam 1967; Van Allen 1983). All, or almost all, the spacecraft launched by the Soviet Union during the next three years evidently carried at least one instrument for observing solar radiations or particles (Guendel 1968). The first solar measurements

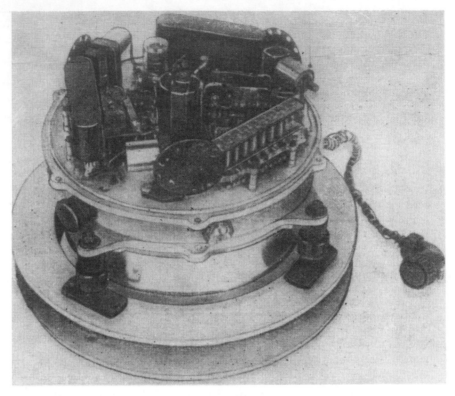

Figure 5.2. The first solar instrument package to be orbited—the soft X-ray and Lyman-alpha photometers that S. L. Mandel'shtam's group sent aloft in *Sputnik 2*'s payload on November 3, 1957. (Source: Aleksandrov and Federov 1961, 164.)

from a spacecraft were obtained by a team from the Soviet Academy's Radio Engineering Institute in Moscow. Their ion traps aboard *Lunik 2*, launched in September 1959, measured the flow rate of what some theorists were beginning to call the solar wind (see chap. 6).

Later that month, Herbert Friedman and his associates at the Naval Research Laboratory got the first American solar instrument into orbit on *Vanguard 3* (Corliss 1967). However, their X-ray and Lyman-alpha photometers were, like those of Mandel'shtam's group, swamped by electrons from the Van Allen radiation belts (Friedman 1960). His team's first successful solar measurements from a spacecraft were made with the photometers on *Solrad 1*, which monitored fluctuations in solar X-ray and Lyman-alpha intensities from July into November 1960 (Kreplin, Chubb, and Friedman 1962). Soon, despite their later start, American scientists were setting the pace in the use of spacecraft for observing solar phenomena. They

Table 5.1
Spacecraft with Instruments That Successfully Observed Solar Electromagnetic Radiations, 1960–1975

	Number Launched																
	1960	'61	'62	'63	'64	'65	'66	'67	'68	'69	'70	'71	'72	'73	'74	'75	Total
American spacecraft with																	
1–3 such instruments	1	1	0	1	4	5	2	6	3	10	2	1	0	1	2	5	44
4+ such instruments[a]	0	0	1	0	0	1	0	2	0	2	0	1	0	1	0	1	9
American total	1	1	1	1	4	6	2	8	3	12	2	2	0	2	2	6	53
Soviet spacecraft with																	
1–3 such instruments	2	0	0	0	2	2	0	1	2	1	2	0	3	1	2	2	20
4+ such instruments	0	0	0	0	0	0	0	0	0	0	0	0	0	0	0	0	0
Soviet total	2	0	0	0	2	2	0	1	2	1	2	0	3	1	2	2	20
All other nations' spacecraft with																	
1–3 such instruments	0	0	1	0	0	0	0	1	1	0	0	1	2	0	2	2	10
4+ such instruments	0	0	0	0	0	0	0	0	0	0	0	0	0	0	0	0	0
All other total	0	0	1	0	0	0	0	1	1	0	0	1	2	0	2	2	10
Grand total	3	1	2	1	6	8	2	10	6	13	4	3	5	3	6	10	83

Source: Hufbauer 1990.

[a]The spacecraft carrying four or more such instruments were NASA's *Orbiting Solar Observatories 1–8* and *Skylab*.

consistently put more solar instruments into space than did Soviet scientists (see table 5.1 and, e.g., Lüst 1967). Moreover, to judge from their research (see, for instance, chap. 6), their instruments generally had greater capabilities than did the Soviet instruments.

American leadership in studying the sun from space was a testament both to the prominence of American scientists in the making of space policy and to the prowess of American engineers—especially in electronics and computing—in implementing that policy. The United States put more plentiful resources and expert manpower into the selection, development, production, and deployment of space-based solar instruments than did the Soviet Union. American scientists, for instance, were able to do much more by way of testing prototypes than were their Soviet counterparts. They made frequent use of high-altitude aircraft (e.g., chap. 7), stratospheric balloons (e.g., Newkirk and Bohlin 1965), and sounding rockets (Corliss 1971; Ezell 1988b) for trying out the predecessors of solar instruments that were later placed in orbit.

The scientists who used spacecraft as solar observing platforms devoted primary attention to emissions that were either totally blocked or much attenuated by the earth's atmosphere. One important class of such emissions comprised particles emanating from the sun's atmosphere—the solar wind from the corona and solar cosmic rays from flares. Since the next chapter focuses on space-age investigations of the solar wind, it suffices here to mention two features of this research. First, virtually all the scientists working in this area were new to solar physics, because the instruments used for studying solar particles came from different instrumental traditions than did those used in the field's older areas. Second, these newcomers obtained a sizable fraction—about 40 percent—of the spacecraft berths allocated to solar instruments. One reason for their success was that observations of solar particles had much to reveal about the sun and its influence in the solar system (see chap. 6). A second reason was that their instruments, besides being relatively cheap and light, could double as tools for investigating conditions near the planets and in interplanetary space.

The other important class of solar emissions observed from space comprised the sun's ultraviolet and X-ray radiations. A fair number of the scientists behind the solar-radiation instruments placed on spacecraft were, like those behind the solar-particle instruments, newcomers to solar research. But the dominant figures here were either physicists who had been using rockets as solar observing platforms or ground-based solar astronomers who were attracted by the prospect of getting an unobstructed view of the sun. Such scientists were especially successful in the United States.

An enthusiastic pragmatist—John C. Lindsay (1916–65)—played the lead role in initiating NASA's solar observatory series (Biography of Lindsay 1962; Bester 1966). An experimental physicist by background, Lindsay (fig. 5.3) got involved in solar research in 1955 when he joined Friedman's team at the Naval Research Laboratory. He participated in Friedman's campaign to observe X rays from flares with rocket-borne instruments (see chap. 4). He also served as project manager for the solar photometers that Friedman's group developed for the Vanguard program. Then in the fall of 1958, like Newell and many others associated with the Naval Research Laboratory's rocket program, Lindsay transferred to NASA. He went to the agency's Goddard Space Flight Center in Greenbelt, Maryland. His first major assignment was as project manager for *Explorer 6,* which was launched in August 1959.

Meanwhile, impatient to observe flares from satellites, Lindsay was urging that NASA begin its program of solar research with a low-budget spacecraft (Bester 1966; U.S. Comptroller General 1964; Goldberg 1981, 1986b; Ezell 1988a). In doing so, he was bucking Newell and the agency's other science planners, who were entertaining grandiose ideas for the first solar observatory in space. Nonetheless, he managed by June 1959 to win over

Figure 5.3. John Lindsay, the initiator of NASA's Orbiting Solar Observatory program. (Source: Bester 1966, facing p. 161.)

Ball Brothers Research Corporation of Boulder, Colorado, a young firm that specialized in producing pointing controls for American sounding rockets. He also won some solar physicists over to his pragmatic approach—notably Michigan's Leo Goldberg, whose postwar enthusiasm for space research (see chap. 4) had been rekindled by *Sputnik*. By August he had made sufficient progress that his spacecraft was included in NASA's official ten-year program. By November he had secured a contract for Ball Brothers to produce the satellite. And on March 7, 1962, about three years after he began agitating for a satellite dedicated to solar observing, *Orbiting Solar Observatory (OSO) 1* (fig. 5.4)—the first spin-stabilized spacecraft with pointing control and several solar instruments—was in orbit (NASA 1965a).

By this time Lindsay had transformed his single mission into an ongoing program (Bester 1966). His success had been made possible by two allies. One was Laurence T. Hogarth, a suave yet tough manager who served as Lindsay's administrative counterpart. The other was the Astronomy Subcommittee of NASA's Space Sciences Steering Committee (Newell 1980). Working with these allies, Lindsay had secured funding for a second solar observatory that would include instruments from such respected scientists as Richard Tousey of the Naval Research Laboratory and Leo Goldberg, who was now with Harvard College Observatory (NASA 1962b). He had also obtained approval for preliminary planning of an Advanced Orbiting Solar Observatory. Thus in the spring of 1962, Lindsay had every reason to be optimistic about NASA's main program of solar research.

Things continued to go well for Lindsay and his program during the next year and a half (Bester 1966; Ezell 1988a). NASA recognized the success of *Orbiting Solar Observatory 1* by giving him its highest individual award. Meanwhile, with Hogarth riding herd, Ball Brothers and the principal investigators readied the second orbiting solar observatory and its instruments for launch. In addition, the Space Sciences Steering Committee's new Solar Physics Subcommittee, which included solar physicists from outside as well as inside the agency (NASA 1962c), approved payload proposals for the third, fourth, and fifth satellites in the series. And after competitive design studies, NASA awarded a contract to Republic Aviation Corporation for detailed planning of the Advanced Orbiting Solar Observatory.

From early in 1964, however, Lindsay's program ran into one difficulty after another (Bester 1966). The U.S. Comptroller General (1964) criticized NASA for wasting $799,000 on the solar observatory program by forcing its pace. Much worse, in April, when the second observatory was being mated to its third-stage rocket at Kennedy Space Flight Center, static electricity ignited the engine. The satellite was badly damaged, and three engineers were killed. Then in February 1965, when the rebuilt satellite was orbited, electrical arcing shut down its most sophisticated instrument—Goldberg's ultraviolet spectrometer-spectroheliograph. A few months later, rumors began circulating that NASA was about to raid the budget for the Advanced Orbiting Solar Observatory (Karth 1965). Then in August, the third solar observatory failed to make orbit because the third-stage rocket fired pre-

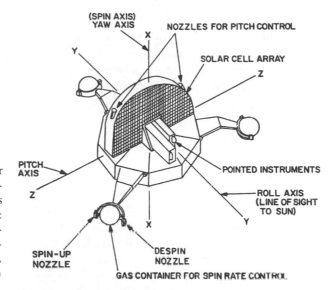

Figure 5.4. The first Orbiting Solar Observatory: *Top*, diagram indicating the satellite's control systems and pointed instruments. (Source: NASA 1962b, 4). *Bottom*, employees of Ball Brothers Research Corporation at work on the spacecraft. (Source: Bester 1966, facing p. 160.)

maturely. Perhaps it was not a coincidence that the beleaguered Lindsay was felled by a heart attack in late September 1965. Not long afterward, NASA canceled the Advanced Orbiting Solar Observatory program.

The case for using spacecraft as solar observing platforms was so compelling that NASA's solar program soon recovered from all these setbacks. Over the next decade, NASA successfully launched the replacement *Orbiting Solar Observatory 3* and five further satellites in the series. In doing so, the agency gave scientific teams from fourteen institutions in the United States, Britain, France, and Italy opportunities to observe the sun with ever greater photometric, spectral, spatial, and temporal resolution. The program did not stop with collecting data. As the quality of the results improved, NASA devoted greater attention to fostering their use by solar physicists who were not direct participants in the missions—for example, by sponsoring special user-oriented conferences (Glaser 1969). Hence NASA's solar observatory series ultimately compiled a record that surpassed all Lindsay's expectations.

In the meantime, NASA had embarked on the most ambitious solar mission that has ever been sent aloft—what came to be known as the *Skylab* Apollo Telescope Mount. Before his death, Lindsay and others had discussed the possibility of putting a manned solar observatory into orbit (Bester 1966). They were attracted to the idea partly because astronauts would have the versatility both to take advantage of unexpected observational opportunities and to handle unanticipated difficulties. They were also stimulated, however, by the very challenge of giving astronauts an integral role in solar research. Notwithstanding this interest, Lindsay had managed only to arrange for Ball Brothers to explore how a manned spacecraft could, despite crew movement, achieve the requisite pointing stability.

The prospects of a manned solar observatory improved markedly in December 1965 when NASA's Office of Space Science and Applications took up the cause (Newkirk, Ertel, and Brooks 1977; Compton and Benson 1983). Homer Newell, the office's chief, had been spurred into action by the agency's decision to kill the Advanced Orbiting Solar Observatory. He immediately set about recruiting the disappointed principal investigators for an Apollo observatory mission that, if approval could be won, would complement the Apollo lunar program. In March 1966, having gotten a favorable response from the scientists, he requested that such a program be established. Newell's initiative aroused the competitive instincts of George E. Mueller, who, as chief of NASA's Office for Manned Space Flight, was responsible for the new Apollo Applications Program. Mueller emerged triumphant from the ensuing turf war. In July 1966 NASA announced that the Apollo Telescope Mount would be managed by the Marshall Space Flight Center in Huntsville, Alabama. Rein Ise, a seasoned Marshall engineer, was charged with organizing a project office there. And the four scientists who had been negotiating with Newell's deputies—Leo Goldberg of Harvard College Observatory, Richard Tousey of the Naval Research Laboratory (NRL), Gordon Newkirk of the High Altitude Observatory (HAO), and

Ricardo Giacconi of American Science and Engineering in Cambridge, Massachusetts—were notified of their selection as principal investigators.

They were also informed that their Apollo Telescope Mount (ATM) instruments must be ready for launch by late 1968. The instrument teams, to judge from a meeting of NASA's Solar Physics Subcommittee in September 1966, were not at all happy with this deadline or the constraints on the mission resulting from the presence of the astronauts:

> The three experimenters present, Harvard, HAO, and NRL, expressed concern over tight delivery schedule (launch December 68) compromising experiment design. Also general agreement was reached that OSO measurements were advantageous over the [brief] ATM measurements. ATM is not a substitute for OSO. Also great dissatisfaction was expressed with proposed hard-mount for ATM experiments which requires no movement by the 2 or 3 astronauts for 20 minutes at a time. Urine disposal appears to form a cloud around ATM at 200 miles altitude which is of great concern. (Davis and Ostaff 1966)

During the next three years, such concerns were met by refinements in spacecraft design,[2] by extension of the mission from a single crew with a two-week observing period to three successive crews with longer observing periods, and by postponements of the scheduled launch date to mid-1972 (Newkirk, Ertel, and Brooks 1977; Compton and Benson 1983). All the while, the principal investigators' teams were hard at work on a set of telescopes that were more sensitive than any solar instruments yet orbited.

As they solved their most difficult design problems, the principal investigators paid increasing attention to operational issues. They wanted sufficient flexibility to ensure a quick response to flares and other transient solar phenomena. Accordingly, they arranged for the astronauts to get enough training in solar physics to take observational initiatives. They also persuaded NASA to give their teams fairly direct access to the astronauts during the mission. Besides flexibility, the scientists wanted a means of avoiding disputes over observing priorities. They achieved this objective in 1971 by agreeing, in emulation of the instrument teams for *Orbiting Solar Observatory 6*, upon a set of coordinated observing routines. The passage of time was, meanwhile, accompanied by a significant turnover in scientific personnel. Goldberg received an offer to serve as director of the Kitt Peak National Observatory in Arizona. Unwilling to forgo this opportunity, he passed the job of principal investigator for Harvard's instrument on to Edmund M.

2. Pointing control was achieved by means of gimballed mounting and gyroscopes; wastes were partly stored for subsequent analysis and partly recycled (Compton and Benson 1983).

Figure 5.5. *Skylab* atop its *Saturn V* en route from the Vehicle Assembly Building to the launch pad at Kennedy Space Center on April 16, 1973. (Courtesy of NASA History Office.)

Reeves. Likewise Newkirk, who was appointed director of the High Altitude Observatory, arranged for his colleague Robert M. MacQueen to take over as principal investigator.

On May 14, 1973, NASA finally launched the unmanned *Skylab* with the Apollo Telescope Mount (fig. 5.5). The four main solar instruments alone represented a major investment—$40.9 million for the Naval Research Laboratory's ultraviolet spectrograph and spectroheliograph, $34.6 million for Harvard's ultraviolet spectroheliometer, $14.7 million for the High Altitude Observatory's white-light coronagraph, and $8.3 million for American Science and Engineering's X-ray telescope (Compton and Benson 1983). Hearts must have sunk when the postlaunch checkout revealed that the micrometeoroid shield, which was also intended to serve as a sun screen, and two of the spacecraft's six solar-power panels were not functioning properly. During ascent the shield had broken loose, taking one of the solar panels with it and jamming the other. The launch of the first crew was postponed while NASA engineers diagnosed the problems, devised a remedial strategy, and rehearsed the astronauts in its implementation. Then on May 25, Charles Conrad, Jr., Joseph P. Kerwin, and Paul J. Weitz were sent up to restore the ailing spacecraft to working order. They replaced the

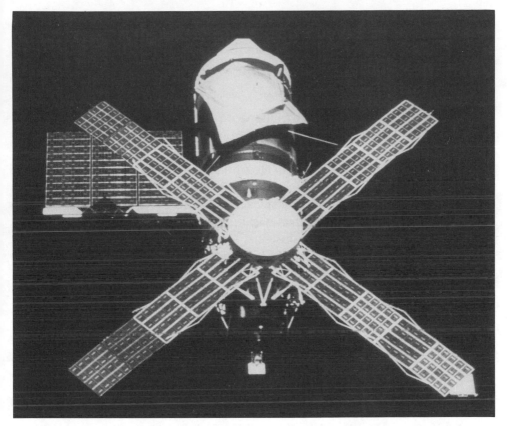

Figure 5.6. *Skylab* after repair; note the parasol atop the workshop and the surviving workshop solar panel in extended position; also note the Apollo Telescope Mount with its four solar panels. (Courtesy of NASA History Office.)

micrometeoroid shield with a special parasol that soon reduced temperatures to a livable level. They also freed up the jammed solar panel, thereby increasing the power supply to the range needed for sustained operations (fig. 5.6).

Once repaired, *Skylab* was a superb solar observatory (Eddy 1979; Compton and Benson 1983). Over the next eight months, the first crew and two follow-up crews worked in close coordination with the ground-based instrument teams, which in turn were backed up by a global observing network. All together the astronauts devoted 953.3 hours to obtaining 127,047 photographs of the sun as well as myriad related electronic data. In the ensuing years, solar physicists working individually and at special Skylab workshops have used these records in studies of coronal holes (see chap. 6) and numerous other phenomena. Thus, contrary to the dire expectations of the Solar Physics Subcommittee back in 1966, the Apollo Telescope Mount was a success—partly because the solar physicists obtained a major say in instrument design and operation and partly because the astronauts performed so capably as repairmen and observers.

Ground-Based Solar Observatories

Besides funding solar research from spacecraft, generous budgets after *Sputnik* provided solar physicists with numerous opportunities to make the case for improving ground-based observational capabilities. They were quick to argue that, thanks to advances in electronics and optics, they could significantly enhance resolving power and hence follow up on prior studies of optical, magnetic, and spectroscopic features. Moreover, as the solar data acquired in space became increasingly precise, they insisted that concurrent ground-based observing was essential for interpretation. And as manned missions increased in frequency, they pointed out that a reliable flare patrol would help protect astronauts from intense solar emissions.

Armed with such arguments, solar physicists obtained support for several new telescopes between 1957 and 1975. Robert R. McMath, one of the pioneers in solar cinematography (see chap. 3), led the campaign for the largest of these new instruments (McMath and Pierce 1960; Petrie 1962; Kloeppel 1983; DeVorkin 1989c). During the mid-1950s, he chaired a National Science Foundation panel that articulated the need for a national optical observatory and supervised the search that settled on Kitt Peak, Arizona, as the best site. He persuaded the panel, which included Goldberg, that an immense solar telescope should be included among the new observatory's instruments. Although the telescope was funded in the aftermath of *Sputnik*, McMath did not live to see its dedication in November 1962. His role was noted in President John F. Kennedy's message to the celebrants:

> The great new solar telescope at the Kitt Peak National Observatory in Arizona is a source of pride to the nation. The largest instrument for solar research in the world, it presents American astronomers with a unique tool for investigating the nearest of the stars, our sun. This project is of exceptional interest to all our citizens, for the observatory is financed by the Federal Government through the National Science Foundation. . . . Bold in concept and magnificent in execution, the instrument is the crowning achievement of the career of the late Robert R. McMath, builder of solar telescopes, for whom it is named. (Kloeppel 1983, 69)

The $4 million McMath Solar Telescope (fig. 5.7) was indeed gargantuan. Its three mirrors, including the 150-cm objective, directed the sun's rays along a

Figure 5.7. Kitt Peak's McMath Solar Telescope some months before its dedication in November, 1962; the tractor in the middle ground gives some idea of scale. (Courtesy of National Optical Astronomy Observatories.)

244-m path before forming a 90-cm solar image in an underground observation room.

Although other teams building solar telescopes between 1957 and 1975 worked on a smaller scale than did McMath and his colleagues, several ended up with instruments that produced images of equal or higher resolution. Those who did so all found one way or another to avoid the air movements along the optical path that often blurred the McMath Solar Telescope's image (Dunn 1985). Some groups built telescopes with very short optical paths (e.g., Kiepenheuer 1966; Zirin 1970; Rösch and Dragesco 1980). Others constructed telescopes with evacuated tubes (fig. 5.8) (e.g., Dunn 1964, 1969; Livingston et al. 1976). A few of the telescopes, besides having more sophisticated designs, had the added advantage of sites with less atmospheric turbulence than Kitt Peak and hence better "seeing"—for example, the Pic du Midi in France, Capri off the coast of Italy, and Big Bear Lake in the southern California mountains (Zirin and Mosher 1988).

Figure 5.8. Sacramento Peak's Vacuum Tower Telescope in the morning sun and (inset) its dedication plaque. (Courtesy of the National Solar Observatory/Sacramento Peak.)

In addition to building new telescopes, solar physicists were upgrading the auxiliary instrumentation of existing telescopes. Step by step, they replaced the weakest components in their apparatus in an unceasing quest for higher resolution. They narrowed the passbands of monochromatic filters, substituted electronic subsystems for electromechanical components, and computerized control and data systems. The labors of Robert F. Howard (b. 1932) on Mount Wilson's magnetograph (see chap. 4) exemplify this process of refinement (Howard 1974, 1976). In 1959, working with the instrument's inventor, Horace Babcock, he installed a new cathode-ray tube for displaying the solar magnetic fields. Three years later he replaced the spectrograph's grating so as to improve the spectral resolution of the input signal. In the mid-1960s he increased the spatial resolution to 11,000 sectors on the solar disk, installed a guider with a digital logic system for controlling the step wise scan of these sectors, developed an electronic subsystem that could determine both the magnetic field and rotational velocity of each sector and digitize the results for storage on computer tape, and devised a servo plotter for producing magnetograms. About 1970 he tackled the optics, replacing the telescope's 30-cm objective lens with a cemented triplet achromatic lens and the spectrograph's feed lens with an apochromatic lens. Finally in 1974, no longer satisfied with his guidance and data-handling systems, he computerized his operations. All these modifications were expensive. Without funding from the Office of Naval Research, the Air Force Cambridge Research Laboratories, the National Science Foundation, and NASA, Howard could never have made most of them.

A Subterranean Solar Detector

Although solar physicists energetically pursued opportunities to acquire new telescopes and auxiliary apparatus, they evinced no interest whatever in establishing subterranean facilities for solar research. Yet during the 1960s a small but growing number of nuclear scientists and theoretical astrophysicists thought that just such facilities would enable direct observation of neutrinos from some of the nuclear reactions powering the sun (Reines 1967; Bahcall and Davis 1982; Pinch 1980, 1985, 1986; Bahcall 1989). Two key advocates of this idea were Raymond Davis, Jr. (b. 1914), a nuclear chemist at Brookhaven National Laboratory, and John N. Bahcall, a young theoretical astrophysicist at the California Institute of Technology. If the sun produced enough neutrinos, Davis first suggested in 1955, he could detect these particles—which theory said must be traveling outward from the sun's energy-generating core at the speed of light—by their ability to transmute the nuclei of chlorine 37 into argon 37. Some seven years later he persuaded Bahcall, an expert on stellar interiors, to undertake the difficult task of

Figure 5.9. The neutrino-trapping tank in the Brookhaven-Homestake Solar Neutrino Observatory in a mine chamber well over a kilometer beneath Lead, South Dakota. (Courtesy of the Brookhaven National Laboratory.)

predicting the flux of solar neutrinos. Bahcall's prediction was high enough to justify serious attempts at detection.

In the mid-1960s, nuclear scientists managed to secure funding from the Atomic Energy Commission for solar-neutrino detectors deep enough in American and South African mines to be shielded by the overburden from cosmic rays. The largest and most expensive ($600,000) was that installed by Davis in South Dakota's Homestake Mine (fig. 5.9). In 1967 he began measuring the rate at which solar neutrinos were transmuting the chlorine in some 400,000 liters of cleaning fluid into argon. The rate turned out to be so much lower than Bahcall's predicted flux that it was near or below the threshold of detection. Davis was delighted. Nothing is quite so satisfying to an experimentalist as coming up with a result that challenges theory. By contrast, Bahcall was disappointed by the failure of his prediction. He soon concluded, however, that Davis's null result could well indicate the need for a major revision in stellar or neutrino theory (Bahcall 1969). The discrepancy between observation and theory was in any case sufficiently intriguing to warrant round after round of refinements in Davis's subterranean solar-neutrino detector.[3]

Between *Sputnik* and the mid-1970s, therefore, the American government underwrote a major expansion of patronage for solar research. Much of the new money went into the development and exploitation of solar observing for spacecraft, but substantial amounts were also expended on ground-based observatories and a subterranean neutrino detector. Meanwhile, to foster the interpretation of the data acquired, increasing funds were also going to support theorists and their computational needs. One consequence of the expanded patronage was that the solar physics community roughly doubled in size, reaching about 150 established investigators in the United States and some 300 worldwide (Smith et al. 1975). Another, to which we now turn, was the emergence of subdisciplinary institutions.

Subdisciplinary Trends, 1965–1975

As their numbers grew, solar physicists came to think they needed more opportunities for communicating with one another about practical concerns

3. The rate Davis found was well below the detection threshold of the solar-neutrino detectors of the other American groups, so they soon abandoned their efforts. In 1987 a Japanese scientific team finally provided independent confirmation of Davis's measurements with a different kind of solar-neutrino detector. At present, several other new types of solar-neutrino detectors are under development (Bahcall, Davis, and Wolfenstein 1988; Bahcall 1989).

and scientific problems. They found it increasingly difficult to get as much time and space as they wanted in the main astronomical societies and journals. This difficulty made them ever more aware of the limitations of existing institutions. The triennial meetings of the International Astronomical Union's solar commissions were too infrequent, and the focus on solar statistics of the *Quarterly Bulletin on Solar Activity* was too narrow, to fulfill their needs for mutual access. From the mid-1960s, therefore, solar physicists formed several subdisciplinary institutions—for example, the Solar Physics Division of the American Astronomical Society, the Joint Organization for Solar Observations in Europe, and most notably the journal *Solar Physics*. In doing so, they were participating in a much broader process of subdisciplinary formation that occurred throughout the republic of science during the bountiful times following *Sputnik*.

Organizing

Henry J. Smith, the chief solar physicist at NASA headquarters, initiated the train of events that gave rise to the American Astronomical Society's Solar Physics Division. He proposed at a meeting of NASA's Solar Physics Subcommittee in April 1965 that NASA-funded solar physicists meet periodically to discuss common problems (NASA 1965b). In the ensuing discussion, Leo Goldberg (fig. 5.10) offered an alternative—that the American Astronomical Society sponsor an annual solar meeting, "perhaps as a first step in establishing a section for Solar Physics." Afterward Goldberg considered how, as the current president of the American Astronomical Society,[4] he might follow through on his suggestion. To raise the issue directly with the Society's council would have been controversial, since many councilors were concerned about the threat that rampant specialization posed to astronomy's unity. Accordingly, Goldberg (1965a) asked John W. Firor, Jr., the director of the High Altitude Observatory in Boulder, to take the initiative.

That July, Firor (1965) presented the council with a proposal that an annual solar meeting be held under the auspices of the American Astronomical Society. He opened by drawing the council's attention to the rapid growth of solar research: "One of the fastest growing fields of astronomical research is the study of the sun. The advent of new techniques of observing the sun—on the ground, from balloons and rockets, from space; in optical and radio wavelengths; and with fast particles and magnetic fields—has brought new people and new enthusiasm to solar research." This growth, he suggested, had created a need for special solar meetings. Although this need could be met by the American Geophysical Union or some other organization, he believed that

---

4. At this time Goldberg was also president of the International Astronomical Union.

if the AAS takes the lead at this time in bringing together the varied aspects of solar research, the traditional and valuable connection of solar studies to the rest of astronomy will be symbolized and encouraged. A meeting devoted to solar astronomy may appear as a step away from the desired unity of astronomy as a subject, but specialization of meetings is an inevitable consequence of the growth of science, and no amount of nostalgia will bring back the situation in which a large fraction of the attendees at a general AAS meeting is competent to discuss most of the contributions. In any case, the gain from bringing into our Society those now studying the sun with new techniques should outweigh the advance in specialization represented by a solar meeting.

He went on to propose that the annual solar meetings be held in the western United States, because of the "considerable concentration" of solar research there, and that a start be made with a three-day meeting in Boulder in October 1966.

The American Astronomical Society Council would not be rushed. It

Figure 5.10. Leo Goldberg about 1963, with a model of the second Orbiting Solar Observatory, for which his Harvard College Observatory group was developing an ultraviolet spectrometer. (Courtesy of W. H. Parkinson.)

approved the meeting in Boulder but declined to make any long-range commitments (Goldberg 1965b). The solar meeting in Boulder was "a great success, and it was the unanimous opinion of those present that another such meeting should be held within a year or so" (Goldberg 1966). When asked once again to approve annual scheduling (Goldberg 1967), however, the council refused to do more than endorse a second solar meeting in Tucson in February 1968. The Tucson meeting drew over two hundred scientists (Firor 1968). Later that year the council yielded to the advocates of specialized meetings, whose ranks had grown to include the planetary astronomers and high-energy astrophysicists. It approved the formation of subdisciplinary divisions within the American Astronomical Society, giving them the right to organize separate meetings under the society's auspices (Goldberg 1968; Schwarzschild 1969; Tatarewicz 1990).

Planning for the Society's Solar Physics Division commenced at a third special solar meeting in Pasadena in February 1969 and culminated with a founding meeting at NASA's Marshall Space Flight Center in November 1970 (Abstracts of papers 1971; AAS Division 1972). The solar physicists attending the meeting in Huntsville adopted a constitution and elected officers. During the next few years, as membership climbed above 180, the Solar Physics Division's main activity was organizing solar meetings (Solar Physics Division 1973; Sturrock 1974). It did, however, take on a few additional responsibilities. In December 1973, for instance, it sponsored a symposium on *Skylab's* coronal observations at a regular meeting of the American Astronomical Society. And the following year it started raising funds for a Hale Prize Lectureship, to recognize outstanding contributors to solar physics.

Across the Atlantic, meanwhile, the Europeans were also forming solar associations. Organization seemed essential for acquiring instruments that would keep them competitive with the Americans. K. O. Kiepenheuer took the lead at the International Astronomical Union's meeting in 1967 (Kiepenheuer 1970; De Jager and Maltby 1970; De Jager 1970). He polled several of his European colleagues about their interest in a solar observatory that would surpass the best American facilities and, finding them receptive to the idea, called a planning meeting at the Fraunhofer Institute in Freiburg, West Germany. The eleven solar physicists from six countries who attended in July 1968 agreed to undertake a cooperative search along the Mediterranean seaboard for a site with exceptional seeing. At a second meeting in March 1969, twenty-one representatives from eight countries worked out the details for site testing in Sicily and Portugal. Nine months later the group reassembled in Catania, Sicily, to discuss initial results and plan the survey's next stage. In two years, therefore, Kiepenheuer had organized a thriving cooperative enterprise by providing his European solar colleagues with a strategy for outdoing the Americans.

All was going so well, in fact, that Kiepenheuer and nine of his colleagues—two each from France, Germany, and Italy, and one each from the Netherlands, Norway, Sweden, and Switzerland—took steps to formalize the venture (Statement of intentions 1970). At the Catania meeting, they constituted themselves as a Provisional Board for the Joint Organization for Solar Observations and elected Kiepenheuer president and Cornelis de Jager of Utrecht secretary. They envisioned the completion of site testing in 1973 at a cost of $450,000 and the construction of the joint observatory by 1976 at a cost of about $900,000. Subsequent progress was slower than anticipated, partly because the board was unable to arrange major funding and partly because, disappointed by the tests at the Mediterranean sites, it broadened the search to the Atlantic Ocean, particularly the Canary Islands (Kiepenheuer 1974, 1975). Still, by 1975, when the search narrowed down to Tenerife and La Palma in the Canaries, the Joint Organization for Solar Observations had secured the ongoing participation of solar physicists representing Spain, Austria, Greece, and Israel.

Besides searching for an observatory site, the Joint Organization of Solar Observations took the initiative in forming a general association of European solar physicists. Kiepenheuer and especially de Jager (1973b) came to think that support from the broader solar community would be essential for fund-raising. They arranged for the provisional board to meet in March 1974 with the three-year-old Committee of European Solar Radio Astronomers (Fokker 1973). The two groups appointed a committee to plan a general meeting for European solar physicists (Reijnen 1974). Held in Florence a year later, the first European Solar Meeting attracted 223 participants (Fokker 1976). Those attending a session devoted to means of promoting solar research supported the organization of a Solar Physics Section in the Astronomy and Astrophysics Division of the European Physical Society. Thus, in Western Europe as in the United States, solar physicists stood ready to join into new subdisciplinary ventures.

Publishing

This same spirit was evident in the publication of solar research. The journals that had traditionally carried the most important solar papers—the *Astrophysical Journal* in the United States and the Royal Astronomical Society's *Monthly Notices* in Britain—continued to be important outlets for new work. But competition to get into these journals was getting more severe. On the watch for additional outlets, solar physicists placed much of their work in two other periodicals. Those who were primarily interested in solar emissions with pronounced terrestrial effects favored the rapidly expanding *Journal of Geophysical Research.* Those interested in optical studies of the sun leaned toward *Solar Physics: A Journal for Solar Research and the Study of Solar Terrestrial Physics,* which first appeared in 1967. This journal's press run climbed to 1,500 copies and the annual output to six thick volumes within five years of its debut. Indeed, *Solar Physics* swiftly evolved into the central forum of the international solar physics community (Hufbauer 1989).

The founder of *Solar Physics* was Utrecht's Cornelis de Jager (b. 1921). About 1960, the Dutch publisher Anton Reidel was seeking ways to turn his family's small printing firm into a full-fledged publishing company. He recognized that the space boom provided an opportunity for a new journal, but an editor had to be found. Reidel asked de Jager whether he would take on the job. De Jager, thinking this could be an excellent way to acquaint himself with the trends in space research, was interested, but he had two conditions. He wanted to limit the journal to invited review papers, and he wanted the assistance of a distinguished editorial board. Reidel readily agreed. The first issue of *Space Science Reviews* appeared in June 1962. Impressed with the success of this journal, Reidel asked de Jager some three years later whether he might consider editing another periodical as well. De Jager thought such a subdisciplinary journal might succeed in solar physics. He felt too busy, however, to take on a second journal. Besides his editorship of *Space Science Reviews,* he was orchestrating a rapid expansion of Utrecht's astronomy department, lecturing once a week in Brussels, serving on several international committees, and, remarkably, pursuing his own program of solar research. Reidel persisted. De Jager, seeing that he would simultaneously buttress Utrecht's position within solar physics and help the field consolidate its position within astronomy, yielded to Reidel's entreaties.

His first step was to canvass fifteen solar physicists around the world about the journal's prospects and organization. Most responded with enthusiasm. Indeed, the new president of the International Astronomical Union's Commission 10 for solar activity—Zdenek Svestka (b. 1925), a Czech specialist on solar flares—reported that he had tried to organize just such a journal at the union's 1964 congress in Hamburg. Some respondents, however, raised a serious problem. They feared that papers of broad significance, especially theoretical studies of the solar atmosphere, would no longer be seen by stellar physicists. In March 1966 de Jager responded to this concern in a circular letter inviting thirty-seven colleagues to join *Solar Physics'* editorial board. He granted that general journals might be more suitable for some theoretical work. Nonetheless, he was confident that *Solar Physics* would be making a real contribution if it brought together work that focused on the sun, particularly on solar activity.

Most of de Jager's colleagues welcomed his invitation. Moreover, Svestka, who had been invited to share in the journal's direction, joined enthusiastically in the work of shaping the new periodical into a truly international forum. In their first circular to potential contributors, de Jager and Svestka (1966) made no concession to the idea that *Solar Physics* was inappropriate for articles that might be of interest to stellar physicists. Instead, they promised that "SOLAR PHYSICS will deal with all aspects of solar problems, from the interior to the solar wind in interplanetary space. It will also deal with sun-earth relations [in those] cases where the emphasis is solar, for instance when the earth's atmosphere or magnetic field is used as a spectrum analyser for the solar electromagnetic or particle spectrum. SOLAR PHYSICS will contain theoretical, observational and instrumental papers, and will refer to ground-level observations as well as to those made from balloons or space vehicles." Two months later, Reidel began advertising for subscribers. The response was so good that the firm decided on a press run of 1,000 copies.

The first issue of *Solar Physics* appeared on schedule in early 1967. By this time the editorial board, including the chief editors, numbered forty-one scientists from twelve nations. Speaking for the board, de Jager and Svestka (1967) expounded on the journal's rationale in their preface. Contemporary scientists were barraged, they observed, with an "ever growing stream of publications . . . scattered over many different journals." In a "well-defined" field like solar physics, such a scattering of research was unnecessary. The new journal would ease the work of solar physicists by providing them with a "vehicle where results . . . could be found together." The ultimate goal was to "establish a closer cooperation between all those concerned with solar research." In short, the editors and their board believed that *Solar Physics,* by serving the field as a forum, would promote research and foster cooperation.

To the delight of the publisher and the editors alike, *Solar Physics* flourished. Circulation grew so that Reidel increased the press run to 1,200 copies in 1969 and to 1,500 copies in 1971. Meanwhile, de Jager and Svestka (fig. 5.11) were under increasing pressure from submissions. As early as the spring of 1967, they could see that eight issues a year would be needed to keep up with the influx of publishable manuscripts. Half a year later, they realized that they would have to shift to monthly publication in 1968. Even monthly publication did not suffice. By November 1968 de Jager and Svestka had to announce a 34 percent increase in the number of pages per issue. In July 1969 they coupled their next announcement of an increase in issue size with a report that referees were being asked to be less lenient in judging manuscripts. The referees were somewhat tougher, boosting rejec-

Figure 5.11. Cornelis de Jager (right) and Zdenek Svestka, coeditors of *Solar Physics,* about 1970. (Courtesy of Zdenek Svestka.)

tions to a modest 11 percent by 1971. Nonetheless, de Jager and Svestka soon had to increase issue size yet again. The consequence was that the annual number of volumes climbed to six and the cost of an individual subscription to just under $100 a year. In July 1972, anxious to avoid any further increases, the editors discussed the problem of runaway growth. Their proposed solution was to give short articles precedence over those longer then ten printed pages.

That fall ten solar physicists at Sacramento Peak Observatory, believing that *Solar Physics* had already become too large and expensive, urged de Jager and Svestka to reconsider the problem. They suggested several means of reducing costs to individual subscribers—introducing page charges; increasing the differential between institutional and individual subscriptions; raising refereeing standards to the level of the *Astrophysical Journal;* limiting free distribution to the most active referees; and printing on cheaper paper. Several of the rebels argued that page charges would be the best remedy— such fees would not only fund lower subscription prices but also discourage verbosity. To ensure that their views were not dismissed, this group informed de Jager and Svestka that the matter would be raised at the January 1973 meeting of the American Astronomical Society's Solar Physics Division. De Jager was swift to respond. He too was concerned about the journal's cost to individual subscribers. He and Svestka feared, however, that introducing page charges would turn *Solar Physics* into an American journal, since American institutions were the only ones willing or able to pay such fees. In concluding, he promised to discuss the matter at length with Svestka and the publisher.

Although de Jager and Svestka obtained a 40 percent reduction in the price of individual subscriptions by February, the concerns about *Solar Physics'* practices, once having been articulated, were not so easily laid to rest. In particular, the discussion of the journal at the Solar Physics Division's meeting seems to have engendered further criticism of its standards. Following up on the discussion, Gordon Newkirk, the division's outgoing chairman, a member of the editorial board since 1968, and Firor's successor as director of the High Altitude Observatory, asked a stellar theorist for an evaluation of *Solar Physics*. He was advised that though the journal carried many fine papers, its acceptance threshold was still well below that of the *Astrophysical Journal*. Newkirk sent a circular letter to the editorial board's members, reporting his concern that some solar as well as nonsolar colleagues were of the opinion that "we include too many papers which are not worthy of publication in a refereed journal" (Newkirk 1973). He urged that those who shared his concern about the journal's future join him in asking the editors to convene a board meeting at the International Astronomical Union's congress in Sydney that August.

De Jager and Svestka soon called the meeting. In doing so, they expressed their concern that several board members thought some bad papers were appearing in *Solar Physics*. They intended to formulate better guidelines for referees. They also believed that steps should be taken to strengthen the board. Perhaps, they suggested, its members should serve staggered terms, retiring after four years unless they were reelected by the board. Through such elections the board could be continually rejuvenated with young solar physicists. The editors-in-chief, they also suggested, should serve staggered eight-year terms. De Jager, indeed, volunteered to step down at the end of 1974 and Svestka, four years later.

At the editorial board's meeting in Sydney, de Jager first discussed the finances of *Solar Physics* with the seventeen members present. Then he reported the startling fact that Newkirk's circular letter had caused the rejection rate to jump from 12 percent to 30 percent and remain there. Those present agreed that a rate of 25 percent should be regarded as the journal's new norm. They also agreed that referees should be asked to judge papers on their originality, brevity, and physical coherence. And after persuading de Jager to remain at the helm, they reached a consensus that board members should be elected for staggered terms. The editorial board's first regular election was held during 1974. Over 90 percent of the board members joined de Jager and Svestka in balloting for the eleven positions that would fall vacant at year's end. The upshot was indeed rejuvenating—while the board as a whole had a median age of fifty-four just before the election, the eleven new and reelected members had a median age of forty.

Between 1966 and 1975, therefore, de Jager and Svestka presided over the development of *Solar Physics* into a thriving international forum for the solar physics community. One indication of their success was the journal's large circulation. Another was its publication of over two hundred articles a year on all aspects of solar physics. The most telling indications, however, were the row over standards and the institution of regular elections for editorial board members. These episodes revealed that leading solar physicists had come, as de Jager and Svestka hoped from the outset, to regard the journal's performance as a collective responsibility.

Lean Times, 1975–1990

In the early 1970s, after treating science munificently for a decade and half, American politicians and bureaucrats slowed the growth of scientific budgets dramatically. They turned parsimonious for two main reasons. The costs of the Vietnam War, the Arab oil embargo, and social and environmental programs raised doubts about the desirability of continuing the expansion of federal patronage for science and technology. And the growth of opposition among scientists to America's foreign and military policies engendered questions about the idea that scientific spending promoted national security. Conservatives and moderates were increasingly skeptical that support for pure science was essential for maintaining American military superiority. Exploiting this skepticism, Senator Mike Mansfield and other liberals who wanted to reduce the military's presence in American society waged an effective campaign to limit the armed forces' patronage of research and development to projects having a demonstrable value to their missions (Nichols 1971).

As American leaders curbed scientific spending between 1970 and 1975, most of their counterparts around the globe followed suit. To be sure, the governments of Japan and a few other countries recognized the advantages of adhering to expansive science policies. Most governments were relieved, however, that fewer sacrifices would be required to maintain a respectable presence in science. Consequently, the leveling off of scientific budgets in the United States inaugurated an era of limits in world science that is likely to persist for a long time to come.

The Solar Physics Community's Struggle against Its Insularity

Like other scientists, solar physicists saw by the mid-1970s that lean times were upon them. One harbinger of future stringency was the High Altitude Observatory's closing of its Climax station (White and Jefferies 1973). Another was NASA's termination in September 1973 of the ninth and tenth Orbiting Solar Observatories (Ezell 1988b). Yet another was the U.S. Air Force's decision to abandon the Sacramento Peak Observatory (Alvensleben 1975). Although two major missions—*Skylab* and *Orbiting Solar Observatory 8*—softened the transition, solar physicists could see that maintaining current support levels would be difficult.

American solar physicists, who had fared very well in the halcyon years following *Sputnik,* were worried. Their concern gave the Solar Physics Division's chairman in 1975—Robert W. Noyes of the Harvard-Smithsonian Center for Astrophysics—an opportunity to fight solar separatism that he had wanted for a decade.[5] Noyes made a persuasive case that the division should hold most of its meetings with the American Astronomical Society (House 1976). He also established a committee to suggest strategies for reducing solar physics' insularity and consequent vulnerability (Bohlin 1976; Dupree et al. 1976). It was not by chance that the committee's chair—Noyes's Harvard colleague Andrea K. Dupree—and six members shared his opposition to solar separatism.

Dupree's committee (Dupree et al. 1976) issued a strongly worded report. It portrayed "the current situation" as grave:

> Solar physics, which once was an integral part of astrophysics and the astronomical community, appears now to be a distinctly separate and isolated field of astrophysical research. Communication and cross-fertilization among the subdisciplines of astrophysics has declined. The astronomical community is largely unaware of and maybe indifferent to current research in solar astrono-

5. Back when Leo Goldberg and John Firor opened the campaign for regular solar meetings under the auspices of the American Astronomical Society, Noyes (1965) had criticized this proposal. "I would be somewhat concerned," he wrote Goldberg, "over any change which would tend to divide the solar astronomers from the rest of the astronomical community. I think there is already too much tendency for solar physics to be insular, due I suppose to the isolation of some of the observatories (Sac Peak is a prime example) and the complexity of the detailed analysis required by the detailed nature of solar observations. I do not think it is good, for instance, for an astronomer to spend his life studying oscillations in the solar atmosphere to the exclusion of all other astronomy! I fear that either creating a separate solar astronomy subsection or devoting one meeting of the AAS each year to solar astronomy would work in just that direction. The solar people would come only to 'their' meeting, would see only their solar colleagues and would be seen by no others."

my. . . . [A]dverse effects . . . could result from the loss of interest and hence support from the astronomical community. Lack of support for solar physics on a national and local level can endanger funding as well as encourage a further decrease in faculty positions in solar physics. Few students are then produced or even exposed to the problems and potential in the study of the sun.

The result was that solar physicists were almost alone in realizing that their subject was "a significant, active, and vital field [with] numerous substantial, exciting, and unsolved problems [that had] broad interdisciplinary extension."

Dupree and her colleagues suggested several remedies. They urged the Solar Physics Division to continue meeting jointly with the American Astronomical Society, to nurture relations with other societies that might be interested in solar phenomena, and to encourage interdisciplinary lectures, symposia, and research projects. In addition, they urged the division's members "to act individually" on solar physics' behalf. Here they stressed the importance of promoting faculty appointments in the field so that university astronomers and students would be exposed "to active programs of solar research." Furthermore, they emphasized the desirability of placing "solar papers of general interest" in the *Astrophysical Journal,* proposing that only papers "of strictly solar interest" should be given to *Solar Physics.*[6] They also exhorted their peers to search out opportunities to apply their techniques to nonsolar problems, write semipopular papers, give lectures, issue press releases, and serve on advisory committees.

Many American solar physicists apparently saw merit in the case for strengthening their field's ties with the rest of astronomy. The division started appointing members to act as liaisons to neighboring subdisciplines (Bohlin 1977). Moreover, the division's Hale Committee arranged for the prize's first recipient—the prominent theorist Eugene N. Parker (b. 1927) of the University of Chicago (see chap. 6)—to give his lecture to the society as a whole (Bohlin 1978). Besides seeking a greater presence within the discipline of astronomy, solar physicists updated George Ellery Hale's arguments for their field's significance. Their campaign, which opened at about the same time in Europe (e.g., Schröter 1974; Zwaan 1974; Pecker and Thomas

6. The number of articles published by *Solar Physics* fell dramatically in the late 1970s. But, contrary to Hufbauer (1989), only a small part of this decline can be attributed to a shift in solar physicists' journal preferences. Rather, it seems to have been caused primarily by the inordinate delays that resulted from the bankruptcy of Reidel's printer and the difficulties of arranging a suitable successor. Since the early 1980s, *Solar Physics* has managed—despite maintaining a rejection rate of over 25 percent—to publish material at a higher average rate than in the mid-1970s (De Jager 1988).

1976) as in the United States (e.g., Smith et al. 1975; Sturrock et al. 1976), has continued to the present (e.g., Eddy 1979; Noyes 1982; Parker 1985a; Zirin 1988a; Tayler 1989). It has underscored four contributions of solar physics to astrophysics—new instruments, new diagnostic techniques, new interpretive insights, and new observational tests for existing theory.

Among solar physics' advocates, Parker (fig. 5.12) has been particularly active and eloquent. The "range and intensity of solar radiation," according to his summary of a 1975 report to the National Academy's Space Science Board (Parker 1979, 5, 7, 10, 12), "have favoured the development of sophisticated . . . spectroscopic and imaging instruments . . . that have subsequently found application in other areas of astrophysics." In addition, the "sun is the principal laboratory for developing and testing theoretical diagnostic tools with which to determine such physical properties as temperature, density, ionization equilibrium, systematic and random velocity fields, mass loss, heating and radiative losses, and chemical abundances." Of still greater importance, "the sun, our daytime star, is sufficiently near [to reveal] a variety of phenomena that at first sight defy rational explanation . . . but ultimately stimulate the theoretical understanding of new effects . . . elsewhere in physics and astrophysics." Finally, "the sun has been, and continues to be, the 'testing ground' of [stellar] astrophysics, where theories of nuclear energy, convection, radiation transport, and other phenomena may be confirmed or refuted." In short, as Parker (1978, 2, 7) insisted at a semipopular symposium on the "new solar physics," the field was the "mother of astrophysics" because it was "the one area in which hard science—the critical interplay of theory and measurement—can function."

Solar Observational Capabilities: A Mixed Record

Solar physicists have had a mixed record in the struggle to sustain the advance of their observational capabilities. On one hand, they have suffered many disappointments. They have witnessed reductions in the numbers of spacecraft and ground-based facilities engaged in solar observing. They have also witnessed one postponement after another of several projects that would significantly enhance their ability to investigate solar phenomena. On the other hand, they have chalked up many successes. They have sent some remarkable instruments into space and constructed some excellent telescopes at ground-based observatories. Moreover, they have gotten access to extremely sensitive radio telescopes designed for galactic and extragalactic work and found various ways—including setting up instruments at the South Pole—to refine observational knowledge of the sun's vibrational modes. It would be fair to say that, despite many setbacks, solar physicists can now observe the sun at higher photometric, spectral, spatial, and temporal resolutions than ever before.

Figure 5.12. Eugene Parker about 1970; noted for his pioneering theoretical research on the solar wind, he would become one of solar physics' most active champions in the mid-1970s. (Courtesy of Eugene Parker.)

Table 5.2
Spacecraft with Instruments That Successfully Observed Solar Electromagnetic Radiations, 1960–1985

	Number Launched					
	1960–65	1966–70	1971–75	1976–80	1981–85	Total
American spacecraft with						
1–3 such instruments	12	23	9	6	7	57
4+ such instruments	2	4	3	3	1	13
American total	14	27	12	9	8	70
Soviet spacecraft with						
1–3 such instruments	6	6	8	3	3	26
4 1 such instruments	0	0	0	3	0	3
Soviet total	6	6	8	6	3	29
All other nations' spacecraft with						
1 3 such instruments	1	2	7	2	1	13
4+ such instruments	0	0	0	0	1	1
All other total	1	2	7	2	2	14
Grand total	21	35	27	17	13	113

Source: Hufbauer 1990.

The decline in the number of spacecraft used as solar observing platforms since the mid-1970s has certainly been noticeable (see table 5.2). Two developments have combined to cause this decline. First was NASA's decision to develop the reusable Shuttle orbiter into its Space Transportation System (Logsdon 1986; Ezell 1988b). Bringing the Shuttle fleet into service turned out to be much more difficult, costly, and time consuming than its advocates within the agency anticipated. The consequence was not only that funds for space science were tight but also that those missions that won approval were subject to repeated delays (e.g., Smith 1979; Waldrop 1981; R. W. Smith 1989). The second development behind the declining number of spacecraft with solar instruments was strong competition from stellar and extragalactic astrophysicists for access to space. Thanks in part to the achievements of space-based solar observing, these rivals no longer had trouble arguing that instrumental sensitivity and spacecraft pointing were sufficient for research on celestial objects much fainter than the sun (e.g., R. W. Smith 1989). As a consequence, funds that might well have gone to solar missions in the 1960s and early 1970s were allocated to nonsolar astrophysical missions such as the *International Ultraviolet Explorer* (launched January 1978), the *Einstein High Energy Astronomical Observatory* (launched November 1978), the *Infrared Astronomical Satellite* (launched January 1983), and the *Hubble Space Telescope* (launched April 1990).

The prospects that the number of spacecraft with solar instruments will return to the levels of the late 1960s and early 1970s are nil. In fact, since the *Challenger* disaster of January 1986 it has been hard to imagine that the number of such missions will soon return to the reduced levels of the late 1970s and early 1980s. In general, the tragedy tempered the optimism of space-science planners not only in the United States but also in the other spacefaring nations (e.g., Waldrop 1986, 1987; Dickson 1986, 1987; Brown and Giacconi 1987). In particular, the loss of the *Challenger* led to several changes in NASA's plans for solar observing from space. For instance, the agency postponed the launch of Ulysses—a European spacecraft for observing the solar wind from above the sun's poles—for four years (see chap. 6). More serious, NASA abandoned development of a solar telescope with a 130-cm primary mirror that would have provided a sustained view of features less than half the size of those visible for short times from the best ground-based telescopes (Jordan 1984) because, contrary to original plans, the Shuttle would rarely be available for flying the instrument (Zirin 1986; Solar telescope canceled 1986).

The reduction in the number of spacecraft with solar instruments since 1975 has been accompanied by a decline in the number of ground-based solar observing programs. Several routine monitoring programs have, according to a report from the World Data Center in Boulder (Coffey 1986), either ceased operation or lapsed into irregular activity. Two casualties have been particularly noteworthy. At the end of 1976, after 102 years of measuring sunspot positions and areas, Greenwich discontinued its program of daily sunspot photography (Newkirk 1979). Four years later Zurich abandoned its traditional responsibility as the center for sunspot statistics (see below). Two additional ground-based solar programs of great prominence appeared to be in serious jeopardy. In 1976, it took a concerted campaign by European as well as American solar physicists to get funding responsibility for the Sacramento Peak Observatory transferred from the Air Force to the National Science Foundation (Evans 1977). Since then, Sacramento Peak has been operated for the foundation by the same academic consortium that operates Kitt Peak National Observatory (Cram 1981). Its long-range future is not entirely secure, however, because important voices within this consortium want to consolidate solar observing at Kitt Peak as a means of freeing funds for other endeavors (Waldrop 1985). Meanwhile, in 1984, needing funds for the Las Campanas Observatory in Chile, the Carnegie Institution of Washington decided to stop supporting Mount Wilson Observatory. Over the next few years, the burden of responsibility for this historic observatory, including the solar telescopes that Hale, his colleagues, and his successors had used so well for eight decades (Howard 1985), was transferred to the new Mount Wilson Institute, which will have to rely on philanthropy as well as federal research grants to cover operating costs (Second life 1984; Berry 1988).

Although the solar physics community has had many reversals in these lean times, it has also managed to strengthen solar observational capabilities in several directions. Solar physicists have sent dozens of instruments into space that, by and large, were more rugged and precise than earlier instruments. In fact, thanks to projects begun in the United States, Europe, and the Soviet Union during the late 1960s and early 1970s, spacecraft with four or more solar instruments continued to go up at a good pace until the fall of 1978. Since then, although the pace has slowed, several major missions have been totally or mainly dedicated to observations of solar radiations and particles.

The best equipped solar mission since 1975 has been a NASA spacecraft for the study of flares (Covault 1979; Rust 1984). The *Solar Maximum Mission* (fig. 5.13), the agency's last major solar mission launched by expendable rocket, was put into orbit on February 14, 1980—just five years after work began on its payload at eight American and five European scientific centers. At first all seven of the satellite's instruments delivered excellent results. Soon, however, component failures in three instruments hampered their performance. Worse yet, nine months after launch, the spacecraft's pointing system went out of commission (Covault 1981a). Since the backup system could point only within a few degrees of the sun, this failure transformed the satellite from a first-rate flare observatory into modest solar monitor (see chap. 7).

Such a turn of events was not unanticipated. Thinking that Shuttle-borne astronauts should be able to retrieve and repair ailing satellites, NASA had designed the spacecraft with a graspable trunnion pin and modular electronic systems (Covault 1979, 1981b). In February 1982, after the Shuttle was operational, NASA requested permission to divert funds to a Solar Maximum Repair Mission (Beggs 1982; Chaikin 1982; Rust 1984). NASA hoped to demonstrate the Shuttle's versatility, and the solar physicists wanted access to more high-quality data. The responsible congressional committees were concerned that such a diversion would compromise other programs. They insisted that the Department of Defense should bear half the costs of the mission, since the military also had much to gain from the demonstration of the capability of Shuttle crews to retrieve and repair satellites (Kerwin 1982). The necessary arrangements were soon made (Murphy 1982). On April 6, 1984, after two hectic years of preparation, NASA sent Robert Crippen and his crew of four up in the *Challenger* to make the repair (Covault 1984; Chaikin 1984). The astronauts had difficulty getting the

Figure 5.13. Illustration of the *Solar Maximum Mission,* launched on February 14, 1980; note the apertures of the solar instrument on the top panel and the grapple fixture below the leading edge of the right-hand solar power array. (Courtesy of NASA Goddard Space Flight Center.)

satellite into the orbiter's bay, but once successful, they repaired the space-craft's pointing system and coronagraph in less than eight hours (fig. 5.14). The outcome was that—at a cost considerably below that of a comparable spacecraft but considerably above that of the most expensive ground-based solar telescope—the *Solar Maximum Mission* was ready for another five years of operation (Maran and Woodgate 1984; Woodgate 1984; Kerr 1989).

Meanwhile, at the many ground-based solar observatories that have remained open, solar physicists around the globe have continued drawing upon new electronic and computing technology to upgrade their auxiliary instrumentation (Dunn 1985). In addition, Japanese, German, Swedish, Soviet, and French solar physicists have obtained support for several good-sized evacuated solar telescopes (Engvold 1985; Dunn 1985; Schröter 1984; Stenflo 1988). By doing so, they have approached or matched the standard

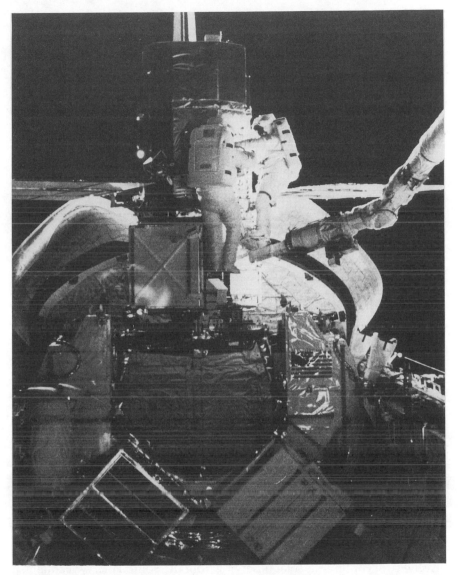

Figure 5.14. Astronauts James D. van Hoften (left) and George D. Nelson replacing the *Solar Maximum Mission*'s coronagraph electronics box in the *Challenger*'s payload bay in April 1984. (Courtesy of NASA Johnson Space Center.)

set by Richard Dunn and his colleagues at the Sacramento Peak Observatory. Indeed, with the completion of the 90-cm telescope at Tenerife in the early 1990s, the French may well surpass this standard.

Ground-based solar physicists have also secured support since the mid-1970s for observing the sun from two surprising sites—Socorro, New Mexico, and the South Pole. They were attracted to Socorro by the Very Large Array, an immense radio telescope that was built primarily for galactic and extragalactic astronomy (Tucker and Tucker 1986). Its use for studying the sun was inaugurated by the University of Maryland's Mukul R. Kundu (b. 1930), an Indian-born radio astronomer who has focused on solar research since his graduate studies in France during the 1950s. He secured observing time by arguing that the Very Large Array's ability to pinpoint the origin of radio emissions from flares would be crucial for interpreting high-resolution optical observations from ground-based facilities and NASA's forthcoming Solar Maximum Mission. He and his colleagues made their first observations at Socorro in July 1977 and May 1978 (Kundu, Schmahl, and Rao 1981). The success of these trials provided adequate justification for Kundu's group and a few others as well to use the Very Large Array as a radioheliograph during the *Solar Maximum Mission's* first year and thereafter (e.g., Roberts and Havlen 1982; Kundu 1982).

Two converging motivations gave rise to solar observing programs at the South Pole (Harvey, Pomerantz, and Duvall 1982; D. H. Smith 1989). In the mid-1970s, the geophysicist Martin A. Pomerantz, director of the Bartol Research Foundation in Swarthmore, Pennsylvania, had the idea of using the American polar station as a base for establishing an astronomical observatory. Stellar astronomers did not think the advantages of observing through the long night of the polar winter warranted going to this trouble, but a few solar physicists saw merit in observing from the South Pole. The reason for their interest was that continuous observations of the sun could be made during the antarctic summer for as many days as the skies remained cloudless. Such observations would be ideal for studying photospheric oscillations, which, if interpreted as seismic signals, could provide fresh insights on the sun's internal structure. The first solar physicist to take an interest in polar observing was Arne A. Wyller, a former Bartol colleague who had become professor of astrophysics at the Royal Swedish Academy of Sciences. In January 1979 the Swedish and Bartol groups demonstrated the feasibility of continuous solar observing from the South Pole. A year later, a French team led by Eric Fossat of Nice joined with the Bartol group in securing an uninterrupted record of solar oscillations for more than five days (Grec, Fossat, and Pomerantz 1983). Since these early successes, solar observing has been a regular activity at the American station during the antarctic summers.

Thus, although solar physicists have had to struggle for support since the mid-1970s, they can investigate many solar phenomena that earlier were beyond their reach. Their instruments in space, while fewer than in the years leading up to *Skylab* and *Orbiting Solar Observatory 8,* are generally more precise, more versatile, and more reliable. They are acquiring a new generation of high-resolution solar telescopes at ground-based facilities. They are upgrading their auxiliary instrumentation as new technologies become available. And to judge from the examples provided by the new solar observing programs at the Very Large Array and the South Pole, they still have opportunities for fresh ventures.

Careers in Solar Physics since *Sputnik,* 1957–1988

The growth in support for solar physics since *Sputnik* has given the field's recruits and practitioners new options for building their careers and going about their research. In comparison with their predecessors, they have had easy access to fellowships, assistantships, regular positions, grants, and contracts. They have also had the use of a broader array of instruments, techniques, and theories for advancing knowledge of the sun. As solar physicists have followed up these opportunities, their career paths and research practices have become increasingly diverse.

The influence of the new patronage on the solar physics community can be gauged by looking at its members who have worked mainly or wholly in this era. Consider, for instance, twenty-five solar physicists under forty in 1957 who were frequently mentioned in a recent poll of the community as being among its leaders in 1980 (see table 5.3). This list reveals that despite the increased participation of women and the citizens of non-Western nations in solar physics, males of European ancestry or birth have remained predominant among the scientists emerging as the field's leaders. In particular, all the persons listed were men, and all but Kundu were born in the United States or Europe. This list also suggests that despite the comparatively small number of universities granting doctorates in solar physics and despite the comparatively large number of newcomers to the field, scientists who wrote theses on solar or stellar subjects have tended to win recognition as its leaders. In fact, 60 percent of those identified as leading solar physicists in 1980 wrote dissertations in solar physics, and 20 percent did so in the closely related field of stellar physics. Finally, the list gives some indication of the effect of post-*Sputnik* patronage on the employment opportunities of leading solar physicists. Almost half of those on the list were at institutions that either did not exist in 1957 or had never before hired a solar physicist. Like their predecessors, some of the leading solar physicists made their marks by concentrating on ground-based telescopes and observational

Table 5.3
Twenty-five Solar Physicists Who Were under Forty in 1957 and among the Field's Leaders in 1980
(in Order of Entering the Field)

Entry		Birth		Doctoral Dissertation		Employing Institution (1980)
Year	Name	Year	Nation	Year	Subject	
1948	C. de Jager	1921	Neth.	1952	Solar physics	Lab. Sp. Res. Utrecht
1949	Z. Svestka	1925	Czech.	1949	Astrophysics	Lab. Sp. Res. Utrecht
1953	R. G. Athay	1923	USA	1953	Solar physics	High Altitude Obs.
1955	E. N. Parker	1927	USA	1951	Astrophysics	U. of Chicago
1955	H. Zirin	1929	USA	1953	Stellar physics	Cal. Inst. Tech.
1956	G. A. Newkirk	1928	USA	1954	Solar physics	High Altitude Obs.
1957	R. F. Howard	1932	USA	1957	Stellar physics	Hale Obs.
1957	M. R. Kundu	1930	India	1957	Solar physics	U. of Maryland
1958	F. L. Deubner	1934	Germany	1969	Solar physics	Würzburg U.
1959	J. M. Beckers	1932	Neth.	1964	Solar physics	Mount Hopkins Obs.
1960	G. E. Brueckner	1934	Germany	1961	Solar physics	Naval Research Lab.
1961	J. A. Eddy	1931	USA	1962	Solar physics	High Altitude Obs.
1961	P. A. Sturrock	1924	GB	1951	Math physics	Stanford U.
1962	R. W. Noyes	1934	USA	1963	Solar physics	Harvard-Smithsonian Ctr. for Ap.
1963	J. W. Harvey	1940	USA	1969	Solar physics	Kitt Peak Obs.
1964	J. M. Wilcox	1925	USA	1954	Nuclear physics	Stanford U.
1965	A. H. Gabriel	1933	GB	1956	Physics	Rutherford-Appleton Lab.
1965	P. A. Gilman	1941	USA	1966	Solar physics	High Altitude Obs.
1965	J. O. Stenflo	1942	Sweden	1968	Solar physics	Zurich Tech. Inst.
1966	A. J. Hundhausen	1936	USA	1965	Solar physics	High Altitude Obs.
1966	G. L. Withbroe	1938	USA	1965	Solar physics	Harvard-Smithsonian Ctr. for Ap.
1968	R. K. Ulrich	1942	USA	1968	Stellar physics	UCLA
1971	J. C. Brown	1947	GB	1973	Solar physics	Glasgow U.
1972	D. O. Gough	ca. 1940	GB	1966	Stellar physics	Cambridge U.
1972	E. R. Priest	1943	GB	1969	Stellar physics	U. of Saint Andrews

Sources: A poll of some twenty solar physicists around the world during 1984–85; annual astronomy bibliographies; biographical directories; dissertation indexes; and in many cases, the scientists themselves.

Note: Year of entry has been defined as the date of the scientist's first article in solar physics.

results. A majority, however, did so in less traditional ways. Some gained recognition through their contributions to space-based solar observing or theoretical solar physics, new specialities that came of age after 1957. And some built reputations by moving from specialty to specialty within the subdiscipline. A closer look at the careers of Newkirk and Stenflo serves to flesh out these generalizations.

Gordon Newkirk

When *Sputnik* went into orbit, Gordon A. Newkirk, Jr., was on the staff of Boulder's High Altitude Observatory (Newkirk 1983b, 1984; Eddy and MacQueen 1986). He had completed his Ph.D. four years before under Leo Goldberg at the University of Michigan with a study of the infrared spectrum of carbon monoxide in the solar atmosphere. He had lost some momentum, however, during a two-year tour of duty in the Army Signal Corps. Moreover, after taking the job at Boulder in late 1955, he had needed to retool in order to participate in the High Altitude Observatory's coronal research program. By 1958, at age thirty, he was hitting his scientific stride. His first significant undertaking was to use coronal data collected at the Climax station during the International Geophysical Year to interpret concurrent radio observations of the corona made at stations around the world.

While still in the early stages of this study, Newkirk embarked on a project that would gain the High Altitude Observatory an important role in the American space program. During his introduction to coronal work, he had realized that even the best mountain stations would never provide views of the *outer* corona comparable to those sometimes captured during total eclipses. The only way to obtain a prolonged look at this coronal region would be to get coronagraphs into space. In order to be competitive when NASA was ready to orbit such instruments, he set himself the goal of photographing the outer corona from a stratospheric balloon. In March 1964, he and his colleagues achieved their goal (Newkirk and Bohlin 1965). Their coronagraph secured pictures of coronal streamers extending beyond five solar radii. About this same time, Newkirk and his former doctoral student John A. Eddy beat out a team from the Naval Research Laboratory in competing for the coronagraph contract for the Advanced Orbiting Solar Observatory. Although NASA soon canceled this satellite, their contract eventually led to the High Altitude Observatory's successful coronagraphs on *Skylab's* Apollo Telescope Mount and the *Solar Maximum Mission*.

In 1970, well before *Skylab* was launched, Newkirk was obliged to relinquish the lead role in the coronagraph project. He had his hands full as Firor's successor at the High Altitude Observatory, which, with its scientific staff of almost twenty solar physicists, was the world's most important all-

Figure 5.15. Gordon Newkirk about 1975, while director of the High Altitude Observatory in Boulder. (Courtesy of the National Center for Atmospheric Research/National Science Foundation.)

around center for solar research. He found that the directorship became ever more absorbing, and stressful, as the funding climate changed in the early and mid-1970s. Nonetheless, by instituting more rigorous personnel practices, he and the observatory's other senior scientists managed not merely to maintain but to improve its standing.[7] Besides his work at the High Altitude

7. Note that five of the solar physicists listed in table 5.3 were employed by the High Altitude Observatory in 1980—Athay, Newkirk, Eddy, Gilman, and Hundhausen.

Observatory, Newkirk (fig. 5.15) had numerous other administrative and committee responsibilities. He served, for instance, as chair of NASA's Skylab Workshop Series (1974–79), president of the International Astronomical Union Commission 10 (1975–79), chair of the Hale Prize Committee (1976–81), and chair of NASA's Working Group for Study of the Solar Cycle from Space (1978–81).

In 1979, after a decade of administrative work, Newkirk resumed fulltime scientific research. His main activity was writing review papers on variations in solar luminosity, an issue his former student Eddy had helped bring to the fore (see chap. 7). Newkirk did not, however, have long to enjoy his return to science. In late 1985, without ever again matching "the excitement of seeing the first out-of-eclipse images of the outer corona" (Newkirk 1984), he died of cancer.

Jan Stenflo

Fourteen years younger than Newkirk, the Swede Jan O. Stenflo was in secondary school at the time of *Sputnik* (Stenflo 1984b, 1986, 1988). He was an avid amateur astronomer with a 15-cm reflecting telescope of his own construction. By 1962, when he entered Lund University, his hope was to become a stellar or galactic astrophysicist. A year later, however, he accepted an offer from Sweden's veteran solar physicist Yngve Öhman to serve as an observer at the Swedish station on Capri. Stenflo got a start in solar physics during his six months of service in the flare-monitoring program there. In his spare time, he measured the brightness of flares and prominences and tried out some new observational techniques. Upon returning to Lund in 1964, he wrote up the results of his study and completed his requirements for graduation.

Although Lund was not at all oriented to solar physics, Stenflo elected to pursue his doctorate there. He was confident that he could do the requisite observational work elsewhere. As circumstance would have it, he ended up doing his observing in the Soviet Union. The preliminary arrangements were made, unbeknown to Stenflo, at a NATO conference on magnetic fields in the cosmos held at Newcastle in the fall of 1964. Thanks to the easing of Cold War tensions, the conferees included two participants from non-NATO countries—Hannes Alfvén from Sweden and Andrei Borisovich Severny from the Soviet Union. Alfvén, whose magnetohydrodynamic theory of the solar cycle was jeopardized by Mount Wilson's claims that the sun's general magnetic field reversed every eleven years, sought to persuade Severny of the unreliability of solar magnetograms on the grounds that they were not resolving the field's fine structure. Severny, the director of the Crimean Astrophysical Observatory, was well enough connected to propose then and there that a collaborative attempt be made to test Alfvén's doubts using his institute's magnetograph. After the conference, Alfvén asked Öhman to recommend someone for the job, and Öhman proposed Stenflo. The result was that Stenflo, backed by the Swedish and Soviet academies, spent the summers of 1965 and 1966 at Severny's institute. He confirmed Alfvén's claim that available magnetographs were not resolving the fine structure of the solar magnetic field, but he concluded that this did not invalidate the observed polarity reversals of the field. Stenflo used this work to take his Ph.D. at Lund in April 1968.

During the next decade, Stenflo moved quickly to the forefront of solar physics. Although Lund was his base, he was often away not only for conferences but also for research stints at Mount Wilson, Boulder, Kitt Peak, Locarno, and the Soviet space center at Kapustin Yar. His most important piece of work was begun at the McMath Solar Telescope at Kitt Peak during September 1971. Assisted by John Harvey, he found indirect yet compelling spectroscopic evidence there for recent suggestions that magnetic fields emerged from the photosphere in intense fluxtubes that covered less than 1 percent of the solar surface. Stenflo and many other solar physicists were soon busy exploring the implications of this new view of solar magnetism.

In January 1972, before he had written up this research, Stenflo was in Moscow with a small Swedish delegation to make arrangements for a joint space project in the Interkosmos series. Severny was promoting the venture from the Soviet side so an agreement was soon concluded. In deciding how to use this opportunity, Stenflo was mindful that budgetary, size, and telemetry constraints ruled out any attempt to put up a solar instrument that would compete directly with those on American satellites. He needed to try something qualitatively new. He opted for measuring the polarization caused by radiative scattering in the solar atmosphere. Over the next three years, he supervised the development of an ultraviolet polarimeter in Sweden, but he and his colleagues were barred by the Soviet military from taking part in spacecraft integration or launch operations. Instead, once the satellite was in orbit, Stenflo would be allowed to journey to the Crimean Astrophysical Observatory to monitor his instrument's performance. Actually, nothing came of this plan because the launch in June 1975 was a failure.

The Soviets soon agreed to produce a new satellite to fly the reserve model of the Swedish instrument. They also agreed—after being shown satellite pictures of Kapustin Yar that had appeared in the Western press—that no further secrets would be lost if Stenflo and two Swedish engineers were allowed to participate in prelaunch integration and mission monitoring at the launch site. Although *Interkosmos 16* made orbit in July 1976, the performance of Stenflo's polarimeter was disappointing. The satellite (fig.

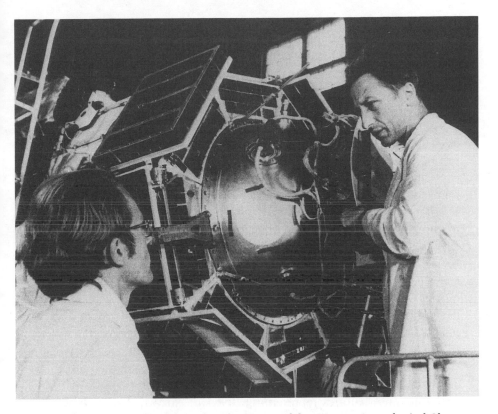

Figure 5.16. Jan Stenflo (left) and Andrei Bruns of the Crimean Astrophysical Observatory in the integration hall at the Soviet launch site of Kapustin Yar on July 24, 1976, three days before the launch of *Interkosmos 16;* note that the Swedish spectrometer polarimeter was in the box between Bruns's hand and head. This scene was shown on Soviet television the day after the launch. (Courtesy of Jan Stenflo.)

5.16), which had been readied for flight without benefit of clean-room facilities, had considerable outgassing. In anticipation of this possibility, the instrument was not unshuttered until the satellite's second day in orbit. Even so, the polarimeter's optical transmission rapidly declined to the point that its data were of marginal scientific value.

In the mid-1970s, Stenflo began applying for professorships in Sweden. On account of the Swedish orientation to galactic astronomy, and perhaps his youth, however, he was passed over on his first four tries. Resilient, he applied in 1979 for professorships in Lund and Zurich. First to respond, the Swiss offered him Max Waldmeier's chair. Stenflo accepted at once. He was attracted by Zurich's emphasis on solar physics, its plans for a comprehensive institute of astronomy, its central location in Europe, and its proximity to the Alps. His most important decision upon taking up duties in the spring of 1980 was that Zurich must abandon the tradition, started by Rudolf Wolf more than a century before, of collating sunspot data from various observatories and issuing reports of the relative sunspot number. A survey of the contributing observatories revealed that the Brussels Observatory had both the motivation and the resources to assume this responsibility. At year's end, the transfer was complete. By taking this initially unpopular step, Stenflo freed up the resources needed for Zurich to move back to the frontier of solar research.

Besides revitalizing Zurich's solar physics program, Stenflo has helped reinvigorate the cooperative project to develop the Large European Solar Telescope (LEST). The Joint Organization of Solar Observations had lost momentum in the late 1970s, partly because of the death of its founder, K. O. Kiepenheuer, and partly because of the general slump in support for science. In the early 1980s, however, Stenflo and others on the board recognized that it was crucial to get off the report-writing treadmill and begin serious fundraising. A formal international organization with its own budget was needed. To this end, they arranged for the Swedish Academy to host an organizing meeting for the LEST Foundation in April 1983 (Stenflo 1984a; Engvold 1985a). The foundation's audacious goal was to construct a 240-cm evacuated or helium-filled steerable telescope with computer-controlled and polarization-free optics having 0.1 arc-second resolution. At this point, only four observatories representing Sweden, Norway, Switzerland, and Israel were ready to commit themselves to the venture. Stenflo, who was elected president, immediately began a campaign for broader participation. That fall he succeeded in recruiting observatories representing Italy and Germany. During the next two years, he succeeded in reaching beyond Europe in his recruiting (Stenflo 1984c, 1986). This prompted a redefinition of the acronym LEST as "Large Earth-Based Solar Telescope." Although the LEST Foundation has grown to nine members, contracted with the Nordic Optical Telescope Association in Denmark for design work, and chosen a site adjacent to the Swedish Solar Observatory on La Palma Island in the Canaries, it has yet to raise funds for construction (Stenflo, Engvold, and Hillerud 1990). Still, with any luck, Stenflo and his colleagues will be operating LEST before the century is out.

The careers of Newkirk and Stenflo illustrate some of the ways solar physicists have capitalized on the support that has become available since *Sputnik*. Both got their start in ground-based optical programs—Newkirk in infrared spectroscopy and coronal research and Stenflo in flare monitoring and magnetographic research. Building on these foundations, they followed

divergent paths. Newkirk worked on coronagraphs for stratospheric balloons and *Skylab* for a dozen years, then focused on administration for a
decade and ended his career as a proponent of new perspectives on solar-
irradiance variability. Stenflo, deftly taking advantage of research opportunities in Europe, the Soviet Union, and United States, has attested to the
abiding vitality of ground-based solar physics by combining observational
skill and interpretive insight in optical studies of solar magnetic fields. And,
in reinvigorating LEST, he has both drawn upon and contributed to solar
physics' long tradition of international cooperation.

Conclusion

Patronage for solar physics grew rapidly from the late 1950s into the early
1970s. Impressed by the military and political implications of *Sputnik*, leaders of advanced nations around the world sharply increased support for
science and technology. Meanwhile, their scientific and technical advisers,
mindful of solar physics' immediate prospects and utility, devoted a share of
the new funds to improving solar observational capabilities. In the 1970s,
changed priorities brought an end to the era of plenty in research and
development. Having fared relatively well during the bountiful times following *Sputnik*, solar physicists have been hard pressed to come up with the
resources needed for new space missions and ongoing programs of research
at ground-based facilities. Still, the current level of worldwide support for
solar work remains well above that of the 1950s.

Thanks to increased support, solar physics has become a substantial enterprise. The number of established solar physicists around the globe has
reached some three hundred. They have created subdisciplinary associations and a journal. They have gained access to particles from the sun's
atmosphere and neutrinos from its interior as well as the entire electromagnetic spectrum. And while doing so, they have attained ever higher photometric, spectral, spatial, and temporal resolutions. But to what effect?
More specifically, how have the various improvements in solar observational capabilities contributed to the progress of solar physics in the past
three decades? The ensuing case studies will suggest answers to these
questions.

PART THREE

Solar Research in the Space Age

An important point in part 2—some would say the most important point is that the space-based observing capabilities developed since *Sputnik* have complemented rather than supplanted ground-based telescopes in solar research. The two chapters in part 3 seek to reinforce this message in the most trenchant way possible. Neither, therefore, focuses on one of the many areas of solar research that had achieved maturity before the late 1950s. In such areas one would not be surprised to find that space-based research has gone hand in hand with ground-based research. Rather, the chapters deal with phenomena—the solar wind and the variability of the solar constant— that remained within the realm of speculation and indirect inference until observed with spacecraft-borne instruments. Although observations from space marked decisive turning points in thinking about these phenomena, the following narratives will show in each case that subsequent research involved an ever more subtle interweaving of data obtained with ground-based as well as space-based instruments.

Besides affirming the continuing importance of ground-based observations of the sun, the two case studies illustrate, yet again, the role of outsiders in enlarging the domain of solar physics. Few of the space scientists behind the observations that conclusively established the existence of the solar wind and solar-irradiance variability thought of themselves as solar physicists at the time. Most, in fact, have never been much involved with solar physics. However, a good number of the scientists following up on these observational breakthroughs have either been solar physicists or come to think of themselves as such. The ensuing reconfiguration of solar physics' domain is, to judge from some comments on the present work, still under way. More than one solar physicist has wondered how a book about the history of solar science could devote so much attention to topics outside its mainstream.

In coping with the many difficulties that confront anyone writing about the history of contemporary science, I have used a variety of tactics. I have relied on review articles, interviews, and citation trails to come up with story lines in areas where the participants often still disagree about how current work will play out. I have consulted review articles, scientific journalism, conference proceedings, and citation records in deciding which among the many participating scientists to highlight. I have depended on biographical directories, interviews and correspondence, a few scientists' archives, and the records in the NASA History Office for crucial background information about the leading figures in my narratives. Finally, I have learned much from the scientists' own attempts to communicate in laymen's terms about how to describe their findings and interpretations without getting lost in technicalities.

CHAPTER SIX

The Solar Wind

1957–1970

During the first half of this century, solar physicists and geophysicists gave increasing credence to the possibility that the sun sent matter as well as light into space. The more they looked into the question, the more it seemed that aurorae and variations in the terrestrial magnetic field were caused by solar corpuscles impinging on the earth's outer atmosphere. Building on this idea, Ludwig Biermann suggested in 1951 that the gas tails of comets were blown persistently away from the sun by an outward flux of solar particles. Six years later, Eugene Parker hypothesized that Biermann's efflux originated as a continuous, high-velocity expansion of the hot solar corona. However, such inferences were bound to remain controversial so long as scientists were unable to make direct observations of what Parker dubbed the "solar wind."

This chapter traces the solar wind's entry into the realm of observational science. After examining the background and early development of Parker's theory, it recounts how Soviet and American space scientists confirmed the solar wind's existence with direct observations of the interplanetary medium. It goes on to describe the exploration of the solar wind during the middle and late 1960s, emphasizing the discovery of a magnetic-sector pattern in the wind and the initial attempts to trace this structure back to the sun. The chapter closes by analyzing the new specialty's development down to about 1970 and by sketching, in the form of an epilogue, the subsequent emergence of a three-dimensional picture of the solar wind.

As might be expected from earlier chapters, the scientists who did the most to inaugurate solar-wind studies did not come to this research with extensive backgrounds in solar physics. Solar physicists had been speculating since the late nineteenth century about the existence and effects of corpuscular emissions from the sun. On account of their strong orientation to optical research, however, they were disinclined to take advantage of the opportunities for direct, sustained observations of solar particles that were opened up by the creation of national space programs. Consequently, the chief protagonists in the emergence of solar-wind research tended to be physicists who were attracted to the challenge of observing charged particles and magnetic fields in interplanetary space. Initially at least, they viewed themselves as space scientists with a special interest in interplanetary phenomena, not as the creators of a new observational specialty within the broad field of solar physics.

As information about the solar wind accumulated, however, space scientists and solar physicists alike came to realize that its interpretation would require them to join forces. The first attempts to correlate space-based observations of the solar wind with ground-based observations of solar features were made during the 1960s. These endeavors yielded some suggestive results that it turned out, heralded the swift emergence during the next decade of a robust model of the wind's three-dimensional structure.

Parker's Hypothesis, 1957–1959

In the late 1950s, Eugene N. Parker (see fig. 5.12) significantly advanced theorizing about the emission of particles from the sun when he hypothesized that the solar corona was expanding at a high velocity into interplanetary space. The roots of this hypothesis were not in solar physics. Rather, they were in the studies of such apparently disparate phenomena as aurorae, geomagnetic disturbances, cosmic rays, and the gas tails of comets.

Parker's Immediate Background

About 1950, when Parker was starting his career, not only solar physicists but also those geophysicists who were interested in the sun's influence on the earth generally believed that the sun sent matter as well as light into interplanetary space (Kiepenheuer 1953). Indirect evidence for this belief had been mounting since the mid-nineteenth century (Maunder 1904b; Chapman and Ferraro 1929; Chapman and Bartels 1940; Schröder 1984). Observers had found that strong aurorae and moderate disturbances in the earth's magnetic field tended to recur in twenty-seven days. This recurrence period, which matched that of the sun's rotation when viewed from the earth, indicated that these phenomena had a solar origin. They had also found that aurorae and geomagnetic disturbances were more intense near the earth's poles. Taken together with the recurrence tendency, this polar property indicated that the agent carrying energy from the sun to the earth was subject to focusing by the terrestrial magnetic field. Solar physicists, auroral physicists, and geomagneticians were wont to presume that this agent must be streams of ions and electrons. They supposed that particular regions of the rotating sun spewed out this "solar corpuscular radiation" for periods lasting up to months, causing the twenty-seven-day recurrence tendency. Numerous attempts to correlate particular solar features with geomagnetic disturbances had failed, however. As of 1950, therefore, the nature of the "M regions," as the German geophysicist Julius Bartels (1932) had named the solar sources of the magnetic storms, was still a conundrum.

Besides thinking that the mysterious solar M regions begat corpuscular streams that in turn generated recurrent aurorae and geomagnetic disturbances, solar physicists and geophysicists believed by 1950 that a weak, continuous flux of charged particles was evaporating from the sun. This hypothesis was consonant with the persistence of faint aurorae at the earth's poles and the incessant minor fluctuations in the earth's magnetic field (Kiepenheuer 1953; Van de Hulst 1953). They surmised that the continuous solar flux gave rise to these phenomena upon striking the earth's field and then spiraling poleward into the upper atmosphere. However, their interest in this comparatively unspectacular form of solar corpuscular radiation was limited to its possible role in the mass and energy balance of the corona.[1]

During the early and mid-1950s, as Parker was moving beyond his doctoral studies, there were two important additions to this matrix of facts and ideas about solar emanations. The first was the inspiration of the German theoretical astrophysicist Ludwig Biermann (fig. 6.1), who had ended up after the war at the Max Planck Institute for Physics in Göttingen. He argued that the gas tails of comets always pointed away from the sun because they encountered the same continuous outflow of solar matter that caused geomagnetic fluctuations and faint polar aurorae (Biermann 1951, 1952, 1953, 1985). He began exploring this possibility when his calculations indicated that, contrary to common belief, solar radiation pressure would exert only a small fraction of the force needed to cause the observed acceleration of knots and other features along the gas tails of comets. His idea was that the charged particles flowing from the sun collided with the ions in a comet's tail, imparting some momentum to them. He estimated that the solar particles would produce the observed antisolar acceleration if their velocity was between 500 and 1000 km/sec and their density between 100 and 1000 ions and electrons/cm^3 at the distance of the earth.

Biermann's basic idea struck his contemporaries as an interesting possibility, but few of those who accepted it were willing to agree that the

1. Solar physicists and geophysicists also believed by 1950 that major flares sometimes blasted very energetic particles—solar cosmic rays (as distinct from galactic cosmic rays)—into the solar system. Research since 1950 on this form of solar corpuscular radiation is not covered in this chapter.

Figure 6.1. Ludwig Biermann in the early 1950s, shortly after he suggested that the continuous solar efflux was responsible for the antisolar orientation of the gas tails of comets. (Courtesy of Mrs. Biermann and Remar Lüst.)

continuous solar efflux could be so intense. For instance, the Swedish theoretical astrophysicist Hannes Alfvén (1957) argued that a much weaker solar efflux would produce the observed acceleration if electromagnetic interactions caused the momentum transfer. Paying little heed to such revisions, Biermann (1957, 110) went beyond his original proposal to suggest that "the solar corpuscular radiation" was identical with "the interplanetary gas." He supposed that the continuous flow of matter from the sun would sweep any stationary gas out of the solar system.

The second important contribution to thinking about solar emanations during the 1950s was made by the veteran solar-terrestrial physicist Sydney Chapman. About 1955, perhaps as a diversion from his labors as president of the International Geophysical Year, he calculated how far the corona extended into the solar system. He found that a very hot corona consisting mainly of electrons and protons would, on account of the efficiency of electrons in conducting heat, extend out beyond the earth's orbit. His confidence in this conclusion was reinforced by various studies of the interplanetary medium's density. Alfred Behr and Heinrich Siedentopf (1953) of Göttingen's Astronomical Institute, for example, had recently argued—on the basis of their measurements of the polarization of the diffuse zodiacal glow seen before dawn and after dusk—that the density of electrons near the earth must be about 600/cm³. In writing up his idea, Chapman (1957, 9) maintained that "the coronal gas" quite possibly "surrounds the earth for long periods."

Parker and His Theory

With the advantage of hindsight, Parker's path to his solar-wind theory is easy to discern (Parker 1985b). He took his doctorate at the California Institute of Technology in 1951 with a theoretical study of possible hydromagnetic effects in the rarefied interstellar medium. His first academic job was at the University of Utah. There, encouraged by the solar theorist Richard N. Thomas and the wide-ranging theoretical physicist Walter M. Elsasser, he extended his domain of competence by developing hydromagnetic theories for the formation of sunspots, the origin of terrestrial and solar magnetic fields, and the acceleration of cosmic rays. By 1955 he was getting a reputation as a theorist of considerable promise. Indeed, Parker was offered research associateships by Walter Roberts at the High Altitude Observatory in Boulder, Colorado, and by John A. Simpson (b. 1916) at the University of Chicago's Enrico Fermi Institute for Nuclear Studies. He opted to join Simpson, who had a large grant from the United States Air Force for research on cosmic rays (Simpson 1985; Needell 1987).

When Parker arrived in Chicago, Simpson's group was already well into an investigation of the sun's ability to influence which galactic cosmic rays reached the earth. They had established that the sun's modulating role was most pronounced at the lower end of the cosmic-ray energy spectrum. One observation—that the energy of the weakest cosmic rays reaching earth varied with the sunspot cycle—suggested that the level of solar activity influenced the ability of cosmic rays to penetrate the interplanetary medium (Meyer and Simpson 1955). Another observation—that maxima in the flux of low-energy cosmic rays had a twenty-seven-day recurrence tendency, as did maxima in geomagnetic activity—suggested that the particles and fields emanating from Bartel's hypothetical M regions might also influence the medium's penetrability. Simpson, in fact, took the initiative in exploring this idea with the Babcocks, who, as we saw in chapter 4, were just then arguing that unipolar magnetic regions in the photosphere were M regions (Simpson, Babcock, and Babcock 1955). To judge from his notes on a letter raising a question about this paper (Davis 1955), his very first assignment for Parker was to consider how solar emanations might affect the interplanetary medium.

Parker swiftly acquainted himself with the latest work on interplanetary conditions. In talks with the German Reimar Lüst,[2] another of Simpson's postdoctoral fellows, he learned about Biermann's idea that an intense outflow of particles from the sun caused the antisolar alignment of the gas tails of comets (Parker 1985b). And on a visit to Boulder during January 1956, he spoke at length with Chapman about the grounds for thinking that the earth was traveling in the extended solar corona (Roberts 1956; Parker 1956; Chapman 1957). At this time Parker was still so preoccupied with explaining the sun's role in modulating the flux of cosmic rays that he paid no heed to the contradiction between Biermann's dynamic explanation of comet tails and Chapman's static theory of the solar corona.

Parker's attention was drawn to this contradiction in mid-1957 during a long conversation with Biermann, who was visiting Simpson's laboratory (Jacobsen and Parker 1973). Having just been appointed to an assistant professorship in the University of Chicago's Department of Physics, Parker may well have been on the alert for problems that would enable him to establish his independence. He came away from the conversation convinced of the cogency of Biermann's case that the continuous solar efflux was strong enough to sweep stationary gas out of the solar system. Yet his prior discussion with Chapman had given him an appreciation for the arguments in favor of the idea that the solar corona extended beyond the earth. He now

2. Reimar Lüst (b. 1923) became director general of the European Space Agency in 1984.

recognized that the two theories were in conflict—either Biermann's continuous efflux would sweep away Chapman's extended corona, or Chapman's corona would block Biermann's efflux. Parker saw that he must choose between, or reconcile, the two pictures of the sun and its relation to the interplanetary medium. In the fall of 1957 he had a promising insight—perhaps Biermann's continuous emission of matter from the sun was nothing other than the expansion of Chapman's solar corona. To evaluate this possibility, he added dynamic terms to Chapman's static equations for the corona and solved for the simple case of a spherically symmetric, nonrotating, nonmagnetic sun. He obtained results that were consistent with Biermann's estimates of the velocity and density of the solar efflux.

In November 1957, trusting in the significance of his work, Parker sent a preliminary report (1958a) to the *Physical Review*. A month later, he displayed the first diagram of his theory (fig. 6.2) at a symposium on the physics of plasmas (gases consisting of charged particles). Early the following year, he submitted a major paper on his theory to the *Astrophysical Journal*. His starting point here was "Biermann's suggestion that gas is . . . streaming outward in all directions from the sun with velocities of the order of 500–1500 km/sec [and] interplanetary densities of 500 ions/cm³." He pointed out that Chapman's hydrostatic equations for the corona did not yield an "equilibrium solution with vanishing pressure at infinity." This failure indi-

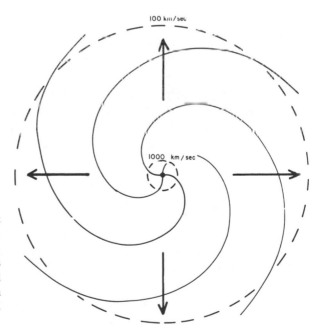

Figure 6.2. Eugene Parker's first sketch of the solar wind; the outward spiraling lines indicate the direction of flow, and the dashed lines show how the wind's speed would affect its angle of impinging upon the earth. (Source: Parker 1958b, 79.)

cated that, in addition to whatever evaporation might be occurring, there must be a continuous "outward hydrodynamic expansion of [coronal] gas." He went on to argue that expansion resulting from a "coronal temperature of 2 or 3 × 10⁶°K over an extended region around the sun would seem to be, then, the simplest origin of the outflowing gas suggested by Biermann." Parker also considered how this solar efflux would affect the magnetic fields in interplanetary space. It appeared that the expanding gas would draw magnetic lines of force out of the corona far into the solar system. Because of the sun's rotation, the resultant interplanetary field would have a spiral pattern in the sun's equatorial plane (Parker 1958c, 664, 666, 671).

Over the next two years, Parker followed up his ambitious paper with many colloquia, lectures, and articles. He soon named the sun's outflowing gas the "solar wind" in order to emphasize its hydrodynamic character (Parker 1958d, 1445, 1958e, 171). He investigated how the typical wind with its associated magnetic field would modulate cosmic rays, how it might produce the polar auroral zones, and how flare-generated gusts in the wind might generate geomagnetic storms. He also inquired how the corona was kept hot enough to sustain the solar wind. This was a serious problem for his theory. According to the values for the wind's velocity and density that he had adopted from Biermann's early papers, the solar wind would carry about a hundred times more energy away from the corona than did electromagnetic radiation. He made the problem somewhat more tractable by following Biermann in reducing the value assigned to the wind's density to 100 ions/cm³. This reduction had the double advantage of lowering the energy requirements by a factor of five and dropping the hypothesized coronal temperature to 2,000,000 K, which was in better agreement with Edlén's coronal line identifications. Still, his theory required a much more effective coronal heating mechanism than any that had been proposed up to that time. Parker's solution to the problem was to suggest that hydromagnetic waves propagating upward from the photosphere dissipated their energy in the coronal plasma by magnetically accelerating the fastest protons encountered there to still higher velocities. Such "suprathermal particle generation," he argued, was "the source of the solar corona and the solar wind" (Parker 1958f, 683).

Parker's theory got a frosty reception (Parker 1985b). The referees for the *Astrophysical Journal* were so critical that he secured publication only by taking the matter up with the editor—his colleague, the preeminent theoretical astrophysicist Subrahmanyan Chandrasekhar. Some of the skeptics, doubting the necessity of invoking an intense efflux of particles from the sun to explain the orientation of comet tails, thought Parker was building on sand. In November 1958, for instance, Chapman (1959, 478) commented at the Royal Society's Discussion of Space Research that "Biermann's proposal [is] not yet fully convincing as the basis for the remarkable implications discussed by Parker. I understand that the cometary evidence is now being re-examined. . . . It is generally agreed that from time to time there is local ejection of matter from the sun, with speed and density perhaps similar to those proposed by Biermann. But it cannot yet be considered certain that such ejection is continual over the whole sun." Other skeptics argued that the solar-wind solution to the hydrodynamic equations for the corona was not unique. Joseph Chamberlain (1960) of the University of Chicago's Yerkes Observatory was especially outspoken. He not only criticized Parker's theory as being arbitrary but also advanced an alternative model that predicted a much lower velocity for the expanding plasma—a solar breeze.

By spring 1959, Parker realized that the controversy engendered by his theory would not be settled until direct observations of the interplanetary medium were made from spacecraft. He was confident that his position would be vindicated. In April, for instance, Parker (1959, 1678) told those attending the Symposium on the Exploration of Space held in Washington, D.C., that

> the most crucial piece of evidence [for the solar-wind theory] will come not too long from now through observations that are being planned by Professor Rossi at MIT. Rossi proposes to measure directly the gas blowing outward from the sun, by a plasma probe flown in a vehicle included in the NASA program. The apparatus will determine the density and energy spectrum of both ions and electrons by a sophisticated form of ion trap. The observations will tell us in detail the solar wind velocity, density, temperature, and time variations. We should [then] be able to proceed with a more elaborate theoretical model than the spherically symmetric one already treated.

Space observations turned out to be every bit as important as Parker predicted. But they were much more difficult to obtain than he or anybody else anticipated.

The First Direct Observations of the Solar Wind, 1959–1962

While Parker and other theorists were debating the existence and character of the solar efflux, several scientists were preparing to make direct observations of the plasma in interplanetary space. Their attitude toward the contending theories was essentially practical. Mainly experimental physicists by background, they were not inclined to endorse one model or another before the measurements were in. Rather, they kept the competing theories in mind when designing their instruments and dramatized the significance of their results by reference to the ongoing controversy. Soviet scientists were the first to measure the solar efflux. But two American teams made the observations that proved decisive for Parker's theory.

Lunik 2's *Observations*

Konstantin I. Gringauz, a radio physicist with the Soviet Academy's Radio Engineering Institute in Moscow, led the team that carried out all the early Soviet investigations of the interplanetary plasma. He and his colleagues got their start in space research by designing instruments for geophysical rockets (Gringauz 1987; Behn 1968; Guendel 1968). In May 1957 they began using rocket-borne radio transmitters to study how the density of electrons in the ionosphere varied with altitude. In August of the same year, they extended the scope of their research by using rocket-borne ion traps to investigate the altitude dependence of ions in the upper atmosphere. This second line of research prepared the way for their participation in the Sputnik program. In fact, three weeks before the launch of *Sputnik 1* on October 4, 1957, Gringauz and a colleague reported how two traps mounted on booms extending from opposite sides of a satellite could be used to measure ion concentrations along a spacecraft's orbit (Gringauz and Zelikman 1957). Eight months later, *Sputnik 3* carried just such traps into space.

Sometime in 1958, probably before *Sputnik 3* was launched, Gringauz's group was chosen to develop ion traps for the initial series of three Soviet lunar probes. This was a major opportunity. The Luniks, as the Soviet probes came to be called, would be the first good-sized spacecraft to venture into the interplanetary realm. The group immediately set about learning what conditions might be encountered en route to the moon. Their literature review, which was carried out before Parker's theory was available in the Soviet Union, indicated that their instruments should be capable of distinguishing between a stationary plasma and fast corpuscular streams. In pursuing this objective, Gringauz and his colleagues (Gringauz et al. 1960; Gringauz 1961) used the same basic strategy from Lunik to Lunik. Four ion traps were embedded at strategic locations on each spacecraft to ensure that, no matter how the spacecraft might be tumbling, at least one instrument would be collecting ions. Each trap (fig. 6.3) had an external grid that was charged to a constant potential ranging between small negative and small positive values—from −10 to +15 volts on *Lunik 1* and *2* and from −19 to +25 volts on *Lunik 3*. If the interplanetary medium should turn out to be stationary, the number of ions collected by the traps would decline from a maximum for the trap with the most negative grid to zero for the trap with

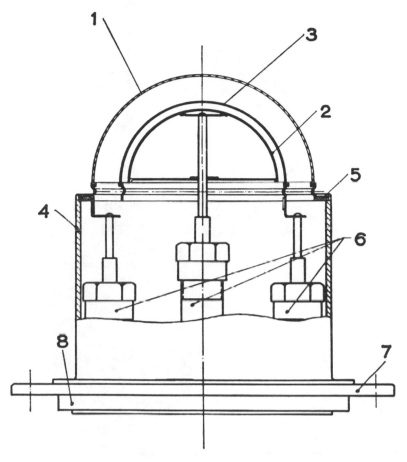

Figure 6.3. Konstantin Gringauz's schematic of the ion trap developed for *Lunik 2;* 1 was the external grid and 3 was the internal grid. (Source: Gringauz 1961, 542.)

the most positive grid. If fast corpuscular streams should be present, however, the ions would enter all four traps at about the same average rate because they would be moving too fast to be deflected by any of the grid voltages. Each trap also had an internal grid charged to −200 volts to repel any photoelectrons that might be produced in the trap by incident radiation.

The Soviet lunar program yielded mixed results (Wukelik 1968; Gringauz et al. 1960). The first mission was a disappointment. Not only did *Lunik 1*, launched on January 2, 1959, miss the moon by some 5,000 km, but its ion traps failed to return data warranting publication. *Lunik 2* (fig. 6.4), by contrast, was a total success. Two days after its launch on September 12, 1959, it collided with the moon. Moreover, all four of its ion traps returned data of good quality during the periods that Soviet ground control

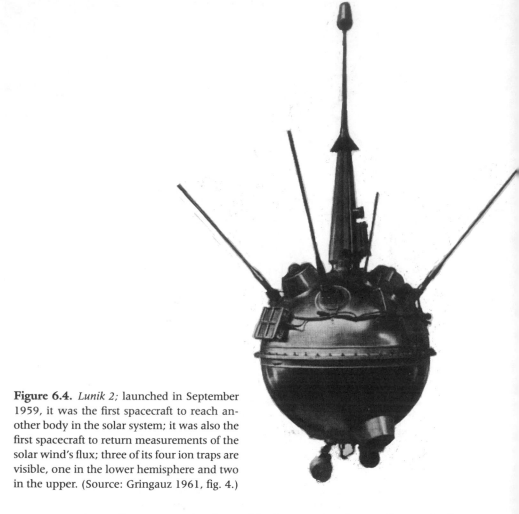

Figure 6.4. *Lunik 2;* launched in September 1959, it was the first spacecraft to reach another body in the solar system; it was also the first spacecraft to return measurements of the solar wind's flux; three of its four ion traps are visible, one in the lower hemisphere and two in the upper. (Source: Gringauz 1961, fig. 4.)

was in radio communication with the spacecraft. *Lunik 3,* which was launched on the second anniversary of *Sputnik 1,* accomplished the impressive feat of photographing the back side of the moon, but the data from its ion traps lacked the completeness of the results from *Lunik 2.*

In February 1960 Gringauz and three colleagues revealed the success of their instrument on *Lunik 2* in a brief paper submitted to the Soviet Academy. They reported that once the spacecraft got beyond the geomagnetic field, the currents received by all four traps traced similar curves. The traps had evidently been in the presence of high-speed ions. The measured flux— the number of ions striking a square centimeter of the collector per second— was approximately 2×10^8. In a concluding assertion of priority, Gringauz and his colleagues remarked that "the corpuscular emission of the sun . . . has thus been observed for the first time in the interplanetary space outside the magnetic field of the earth" (Gringauz et al. 1960, 364).

Soviet astrophysicists were soon commenting on the larger implications of *Lunik 2's* findings. At Gringauz's instigation (Gringauz 1987), Iosef Shklovskii and two colleagues at the Shternberg State Astronomical Institute in Moscow submitted a short paper on Gringauz's results to the leading Soviet astronomical journal in April 1960. After emphasizing the need for a "direct determination of the basic characteristics of the 'solar wind,'" they pointed out that Gringauz's value of the flux was an order of magnitude smaller than Biermann's predicted value. They went on to suggest that the wind's "weakness" was primarily a consequence of its low density, which they estimated at one to two ions/cm^3. And they claimed that *Lunik 2's* results were consonant with the view that "the corpuscular fluxes, with their 'frozen-in' magnetic fields, 'sweep up' the interplanetary gas from the inner portions of the solar system" (Shklovskii, Moroz, and Kurt 1961, 871, 873).

While the article by Shklovskii's group was still in press, the Soviet solar physicist Andrei Severny (1960) gave a preview of the Soviet findings in the West. He did so as a participant in the Fourth Symposium on Cosmical Gas Dynamics, which was held in Italy during August 1960 and attended by Parker, Biermann, and many other theorists interested in interplanetary processes. On at least one occasion during the symposium, Severny reported Gringauz's figure for the solar-wind flux. He also argued that this result, taken in conjunction with recent British investigations of the zodiacal light, indicated that the density of the interplanetary medium must be smaller than 100 ions/cm^3.

Despite Soviet reports about *Lunik 2's* plasma measurements, Western space scientists and astrophysicists paid late and little attention to these first direct observations of the solar wind. One reason was that, despite translation services and international conferences, communications were slow. Another was that, despite the internal grids in the Lunik ion traps, there was some possibility that the Soviet results were compromised by photoelectric currents. Yet another reason was that the Soviet observations provided no data on the speed and direction of the solar wind, parameters needed for understanding the physics of the sun's relation to the interplanetary medium. Perhaps these reasons alone suffice to explain the modest recognition accorded to Gringauz's work, but it is not inconceivable that political rivalries also played a role in inclining scientists in the West to ignore or discount the Soviet achievement.

Explorer 10's *Observations*

By the time of *Lunik 2's* launch in September 1959, American scientists were well along with planning for their own direct observations of the interplanetary plasma. It was the National Academy of Science's Space Science Board that alerted NASA to the research opportunities in this area. At its first meeting in June 1958, the Space Science Board (1958a) divided the work of preliminary planning for the American space program among twelve committees. John Simpson, who was eager to send cosmic-ray detectors above the atmosphere, obtained the chairmanship of the Committee on Physics of Fields and Particles in Space. Within a month, he provided the board with a long list of observations that could, and should, be carried out during 1959–60 (Space Science Board 1958b). Among them was the "Detection of Interplanetary Particles." His chief informant here seems to have been his junior colleague Parker, for he stressed the importance of obtaining "information on suprathermal particles in nature and [the] interplanetary medium." Simpson proposed that measurements of the particles' charge, energy, and mass could be made with ion traps, mass spectrometers, and electronic detectors carried either by satellites with very eccentric orbits or by lunar probes. Such instruments would need to be developed from scratch however, since the particles would not be energetic enough to observed with standard cosmic-ray detectors. Not knowing of any group that was working in this direction, Simpson suggested that the Naval Research Laboratory, the Air Force Geophysics Research Directorate, or the University of Maryland might be persuaded to develop the requisite instrumentation. At its next meeting, the Space Science Board (1958c) agreed that much "exploratory work by experts" would be needed to devise instruments suitable for studying the interplanetary plasma.

At this point Bruno B. Rossi (b. 1905), chair of the Space Science Board's Committee on Space Projects, moved to the fore in advocating investigations of the interplanetary medium. His cosmic-ray group at the Massachusetts Institute of Technology, having been marginalized in nuclear physics by the development of high-energy accelerators, was on the lookout for new research problems (Hirsh 1983). The interplanetary plasma was among the subjects that struck Rossi as offering opportunities for pioneering work. The Committee on Space Projects was in complete agreement. That December, it urged that the "first order of priority" on space probes be "given to the full exploration of the interplanetary plasma" (Space Science Board 1958d). Soon afterward, the full Space Science Board included this recommendation in the long list of desiderata that it sent to the fledgling National Aeronautics and Space Administration.

Homer E. Newell, Jr., and other NASA administrators involved in organizing the agency's space-science program had doubts about the feasibility of Rossi's proposal (Newell 1980). At the time, NASA's capabilities for getting medium-sized satellites into interplanetary space were still very limited. In addition, instrumentation for measuring plasma characteristics had not yet gone beyond the design stage. Unfazed by NASA's lukewarm response, Rossi continued pressing the case for research on the interplanetary medium. His persistence soon paid off (Parker 1959). In the spring of 1959, he obtained a berth for a plasma trap on a satellite scheduled to make an exploratory study of the interplanetary magnetic field some eighteen months later. Rossi followed up on this success with a campaign for berths on subsequent missions. He was optimistic that his group's initial results would be of sufficient interest to justify more sophisticated studies. Indeed, according to Newell (1959b), Rossi believed "that in our competition with the Soviets we might well capture some prizes here, whereas in the other areas [lunar and planetary studies] we would be trailing for some time to come." His reasoning here was that "our superior know-how in instrumentation and telemetering [would offset] our inferiority in rocketry" (Rossi 1960, 598).

In the meantime, Rossi had put his long-time collaborator Herbert S. Bridge (b. 1919) in charge of the group designing the plasma trap (Rossi 1984; Bridge 1987). One of the group's earliest decisions was that it would be folly to seek high accuracy in the initial observations. They were impressed by the disagreements among theorists regarding the plasma's velocity and density. The mean velocity "could be less than 10 km/sec and could exceed several thousand kilometers per second [and] values of the particle density [could] range from less than 1 to several thousands per cubic centimeter." They sought, accordingly, "to devise an instrument capable of giving information . . . on the magnitude and direction of the bulk velocity of the plasma over as wide a range of values as possible" (Bridge et al. 1960, 3053). The resulting instrument was a four-grid ion probe (fig. 6.5) that, with its transistorized electronic system, weighed just over a kilogram. Its key feature was a grid that would be impressed with a positive square-wave of voltage at 1500 cycles per second. This grid would control the incoming flux by alternately accepting all incident protons when its voltage was zero and accepting only those with velocities high enough to overcome the grid's repulsive force when its voltage was positive.

The modulating grid gave Rossi and Bridge's plasma probe two advantages over *Lunik 2's* simpler instruments. First, the grid would ensure that photoelectric currents generated by solar ultraviolet radiation and instrumental background noise did not contaminate the data. Such signals could be filtered out by subtracting the current when the grid voltage was zero from the current when the voltage was positive. Second, the modulating grid would permit determination of the velocities of positive ions, which were presumed to be mainly protons. The electronic subtraction technique would reveal—for each of several positive voltages—the flux of protons

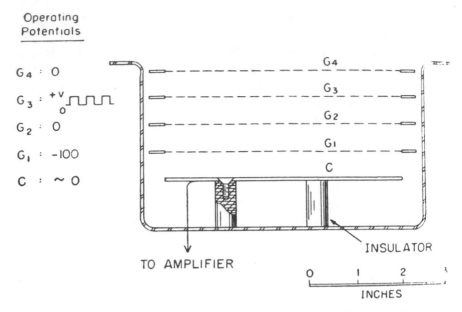

Operating
Potentials

$G_4 : 0$

$G_3 : \overset{+v}{\underset{0}{\text{⊓⊔⊓⊔}}}$

$G_2 : 0$

$G_1 : -100$

$C : \sim 0$

TO AMPLIFIER

INSULATOR

0   1   2   3
INCHES

Figure 6.5. Early schematic of the MIT plasma probe. (Source: Bridge et al. 1960, 3054. © American Geophysical Union.)

moving too slowly to overcome the grid's repulsive force at that voltage. Comparing the results for the different voltages would in turn reveal the proton flux within each of several velocity steps. In hopes of learning whether the solar emissions were breezy or windy, Rossi and Bridge (fig. 6.6) eventually decided to set the probe's voltages to determine the flux in six intervals ranging up to 660 km/sec (Bridge et al. 1962).

In March 1961, some two and a half years after Rossi's group began designing its instrument, NASA launched *Explorer 10*. The satellite's batteries limited data return to sixty hours. During this time, the spacecraft (fig. 6.7) ascended along a highly eccentric orbit to its apogee at 240,000 km. The plan was that the second half of the ascent would be spent observing conditions well beyond the earth's magnetic influence. In fact, although this was not immediately clear, *Explorer 10* never reached the undisturbed interplanetary medium. Rising above the earth's nightside, the spacecraft ended up threading the boundary—soon known as the magnetopause—between the relatively quiet region dominated by the earth's magnetic field and the somewhat turbulent solar wind that was streaming around the teardrop-shaped magnetosphere. The result was that the instrumental readouts were considerably richer, and initially more difficult to interpret, than expected (Bridge 1963).

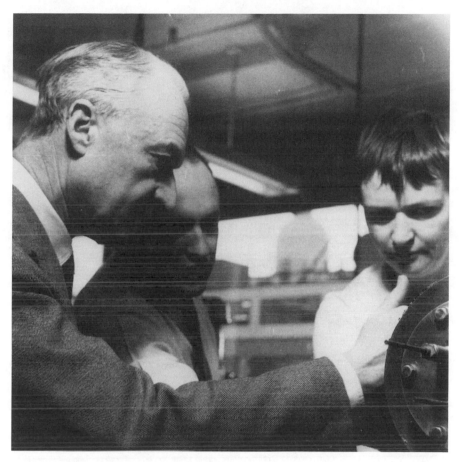

Figure 6.6. Bruno Rossi, Herbert Bridge, and Conrad Dilworth inspecting the MIT plasma probe in 1960. (Courtesy of Herbert Bridge.)

Just three weeks after the mission, representatives of *Explorer 10's* instrument teams reported preliminary results at a NASA press conference. The main news from Rossi and Bridge's group was that direct observation sustained the solar-wind theory. One reporter asked Bridge whether *Explorer 10's* priority was in detecting or in measuring the solar wind. He replied that the spacecraft's plasma probe had provided the first measurement of proton velocities, then he admitted that "the Russians" had reported "protons out there" (NASA 1961a). Soon afterward, Rossi and a member of *Explorer 10's* magnetometer team presented more detailed reports at a meeting of the Particles and Fields Subcommittee of NASA's Space Sciences Steering Committee. Rossi described the plasma observations beyond twenty earth radii. Parker and others on the subcommittee were concerned about the many

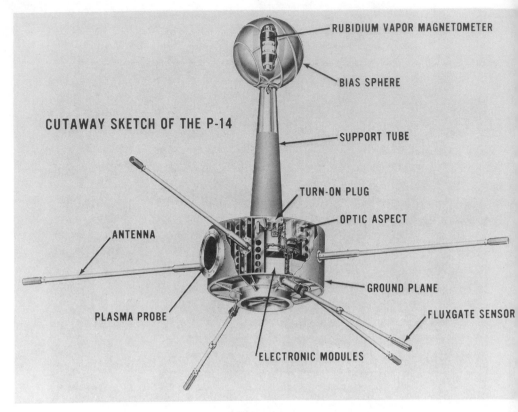

Figure 6.7. Schematic of the P-14 satellite, which was named *Explorer 10* upon its successful launch on March 25, 1961. (Courtesy of NASA History Office.)

abrupt transitions in the data. Doubting that the satellite "ever got completely free of the geomagnetic field," they emphasized "the need for probes on the Sunward side of the Earth where the phenomena should be less complex." The subcommittee thought the time was ripe for "the addition to the space program of some specifically interplanetary missions—separate from the planetary and lunar missions" (NASA 1961b).

In September 1961, *Explorer 10's* scientists presented their first formal reports at the International Conference on Cosmic Rays and the Earth Storm held in Kyoto, Japan. Bridge read a paper on his group's "direct observations of the interplanetary plasma." Without ignoring the irregularities in the data, he focused on the conditions prevailing beyond the geomagnetic boundary. In that region, he reported, the plasma probe found that the total flux was typically about 4×10^8 ions/cm²/sec. The speed of the ions, which were presumed to be protons, ranged from a minimum of 120 km/sec up to the maximum measurable velocity of 660 km/sec with an average of about

300 km/sec. Division of the total flux by the velocity indicated that the plasma density fluctuated between 6 and 20 protons/cm³. The general direction of the flow, as revealed by an analysis of the angular response of the ion trap, was away from the sun. Once having presented his own group's results, Bridge acknowledged that Gringauz's group had earlier obtained a measure of the flux from *Lunik 2* that was "practically identical" to the measure obtained from *Explorer 10*. He pointed out, however, that the Soviet scientists had not reported any "information concerning the directionality of the particles or their energy spectrum" (Bridge et al. 1962, 553, 559).

Explorer 10's plasma observations did much to strengthen the case for the existence of the solar wind. Not everyone, to be sure, was convinced. For instance, some skeptics at the April 1961 meeting of the American Geophysical Union, while granting that *Explorer 10's* data struck "a blow" at their position, insisted that "several more probes" would be needed "to confirm the findings" (Grigg 1961). Parker was confident, nonetheless, that the tide was running in his favor. Elated, he began writing a monograph on his theory. That September at the Kyoto conference, Parker (1962, 563) gave free rein to his optimism in the opening sentences of his paper:

> Recent observations in space (Shklovskii *et al.*, Bridge *et al.*) confirm the earlier inference from the analysis of comet tails (Biermann) etc. that there is a continual outward flow of solar corpuscular radiation from the sun. The origin of this corpuscular radiation, or solar wind, is evidently the hydrodynamic expansion of the solar corona (Parker), though Chamberlain has disagreed with this point of view. Expansion of the quiet-day corona, of $1-2 \times 10^{6}°K$, yields the quiet-day solar wind of some 300 km/sec and 10 protons/cm³ at the orbit of Earth, as observed in space by Bridge *et al.*

For all his apparent detachment, Parker was clearly relishing his triumph over Chamberlain, Chapman, and others who had opposed the solar-wind theory.

One consequence of Parker's vindication was that he was proposed for a full professorship at the University of Chicago. Chandrasekhar (1962), who had earlier doubted Parker's theory, wrote on his behalf that "some of his ideas have received spectacular confirmation. I have in mind particularly his work on the solar wind. . . . When Parker first developed these ideas, they were only one of several competing proposals; and people were generally inclined to look at Parker's ideas with skepticism. But the fact now is the recent satellite observations have vindicated his ideas completely. And in a field as complicated as this, nothing is more successful than success." Thanks, therefore, to *Explorer 10's* observations, both Parker's theory and his career were prospering.

Mariner 2's *Observations*

While the earliest Soviet and American observations of the interplanetary plasma lent powerful support to the argument from comet tails for the existence of the solar wind, the new evidence was not in itself overwhelming. The ion traps on *Lunik 2* had given the approximate magnitude of the wind's flux, but they had done so for less than fifteen hours and revealed nothing about the wind's speed or direction. The plasma probe on *Explorer 10* had provided rough measures of the wind's flux, speed, and direction. But as the analysis of the spacecraft's data proceeded, there was a growing recognition that it had not reached the undisturbed interplanetary medium. To clinch the case for the solar wind, someone had to get an instrument with greater sensitivity onto a spacecraft with greater range. In 1962 just such an instrument was included in the payload on NASA's first successful planetary mission. Its observations were decisive.

The planning that culminated in these observations began in the fall of 1958 at the Jet Propulsion Laboratory in Pasadena, California. At the time, negotiations were under way for the transfer of funding responsibility for the laboratory, which was managed by the California Institute of Technology, from the Army to NASA (Koppes 1982). Its principal activity had been developing military rockets, but the laboratory's leaders saw the impending transfer to NASA as an opportunity to realize their dream of transforming the Jet Propulsion Laboratory into the nation's preeminent center for space exploration (Snyder 1988). Albert R. Hibbs, a theoretical physicist by background, was put in charge of developing the laboratory's strategy for research in space. His Mission Survey Group of about twenty included a subgroup that had spent the preceding two years studying nuclear propulsion for rockets. Conway W. Snyder (b. 1918), the subgroup's leader, was a nuclear physicist who had long been interested in the possibility of space travel. His office mate Marcia M. Neugebauer (b. 1932) was an associate research engineer with a master's degree in physics. Since interest in nuclear rockets had proved evanescent, both threw themselves into the planning effort. In November, indeed, Neugebauer appeared as second author of a report to Hibbs canvassing dozens of problems that might be studied with satellites and interplanetary probes. One proposal was that Parker's theory be tested with measurements of the velocity and density of the plasma flowing from the sun (Newburn and Neugebauer 1958).

Hibbs (1958) soon divided up the work of suggesting instruments for investigating various classes of phenomena. The assignments were wide ranging—Snyder, for instance, got cosmic rays, magnetic fields, and gamma radiation, and Neugebauer got micrometeorites, solar corpuscular radiation, interplanetary gas, and extraterrestrial life! In mid-December Neu-

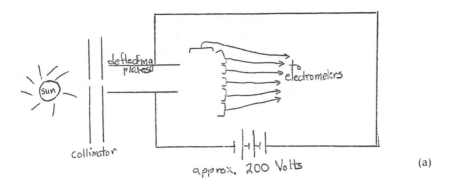

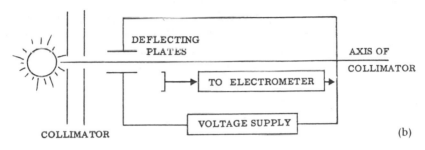

Figure 6.8. Marcia Neugebauer's conceptual sketches of what became the Jet Propulsion Laboratory's plasma spectrometer: (a) A constant-voltage instrument with several charge collectors, each with it own electrometer (December 1958); (b) A variable-voltage instrument with one charge collector and electrometer (January 1959). (Sources: Neugebauer 1958; Jet Propulsion Laboratory 1959, 64.)

gebauer (1958) submitted a short description and sketch of a plasma ana-lyzer. Her idea was to channel solar plasma through a constant electric field that would deflect the protons and electrons—in opposite directions and in proportion to their energies—to an array of current collectors (fig. 6.8a). The proposal was incorporated into the research plan that Hibbs's team dissemi-nated through the laboratory shortly after its formal transfer to NASA (Jet Propulsion Laboratory 1959). A crucial design change that Snyder suggested on the basis of his prior experience with electrostatic detectors in nuclear research did away with the need for the collector array, reducing the instru-ment's intricacy and weight (Snyder 1988). A variable rather than constant voltage would be imposed on the deflecting plates so that a single collector could be used to provide spectra of proton and electron velocities (fig. 6.8b). Plasma analyzers should be included, the Mission Survey Group recom-mended, in the payloads of two missions toward Venus in January 1961.

By mid-winter 1959, Snyder and Neugebauer had decided they wanted responsibility for the plasma analyzers (Snyder 1988). They reasoned that they stood little chance of competing successfully in astronomical, upper-atmosphere, or cosmic-ray research. But since nobody had ever investigated plasmas having the properties anticipated for the interplanetary medium, they supposed they were at no disadvantage there. Snyder and Neugebauer were successful in their bid for the project. Their confidence, however, remained shaky for a time. For instance, upon learning in the spring that both Rossi's group at the Massachusetts Institute of Technology and Michel Bader's group at NASA's Ames Research Center were also at work on plasma instruments, their first inclination was to abandon the effort (Snyder 1986). But once they were acquainted with the designs of the rival instruments, they had no hesitation about arguing for the continuation of their work (Neugebauer 1959a, 1959b). They did consider having Bader's group, which was also pursuing an electrostatic design, study the solar plasma's electrons while they studied its protons. But Hibbs insisted that, as far as possible, only instruments developed by the Jet Propulsion Laboratory should fly on the laboratory's missions. In fact he made sure that Neugebauer and Snyder's plasma analyzers were assigned berths on the two Vega satellites the laboratory was hoping to launch in the first half of 1961 (Hall 1977; Koppes 1982).

Much of the actual work of translating the conceptual design into flight hardware was done by two young electrical engineers—Conrad S. Josias and James L. Lawrence, Jr. (Snyder 1988). The instrument (fig. 6.9), which was named the solar corpuscular radiation electrostatic particle analyzer, would use curved plates with opposite charges, rather than grids, to control the type and speed of the particles reaching the collector. Only particles of

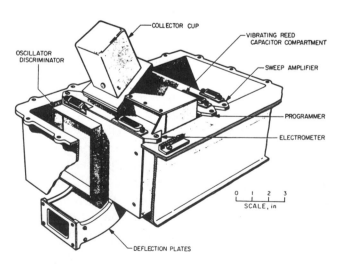

Figure 6.9. Schematic of the JPL plasma spectrometer that flew on *Mariner 2;* the solar wind entered the instrument through the rectangular aperture at lower left. (Source: Snyder and Neugebauer 1964, 92.)

the desired charge and near certain velocities—depending upon the curvature, separation, and voltage difference of the plates—would be registered. Particles with the opposite charge or higher speeds would impinge on the outer plate, while those with speeds that were too low would be drawn into the inner plate. The narrower the spacing between the plates, the greater would be the velocity and directional accuracies. During flight, the flux's velocity spectrum could be measured in a short time by rapidly changing the deflecting voltage. Such electrostatic analyzers, the JPL group argued, could do a much better job of determining conditions in the interplanetary plasma than the ion traps that Rossi's group was developing (Neugebauer 1959b; Josias, Neugebauer, and Snyder 1959).

Notwithstanding their eagerness, Neugebauer and Snyder did not get an early opportunity to fly their instrument (Koppes 1982; Hall 1977; Neugebauer 1982). As of November 1959, it was scheduled for flight on the first Vega in March 1961. But in December 1959 the restructuring of the Vega series into the Ranger series resulted in a delay of several months, partly because the first two Rangers had the difficult technical goal of demonstrating that a long-range mission could be launched from a low "parking" orbit. Finally, in late August 1961, nearly half a year after *Explorer 10's* flight, Neugebauer and Snyder had their chance. An Atlas rocket put *Ranger 1*, which carried six electrostatic analyzers in its payload (NASA 1961c; Neugebauer 1982), into parking orbit. Then, however, the Agena rocket that was to boost the spacecraft into an orbit beyond the magnetosphere malfunctioned. The plasma analyzers never reached the solar wind. Three months later *Ranger 2*, which carried another six analyzers, also failed to get beyond its parking orbit. Through no fault of their own, Neugebauer and Snyder still had no data on the interplanetary plasma.

At this juncture, had the development of the Centaur booster by General Dynamics Corporation been on schedule, Neugebauer and Snyder would have been without immediate prospects for flying their plasma analyzer. The plan had been to send two Mariner spacecraft to Venus with Atlas-Centaur launchers during the summer of 1962 (Jet Propulsion Laboratory 1963a, 1965; Koppes 1982). The Mariners were—in deference to NASA headquarters—to have carried a fair number of instruments from external groups, including two ion traps under development by Bridge's team. One week before the launch of the first Ranger, however, it became clear that slippage in the Centaur's production schedule would force postponement of the Mariner mission until 1964. In response, the Jet Propulsion Laboratory proposed using an Atlas-Agena to propel a modified Ranger spacecraft toward Venus so that the 1962 launch window would not be missed. By mid-October 1961, NASA agreed that the Mariner-R(anger) proposal was the best way to capitalize on the 1962 opportunity.

One consequence of the change to the less powerful Agena rocket was that the Mariner payload had to be lightened. This was the point at which responsibility for interplanetary-plasma measurements was transferred from Rossi and Bridge to Neugebauer and Snyder (Snyder 1986, 1987). In planning for the Mariner missions, Neugebauer and Snyder paid close attention to *Explorer 10's* results. For instance, they oriented the single analyzer that could be included in each Mariner's payload toward the sun. And instead of preparing for the possibility that the solar efflux would be breezy, they had their engineering group set each Mariner's analyzer so that it would measure proton fluxes at ten velocities between 200 and 1250 km/sec (Snyder and Neugebauer 1964).

In late July 1962, less than a year after the Jet Propulsion Laboratory threw its energies into the Mariner-R project and just months after the initial testing of the spacecraft in the space simulator (fig. 6.10), the first Mariner was launched. To everyone's dismay, it went askew and had to be destroyed by the range safety officer. Neugebauer and Snyder wondered whether they were fated to relive the disappointments of *Ranger 1* and *2*. In fact, although there were tense moments, *Mariner 2* was a major success (Jet Propulsion Laboratory 1965). Launched into parking orbit on August 27, 1962, the spacecraft was boosted on its way to Venus two days later. Thereafter— except during the midcourse maneuver and a temporary malfunction of one of the solar panels—*Mariner 2* kept the plasma analyzer pointed to within 0.1° of the sun's center and sent a steady stream of data back to earth. Transmission continued until January 3, 1963, when the spacecraft was 20 days beyond Venus and at a distance of almost 87,000,000 km. Although there were some gaps in reception, especially after December 17, 1962, *Mariner 2's* scientists obtained 104 days' worth of data (Snyder and Neugebauer 1964).

Six weeks after launch, NASA (1962d) sponsored a news conference on the mission's status in Washington, D.C. (fig. 6.11). The preliminary results from the plasma analyzer were given pride of place. After describing the lingering theoretical controversy, Neugebauer emphasized that the initial data established that "the sun sends out this plasma continuously, or for at least as long as Mariner has been observing the phenomenon." Preliminary analysis indicated that the "solar wind has peaks of activity and quiet periods." The velocity usually ranged between 400 and 700 km/sec but occasionally climbed to 1250 km/sec and perhaps beyond. Picking up on these results, the *New York Times* headlined Mariner's disclosure of "a constant 'solar wind' " in its front-page report (Finney 1962).

In late November, another six weeks into the mission, Neugebauer and Snyder (1962) completed their first paper on the plasma observations. It was

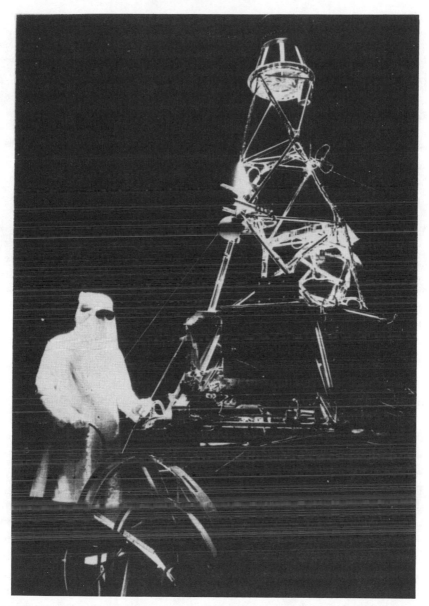

Figure 6.10 The Mariner spacecraft in the Jet Propulsion Laboratory's space-simulator chamber; the spacecraft's thermal behavior predicted on the basis of the simulation tests did not match the data returned by *Mariner 2*, giving rise to questions about the reliability of the value for the solar constant used in the tests (see chap. 7). (Source: Jet Propulsion Laboratory 1963b, 42.)

Figure 6.11. NASA Press Conference on October 10, 1962, to announce the progress of the *Mariner 2* mission; Homer Newell (center right) presided and Marcia Neugebauer (second from left) reported the preliminary results from the solar-plasma instrument. (Source: Newlan 1963, 51.)

based on a preliminary inspection of more than 20,000 velocity spectra taken every 3.7 minutes between August 29 and October 31. Their main points were that the plasma was "always" flowing away from the sun and that it had a flux "in good agreement" with prior reported values and a velocity that agreed "fairly well with the value predicted from Parker's 'solar wind' theory." Neugebauer and Snyder were also able to report two phenomena that were not included in Parker's basic theory. On eight occasions their instrument had recorded the passage of fronts that were faster and denser than the ordinary solar wind. In addition, a fair fraction of their spectra had two maxima rather than the single peak expected for an ordered flow of hot ions. The second maximum indicated that the solar wind contained helium nuclei as well as protons. With twice the mass-to-charge ratio, helium nuclei having the same velocity as the typical proton would behave in the plasma analyzer as if they were protons of twice the typical energy.

This report, which appeared in *Science* in early December 1962, dissolved any remaining doubts about the general soundness of Parker's theory of the solar wind. *Mariner 2's* observations, more precise and prolonged than those from

Lunik 2 and *Explorer 10,* established that plasma was continuously emanating from the sun. But as Parker had forecast back in 1959, the first close look at the solar wind from space did more than confirm his basic model. By measuring variations in the wind's speed and density and by revealing helium's presence in the plasma, *Mariner 2* not only closed the era of the solar wind's discovery but also inaugurated the era of its exploration.

Exploration of the Solar Wind, 1963–1969

Not long after Neugebauer and Snyder's report in *Science,* Parker completed his monograph on the solar wind. He was certain by this time that his basic approach was sound. Just the same, he insisted in the preface that the theory of the solar wind was still "in a rudimentary state." The subject was "so complex [that] real theoretical progress" would depend upon the success of observers in refining and extending knowledge of the wind's characteristics. "We will be surprised," he remarked, "if future observation does not indicate a restatement of many of the small-scale effects that are assumed, and we will be very surprised indeed if observation does not discover a number of complications that have been entirely unanticipated in the present writing" (Parker 1963, vi). This was a safe prediction.

Snyder and Neugebauer, in fact, soon extracted a structured picture of the solar wind from their store of data. In June 1963, Snyder discussed their results at the fourth annual meeting of the International Committee on Space Research in Warsaw. Three months later—in collaboration with Bridge's student U. R. Rao, who had been examining the relation between their data and geomagnetic activity—they submitted a more detailed paper to the *Journal of Geophysical Research* (Snyder, Neugebauer, and Rao 1963). In these accounts they reported conclusive evidence for the old hypothesis that high-speed corpuscular streams from the sun caused the major geomagnetic disturbances that tended to recur every twenty-seven days. First they showed that the solar wind contained recurrent high-speed streams with typical velocities between 600 and 700 km/sec. Then they established that the stream arrivals at earth were correlated with the recurrent peaks in geomagnetic activity (fig. 6.12). Building on this success, they used their data to determine that the average velocity of the solar wind impinging on the earth during any given day was 330 km/sec plus 8.44 times that day's mean geomagnetic index. Snyder and Neugebauer also attempted to identify the solar sources of the high-speed streams—the areas that Bartels had named M regions three decades earlier. It seemed clear that these streams did not arise in a "hydrodynamic expansion of a homogeneous solar corona

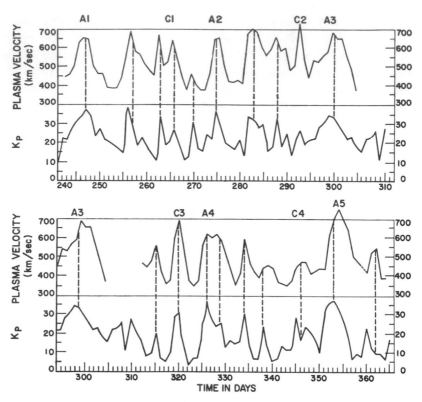

Figure 6.12. Graph indicating the relation between fast solar-wind streams observed by *Mariner 2* and geomagnetic disturbances; Conway Snyder and Marcia Neugebauer labeled two persistent streams having a twenty-seven-day recurrence tendency with A's and C's. (Source: Snyder and Neugebauer 1964, 110.)

but came instead from long-lived local regions in the corona which were abnormal in some respect, perhaps having higher than average temperature." But, they reported, "Attempts to use the measured velocity to extrapolate back to the . . . source points . . . have not yielded useful results" (Snyder and Neugebauer 1964, 110–111).

Eager to disseminate their findings, Neugebauer and Snyder persuaded NASA to fund a Conference on the Solar Wind at the Jet Propulsion Laboratory (Mackin and Neugebauer 1966). The four-day meeting in April 1964 was attended by eighty-three scientists, including the pioneering theorists Biermann, Chapman, and Parker. It opened with reports by Neugebauer, Snyder, and others engaged in interplanetary observing and proceeded to discussions of the solar wind's origin, propagation, and interactions with the earth's magnetosphere, comets, and the moon. Among the highlights of the meeting was Neugebauer's report that the temperature of the protons in the solar wind, which averaged about 150,000 K, increased or decreased along with the velocity. Snyder also announced, after further efforts to trace the high-speed streams back to particular solar features, "that it is still true that the M regions are not visible on the surface of the Sun" (Snyder and Neugebauer 1966, 29). By fostering discussion of techniques, findings, and theories, the conference gave observers and theorists who had been involved in studies of the solar wind a sense of common purpose.

Even as Neugebauer and Snyder's results were being analyzed, dispersed, and absorbed, space scientists were making and planning further observations of the solar wind. Between 1963 and 1970, indeed, they secured berths for solar-wind instruments on about forty spacecraft that reached interplanetary space. NASA, which accounted for over half of these missions, had the most ambitious program. The agency sponsored observations of one or more properties of the solar wind on seven Interplanetary Monitoring Platforms, three Orbiting Geophysical Observatories, six Pioneer and Mariner planetary probes, and two Apollo missions (NASA ca. 1982). Besides its own program, NASA also launched the European Space Research Organisation's *Highly Eccentric Orbit Satellite*, which carried two solar-wind instruments (De Jager 1973a). Meanwhile, starting in 1964, scientists at the Los Alamos Scientific Laboratory were regularly sending up plasma analyzers on the satellites that the Department of Defense and Atomic Energy Commission were using to monitor Soviet compliance with the limited test ban treaty of 1963 (Singer 1965). A detailed knowledge of the background environment was necessary for interpreting the data from the instruments responsible for detecting surreptitious nuclear tests in space. Finally, in the Soviet Union Gringauz and other investigators managed to get instruments for observing the solar wind on many, if not all, of their nation's lunar and planetary probes. However, the Soviet scientists seem rarely to have obtained data that merited rapid or detailed publication.

Magnetic Sectors

One early result from this campaign of exploration was quickly recognized as comparable in significance to Neugebauer and Snyder's work—the discovery by Norman F. Ness (b. 1933) and John M. Wilcox (1925–83) of what appeared to be magnetic sectors in the solar wind. Ness, a solid-earth geophysicist by training, used computer skills he had acquired as a graduate student to get into space science (Ness 1986). His first job was as a research geophysicist at the University of California at Los Angeles, where he helped analyze free oscillations of the earth with a new IBM 709 computer. In mid-1960 he proceeded to Goddard Space Flight Center's Theoretical Division. He was attracted there not only by the opportunity to work with the division's new IBM 7090 transistorized computer but also by a desire to get acquainted with the American space program. The following year, he participated in the analysis of the data collected by *Explorer 10's* magnctometers. This contact with data from space convinced him he did not want to return to Los Angeles and terrestrial geophysics. Declining an assistant professorship at the University of California, he joined James P. Heppner's magnetometer group at Goddard Space Flight Center.

Ness soon obtained two major assignments (Ness 1986). Both were connected with the new Interplanetary Monitoring Platform series that NASA began in response to the urgings of the Particles and Fields Subcommittee (NASA 1961b, 1961d, 1962a). He was put in charge of creating a system that could cope with the avalanche of data these satellites would be generating over the long lifetimes made possible by the transition from batteries to solar panels as power sources. He solved this problem by organizing a data-management system around the new IBM 1410. Ness was also put in charge of the magnetometers for the series' first satellite. He started here by overseeing improvements in the magnetometers that Heppner's team had developed for the *Explorer 10*—one to determine the ambient field's magnitude and two others of a different design to measure its direction (Ness, Scearce, and Seek 1964). He realized, however, that he would need more than accurate instruments to secure reliable magnetic data. The dubious interplanetary-field readings from the magnetometers on *Pioneer 5* and *Mariner 2* had demonstrated the danger of interference from the spacecraft itself. Ness's group circumvented this hazard by excluding all magnetic materials from the vehicle and its payload, by testing for fields generated by the spacecraft's electrical circuits, and when these tests revealed problems, by doubling the length of the booms carrying the directional magnetometers (Ness 1986).

In June 1963, some five months before launch, Ness met John Wilcox in Washington, D.C., at a four-day Symposium on Plasma Space Science (Catholic University of America 1963). An experimental physicist by background, Wilcox was on the lookout for a way to get involved in space research. He had begun his scientific career in Berkeley's Radiation Laboratory, working first in nuclear physics, then in plasma physics (Wilcox 1960). While on sabbatical at Alfvén's institute in Stockholm during 1961–62, he had gotten interested in studying plasmas in space. Once back in California, Wilcox persuaded the Jet Propulsion Laboratory's magnetometer team for the Mariner missions that interpreting their forthcoming data would be easier if concurrent measures of the sun's field structure were made with Mount Wilson's solar magnetograph (Wilcox 1963a). However, he had been frustrated in this undertaking by bad weather and instrument problems. Consequently Wilcox was enthusiastic when Ness sought him out at the symposium to suggest a collaborative investigation of the relation between solar and interplanetary fields. Indeed, they spent the Saturday following the symposium together at Goddard laying the groundwork for concurrent observations from Mount Wilson and the first Interplanetary Monitoring Platform (Ness 1963).

On November 27, 1963, NASA launched the *Interplanetary Monitoring Platform 1* into an eccentric orbit with an apogee of nearly 200,000 km and an orbital period of about four days (Ness and Wilcox 1964; Wilcox and Ness 1965). Initially the spin-stabilized spacecraft spent about three days of each orbit beyond the solar-wind bow shock. As the mission continued, however, the spacecraft's time observing the undisturbed interplanetary medium declined because of the earth's orbital motion around the sun (fig. 6.13). Still, solar-wind observations of high quality were obtained by the

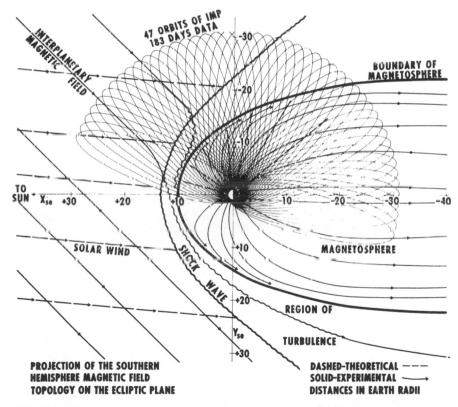

Figure 6.13. Trajectory of the *Interplanetary Monitoring Platform 1*, November 27, 1963-May 30, 1964, with Norman Ness's interpretation of the satellite's magnetic observations; note that after twenty-four orbits the spacecraft was no longer encountering undisturbed solar wind. (Source: Ness 1965, 3002. © American Geophysical Union.)

Goddard magnetometers and a Massachusetts Institute of Technology plasma trap for almost three solar rotations before the interval for interplanetary observing fell to zero in mid-February 1964. Meanwhile, in California Wilcox (1963b) was ensuring the availability of data on solar fields by serving as a Goddard consultant to Robert Howard's magnetograph group at Mount Wilson.

Ness began reporting his group's results in the spring of 1964. At the solar-wind conference in Pasadena that April, he emphasized that the direction of the field was often "at the streaming angle . . . predicted by Parker [for] a solar-wind velocity of approximately 400 km/sec." When the field was not at the streaming angle, it was usually directed at "the antistreaming angle [i.e.,] toward, rather than way from, the Sun." The abruptness and frequency with which the field shifted back and forth between the streaming and antistreaming angles revealed, Ness suggested after discussions and correspondence with Wilcox (1964a), the existence of "filamentary structures" in the solar wind. The presence of these filaments indicated in turn that the wind originated in "individual and discrete sources, either in the photosphere or at greater heights in the solar corona" (Ness 1966, 94, 96, 98).

Soon after the conference, Ness and Wilcox began their detailed investigation of the relation between the interplanetary-field readings and Mount Wilson's solar magnetograms. Wilcox, who had secured a position at the University of California's Space Sciences Laboratory in Berkeley, arranged to spend the summer at Goddard so he could be a full partner in the study. In August 1964, thinking the importance of their findings warranted "a prompt initial presentation" (Ness 1964), they sent a brief paper to *Physical Review Letters*. They opened with a restatement of Ness's preliminary results on interplanetary-field direction. The measurements of the directional magnetometers were "consistent with [Parker's] general model" except for the unanticipated switching of the field back and forth between orientations "parallel or antiparallel to the theoretical angle [135°] proposed by Parker." They backed up this claim with a histogram of the 5.46-minute averages of the field's orientation in the ecliptic plane. They then reiterated Ness's earlier proposal that the field reversals indicated the presence of "magnetic-field filaments" in the solar wind. Breaking fresh ground, Ness and Wilcox also presented evidence that these filaments had a strong twenty-seven-day recurrence tendency and hence a solar origin. In addition, they reported that the interplanetary field's polarity was highly correlated with equatorial photospheric polarity "at a lag of approximately $4\frac{1}{2}$ days." This lag indicated an average solar-wind velocity of about 385 km/sec, in good agreement with the average of 398 km/sec recorded by the plasma trap during the satellite's

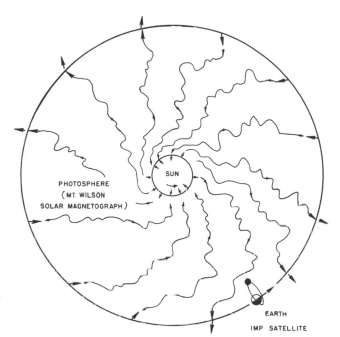

Figure 6.14. John Wilcox's first portrayal of the solar origin of the interplanetary field; note that the arrows indicate the predominant field directions (not the direction of flow). (Source: American Institute of Physics 1964.)

first seven orbits. It appeared, they concluded, that some of the photospheric field lines were "dragged out by the solar wind plasma to become part of the nearby interplanetary magnetic field" (Ness and Wilcox 1964, 462–64).

Ness and Wilcox's paper inspired a press release by the American Institute of Physics (1964). It recounted their confirmation of Parker's model that interplanetary field lines "should leave the sun like streams of water whipping out from a whirling head of a lawn sprinkler, because of the combined effects of the solar wind flow, and the spin of the sun." Accompanied by a drawing (fig. 6.14) provided by Wilcox (1964b), this release soon gave rise to news items in the *New Scientist* and *Sky and Telescope*.

Meanwhile, Ness and Wilcox were pushing ahead with their data analysis. The main burden fell on Wilcox. Ness was terribly busy not only with expounding on the first *Interplanetary Monitoring Platform's* observations of the earth's magnetic tail, but also with monitoring the magnetometers on the series' second satellite and with preparing for the launches of three more satellites. As the analysis proceeded, Wilcox and Ness came to think that the basic units in the interplanetary field were broad sectors, not wispy filaments.

Available records reveal three steps in this significant shift toward a macroscopic view of the magnetic field borne by the solar wind. In November, immediately after one telephone conversation, Wilcox (1964c) sent Ness

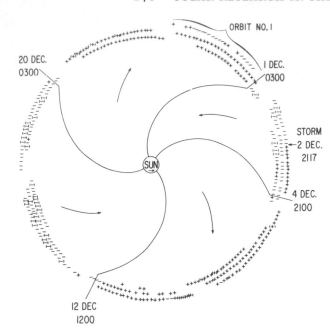

Figure 6.15. Ness and Wilcox's picture of interplanetary-field sectors during December 1963 and January 1964; the outer ring of pluses and minuses was the observed pattern of field polarities. (Source: Ness and Wilcox 1965, 1593. By permission of *Science* © AAAS.)

several color-coded sheets of data. He thought they constituted "a strong suggestion that interplanetary space is divided into seven approximately equal segments." Some four months later, when discussing the relation between the data sets from Massachusetts Institute of Technology's plasma trap and Goddard's magnetometers, Wilcox (1965a) referred to the segments as "sectors" and defined "a super sector [as] two adjoining $\frac{1}{7}$ sectors having the same sense of field." He sent Ness a graph indicating that "the solar wind velocity is 'organized' . . . in terms of the super-sector structure." In mid-March 1965, a month later yet, Ness and Wilcox (1966) submitted a full account of their research to the *Astrophysical Journal.* By this time their focus was squarely on the large-scale structure of the interplanetary field. They referred to Wilcox's super sectors simply as "sectors."

Soon after completing their article for the *Astrophysical Journal,* Ness and Wilcox decided to get out a summary report. It was illustrated with a diagram of the sectors based on three-hour averages of the magnetometer readings that, being much less variable than the 5.46-minute averages, brought out the structure in a very striking way (fig. 6.15). To test the waters, they sent copies to Parker and others for comments. Parker (1965) soon replied that he was "impressed with the remarkably stationary character of the field pattern. [The figure] is nothing less than spectacular!" Pleased, Ness and Wilcox submitted their report to *Science* in mid-April 1965. They first argued that the *Interplanetary Monitoring Platform's* magnetometers had "revealed a regular longitudinal sector structure" in the field during the three solar rotations covered by the mission. Then they announced a few preliminary findings on the sectors themselves. The sectors with fields pointing away from the sun did not differ in any essential way from those with fields pointing into the sun. In both cases, as a sector boundary swept past the earth, the field's direction reversed and its magnitude first climbed above and then fell below its average value. Likewise, the index of geomagnetic activity rose and then declined. These results gave them confidence that their work had "important implications [for] determining the structure of the sun" (Ness and Wilcox 1965, 1592, 1594).

Wilcox and Ness followed up on their report in *Science* with a detailed account of their research in the *Journal of Geophysical Research*. The manuscript had rough sledding because one referee was annoyed that its results had "already been presented in various combinations and permutations in at least . . . four places" (Referee's report 1965). Wilcox (1965b) prevailed by identifying several new points in the manuscript and by stressing the value of a more comprehensive study than could be published in *Science*. His case had merit, for the article was soon recognized as a classic in the solar-wind literature.[3] Wilcox and Ness's discussion of the internal structure of sectors was at the heart of their study. Using plasma measurements, they showed that the solar wind had changed in regular ways as the sectors swept past *Interplanetary Monitoring Platform 1*. The velocity, for instance, had followed the field's magnitude, ascending to a peak then declining in the trailing part of the sectors. All indications, they insisted, pointed to the conclusion that "the sector structure . . . is a fundamental property of the interplanetary medium" (Wilcox and Ness 1965, 5798).

The Quest for Solar Sources

Ness and Wilcox's view of the interplanetary field served as the starting point for many studies examining whether and, if so, how the sector pattern changed with time. This research indicated that sectors were present, in varying numbers, throughout the sunspot cycle. Their view also stimulated attempts to identify the solar sources of the wind streams responsible for recurrent geomagnetic disturbances. Indeed, in their classic paper Wilcox and Ness discussed the source of a stream—velocity 470 cm/sec, outward-directed field—that caused such a disturbance on December 2, 1963. They traced the stream back to a weak magnetic region of positive polarity on a

3. Three solar-wind papers published before 1970 received more than 175 citations during the first decade after their appearance from the sources covered by the *Science Citation Index*—Parker (1958c) received 183 citations; Snyder, Neugebauer, and Rao (1963), 221; and Wilcox and Ness (1965), 201.

solar magnetogram provided by Howard at Mount Wilson. In their view, this "ghost UMR [unipolar magnetic region] is to be identified with Bartels's M region. The weak magnetic fields . . . might be such as not to interfere with the escape of solar wind plasma from the sun." They stressed, however, that further observations would be needed "to determine the generality" of this conclusion (Wilcox and Ness 1965, 5803).

In the years that followed, Wilcox became something of an authority on the origin of sectors in the solar wind. He addressed the issue at length, for instance, in Cornelis de Jager's *Space Science Reviews*. There he made the case that large, long-lived unipolar regions in the photosphere were the ultimate sources of the sectors. The distribution of these regions in the equatorial latitudes, he argued, was mapped out in the interplanetary field's sector pattern. He illustrated how this mapping might occur with a diagram of the sun that portrayed strong solar wind streams issuing from magnetically open areas and sector boundaries over magnetically closed areas (fig. 6.16).

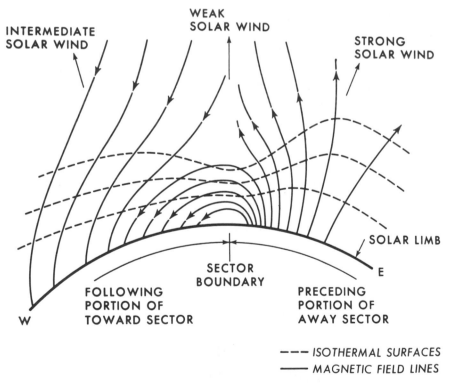

Figure 6.16. Wilcox's view of the origin of fast streams and sector boundaries; note that he regarded magnetically open regions of the sun as the source of strong winds. (Source: Wilcox 1968, 319.)

Notwithstanding his definiteness, Wilcox was careful not to paint himself into a corner. His supposition that "there could be a delay of several weeks" between the appearance of new features in the photosphere and matching changes in the coronal stratum that produced the solar wind gave his entire scheme a good deal of elasticity (Wilcox 1968, 303).

One of the first to follow up these general ideas was Wilcox's doctoral student Kenneth H. Schatten (b. 1944). Imbued with a taste for rigor, Schatten (1986) sought to put more physics into his mentor's phenomenological picture of the solar wind. He set himself the goal of tracing how photospheric magnetic fields evolved into the interplanetary sector structure. He wanted to show that Howard's observations of the photosphere with the Mount Wilson magnetograph had a physical link to Ness's observations of the interplanetary field with the magnetometers aboard *Interplanetary Monitoring Platform 3*. To do so, Schatten developed what he called "the source surface model." He saw the source surface as a thin region in the corona situated about 0.6 solar radii above the photosphere. Various coronal observations as well as his calculations indicated that by this point the photosphere's complex field was smoothed out into a sectorlike pattern that the solar wind then transported outward through the solar system. Completed in 1968, Schatten's thesis both refined and buttressed Wilcox's picture of the origin of the solar wind (Schatten 1968; Schatten, Wilcox, and Ness 1969).

Besides Wilcox and Schatten, several other scientists sought in the late 1960s to trace features of the solar wind back to the sun. Hindsight indicates that Ronald L. Rosenberg (b. 1941) and Paul J. Coleman, Jr. (b. 1932), were especially fortunate in their attempt. In early 1969, Rosenberg was three years beyond an M.A. in physics at Berkeley and in the midst of his first year as a research geophysicist at the Institute of Geophysics and Planetary Physics on the Los Angeles campus of the University of California. He was helping Coleman, who had been involved with space magnetometers since 1960, analyze interplanetary field data collected by ten spacecraft from 1962 into 1968. In particular, they were investigating if the field's polarity depended on whether it was observed from above or below the plane through the solar equator (Coleman 1986).

They found just such a dependence during the six years in question. Above the solar equatorial plane, the field showed "a strong tendency" to point in toward the sun, while below the tendency was outward. Concurrent observations at Mount Wilson revealed the same orientation in the sun's polar fields—inward at the north pole and outward at the south pole. This similarity, Rosenberg and Coleman argued, indicated that "the *dominant* polarity of the interplanetary field in either the northern or southern hemisphere of interplanetary space [was] an extension of . . . the weak dipolar background field of the sun." They anticipated that an important objection to their dipolar model would be Wilcox's evidence that sectors originated in the photosphere's equatorial zone. To circumvent this difficulty, they offered three explanations for the frequent detection of wind streams of "nondominant polarity" on the wrong side of the heliographic equator. One of their ideas was that the solar phenomena giving rise to the wind might distort the "plane separating the two dominant polarity regions," causing high-speed streams of contrary polarity to cross the equatorial plane (Rosenberg and Coleman 1969, 5611, 5618, 5621).

Skepticism greeted Rosenberg and Coleman's model. Wilcox was particularly critical. He believed their statistics, though suggestive, did not establish the reality of a heliographic latitude effect. If the effect should turn out to be real, Wilcox (1969) thought it would be "related to some rather subtle, interesting, and not yet understood mechanism in the development of the photospheric field in equatorial latitudes rather than a direct influence of the sun's polar field." Wilcox felt strongly on this point. He was, for instance, severe in a critique of Rosenberg's attempt to elaborate on the dipolar model of the solar wind, proclaiming that "no conceivable repairs" would make the manuscript "suitable for publication in the Astrophysical Journal" (Wilcox 1970). Rosenberg's article, which soon appeared in *Solar Physics* (1970), would eventually be recognized, however, as a daring forerunner of the three-dimensional pictures of the solar wind that emerged triumphant in the mid-1970s.

The State of Solar-Wind Research about 1970

By about 1970 solar-wind researchers believed the time was ripe for appraising their progress. One sign was the appearance of several review articles and monographs (e.g., Dessler 1967; Wilcox 1968; Brandt 1970; Cowling 1971; Hundhausen 1972). Another was the scheduling of a second week-long conference on the solar wind for March 1971 (Sonett, Coleman, and Wilcox 1972). Such developments attested to the sense of accomplishment felt by those involved in securing and interpreting data on the solar wind. During the preceding decade, they had established the existence of the solar wind, characterized high-speed streams and associated them with geomagnetic disturbances and interplanetary magnetic sectors, and started serious work on the solar sources of various features in the wind's structure.

Throughout these early years of solar-wind research, there was a lively interplay between theorists and observers. The theorists depended on the observers for empirical results that could serve both as points of departure for new inquiries and as checks on one or another line of interpretation. The observers, meanwhile, depended on the theorists for ideas about the kinds of measurements needed and for endorsements of their missions and instruments. At the outset, some theorists—most notably Parker—had the initiative. The chief import of the direct observations of the solar wind from *Lunik 2* (1959), *Explorer 10* (1961), and *Mariner 2* (1962) was to confirm his basic theory of the solar wind. The space scientists did have the pleasure, however, of reducing theoretical estimates of the solar wind's flux and density by two orders of magnitude. With the *Interplanetary Monitoring Platform 1* (1963–64), the initiative passed to the space scientists. Thereafter, they and the theorists who had ready access to their data led the way in advancing knowledge of the solar wind.

Although space scientists assumed the lead in the study of the solar wind in the mid-1960s, their endeavors increased rather than reduced the value of ground-based programs for observing related phenomena. For instance, Snyder and Neugebauer transformed the geomagnetic index into a solar-wind velocity indicator when they found a linear relationship between the two parameters. Again, Ness and Wilcox enhanced the importance of Mount Wilson's magnetographic observations of the sun when they sought to link solar-wind and photospheric field patterns. These examples of a mutual reinforcement of space and ground-based observing programs were harbingers of much stronger links that would be forged in the decade to come.

This chapter's narrative has focused, after Parker, on seven scientists who were especially successful between 1959 and 1970 in using the opportunities provided by the Soviet and American space programs for making direct observations of the solar wind. Yet, the successes of Gringauz, Rossi, Bridge, Neugebauer, Snyder, and Ness were fundamentally the successes of the institutions that employed them—the Soviet Academy's Radio Engineering Institute, the Massachusetts Institute of Technology, the Jet Propulsion Laboratory, and NASA's Goddard Space Flight Center. The very ability of these scientists to develop their plasma probes and magnetometers, get these instruments into space, and retrieve the resulting solar-wind data depended on the backing of these institutions and the access to the Soviet and American space programs that this backing gave them. Ness's collaborator Wilcox was something of an exception, since he did not participate in space observations. Still, like virtually everyone else in the field, he was dependent for solar-wind data on the institutions and enterprises spawned by the superpower space programs.

As solar-wind research turned into an ongoing activity, a vigorous solar-wind subcommunity emerged at the intersection of the solar and geophysical communities. As of 1970, this subcommunity consisted chiefly of Americans, with a sprinkling of Europeans. Their most important patron was NASA; their most common gathering places were the meetings of the American Geophysical Union and the International Committee on Space Research; and their most frequented periodicals were the *Journal of Geophysical Research, Solar Physics,* and the *Astrophysical Journal.* Proud of their field's progress since *Mariner 2,* solar-wind physicists expected that its advance would continue during the 1970s as further refinements were made in their instrumentation and more distant and long-lived missions were undertaken. They were not to be disappointed. Indeed, as the result of unanticipated successes in correlating solar-wind observations with coronal and geomagnetic observations, a third dimension was soon added to their picture of the solar wind's structure.

Epilogue

Through the 1960s, solar-wind physicists paid little heed to the problem of the wind's three-dimensional structure. Their tendency to avoid the issue was a prudent response to the fact that all spacecraft had remained near the ecliptic plane, which intersects the plane through the sun's equator at a mere 7°. They knew, to be sure, that the sun's magnetic field and the corona had different geometries at the solar equator and poles, but no one was able to translate such knowledge into a convincing picture of the solar wind's large-scale organization. As of 1970, therefore, many solar-wind physicists took an agnostic stance on the question of the wind's topology. Others presumed that the wind's properties depended on heliographic latitude. Still others supposed that Ness and Wilcox's magnetic sectors extended poleward from the ecliptic like the sections of an orange. Whatever their stance, they generally thought this speculative issue would remain unresolved until spacecraft were sent far above or below the ecliptic plane. Hence it came as a surprise to all concerned that despite the continuing lack of direct observations at high heliographic latitudes, they reached a consensus regarding the wind's three-dimensional structure between 1970 and 1977.

The story is complex, involving the convergence of three lines of research in two stages. One line was a series of tenuously related attempts between 1970 and mid-1975 to develop models in which a warped neutral sheet divided the solar wind into two hemispheres of opposite polarity. The scientists who proposed such two-hemisphere models—Ronald Rosenberg (1970), Leverett Davis, Jr. (1972), Michael Schulz (1973), John Wilcox's group (Svalgaard, Wilcox, and Duvall 1974), Russell Howard and Martin Koomen (1974), Hannes Alfvén (1975), Eugene Levy (1975), and Takao Saito (1975)—did so in the course of trying to resolve one puzzle or another in coronal, solar-wind, cosmic-ray, or geomagnetic physics. Having different concerns, they did not use the same terms or emphasize the same features when describing their models. Yet it gradually became clear that all their models had much in common. By 1975–76, the proponents of two-hemisphere models had sufficient confidence in their approach that they were taking the trouble to work up three-dimensional representations of the neutral sheet (fig. 6.17).

Meanwhile, an independent line of research led to the identification of the sources of fast streams in the solar wind—Bartels's long-sought M regions. Back in the mid-1950s, the Babcocks had suggested that fast streams might be produced by the unipolar regions in the solar photosphere mapped by their magnetograph (see chap. 4). A dozen years later, John Wilcox had increased the plausibility of this idea by tentatively tracing magnetic sectors and their associated streams back to unipolar regions and by portraying these regions as areas where the field lines ran freely out into space (see fig. 6.16). By the early 1970s, therefore, solar-wind theorists were becoming accustomed to the hypothesis that streams originated in magnetically open areas in the corona (e.g., Hundhausen 1972). Allen Krieger, Adrienne Timothy, and Edmond Roelof (1973) put this view of M regions on a firmer footing when, by comparing a detailed X-ray photograph of the corona (fig. 6.18) with concurrent solar-wind measurements, they identified a dark coronal hole as the source of a fast stream (fig. 6.19). Building on this achievement, solar and solar-wind physicists hastened to monitor coronal holes with telescopes on Orbiting Solar Observatory 7, Skylab, and Kitt Peak and to collect concurrent stream data with Interplanetary Monitoring Platform 6, 7, and 8. Soon, these investigators were reporting a high, but not perfect, statistical relation between holes and streams (e.g., Neupert and Pizzo 1974; Krieger et al. 1974).

Between mid-1975 and 1977, these two lines of research converged and joined with yet another line of research. First, scientists at the NASA-sponsored Skylab Solar Workshop on Coronal Holes began interpreting the hole-stream relationship with the aid of two-hemisphere models. They found that if in accord with this model they defined coronal holes to include polar as well as midlatitude holes, all the fast streams observed during Skylab's flight could be traced back to specific coronal holes (Hundhausen 1977; Zirker 1977). Shortly afterward, Edward Smith's team behind the magnetometers on Pioneer 11 gave weighty support to the basic two-hemisphere model. They reported that—as Schulz had predicted using his version of the

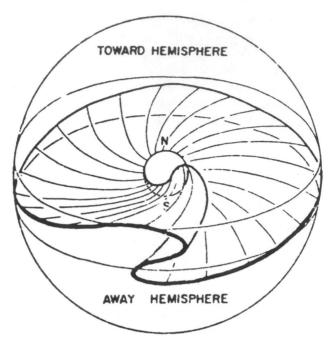

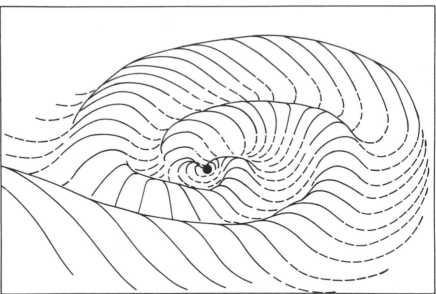

Figure 6.17. Early three-dimensional representations—*top*, by Takao Saito; *bottom*, by Leif Svalgaard and John Wilcox—of the warped neutral sheet dividing the solar wind into two hemispheres. (Sources: Saito 1975, 49; Svalgaard and Wilcox 1976, 766. By permission of *Nature* © 1976 Macmillan Magazines Ltd.)

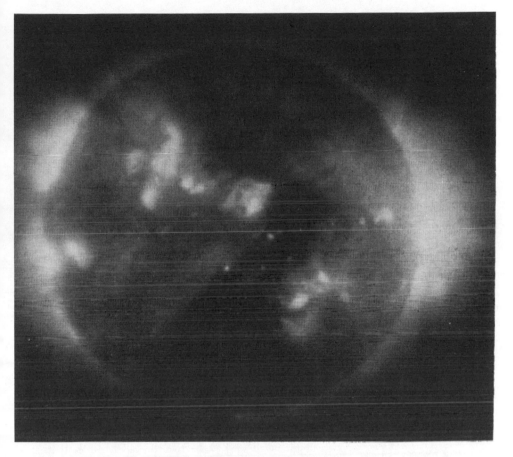

Figure 6.18. Photograph of the corona taken on November 24, 1970, by a rocket-borne prototype of American Science and Engineering's Skylab grazing-incidence X-ray telescope; note the dark coronal hole slanting northward up to the solar equator. (Source: Krieger, Timothy, and Roelof 1973, 507.)

model—the magnetic-sector pattern disappeared when the spacecraft rose above 15° N in heliographic latitude (NASA 1976; Smith, Tsurutani, and Rosenberg 1978). This announcement signaled the triumph of two-hemisphere modeling (e.g., Alfvén 1977; Svalgaard and Wilcox 1978).

Research after 1977 focused on improving the two-hemisphere model's verisimilitude, exploring its implications, and finding its limits. As this work progressed, the hypothetical vantage point for visualizing the wind and related phenomena shifted decisively from the earth to the sun, increasing

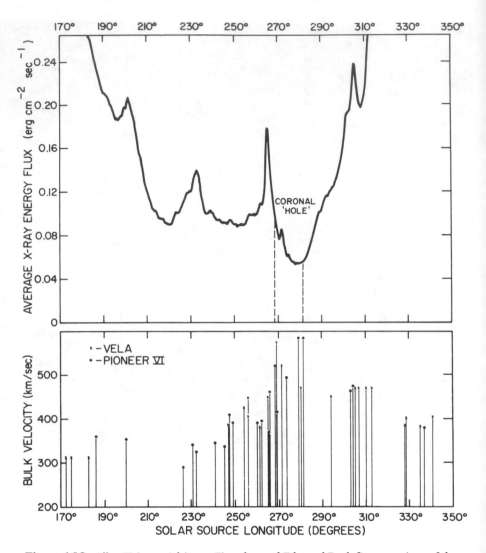

Figure 6.19. Allen Krieger, Adrienne Timothy, and Edmond Roelof's comparison of the coronal intensity along the solar equator (top curve) with the velocity of the solar wind emanating from the corresponding longitudes; note the coronal hole's association with a fast wind stream. (Source: Krieger, Timothy, and Roelof 1973, 519.)

the currency of the term "heliosphere" for the realm extending from the sun out to the distant boundary where the solar wind can no longer sweep the interstellar wind before it (Simpson 1989). In addition, solar-wind physicists came to think that their basic model was too simple for the three years around the sunspot cycle's maximum when the solar field was reversing. Indirect radio-scintillation techniques for measuring the solar wind's velocity at all heliographic latitudes indicated that the high polar wind speeds that were typical during most of the cycle abated during maximum (Rickett and Sime 1979; Coles et al. 1980). Likewise, techniques for using coronal and photospheric data to locate the neutral line where the "heliospheric current sheet" departed from the sun showed that the picture became very confused during maximum. It was ultimately Wilcox's group at Stanford University that devised a convincing means of tracing the solar wind's structure through solar maximum (Hoeksema, Wilcox, and Scherrer 1982; Hoeksema 1984). In their view, the main neutral sheet is highly mobile during this phase of the sunspot cycle, secondary sheets are often present, and average speeds show little dependence on heliographic latitude.

Notwithstanding the plausibility of the present picture of the heliosphere's evolution through the sunspot cycle (e.g., Holzer 1989), a significant obstacle still stands in the way of its unqualified acceptance—the solar wind's speed and polarity have yet to be *directly* observed at heliographic latitudes greater than 16°. The quest for a mission away from the ecliptic plane that could provide such observations has been frustrating. In the early 1970s, the mission's proponents were hoping for launch in May 1974. But three underlying problems—the comparatively small constituency for heliospheric physics, the tight budgets for space science, and the technical problems of the Shuttle and Shuttle-compatible boosters—have forced repeated delays. Known originally as the Out-of-the-Ecliptic Mission, then the Solar Polar Mission, then the International Solar Polar Mission, it is now dubbed Ulysses. The spacecraft was successfully launched in October 1990. If all goes as planned before and after *Ulysses* uses a gravitational assist from Jupiter to enter a polar trajectory over the sun, solar-wind physicists should be debating how well the modified two-hemisphere model stands up to direct observations before the decade is out.

The advance in understanding of the solar wind's topology since 1970 attests to the puissance of contemporary solar science. In particular, the story illuminates how a lack of direct observations of a phenomenon can be overcome with the aid of theoretical models and indirect observational techniques. In this case, two-hemisphere modeling first served as a framework for interpreting several findings—magnetic sectors in the solar wind, the latitude dependence of the wind's polarity, the hole-stream relationship, and the ascent of *Pioneer 11* above the magnetic-sector pattern. Once accepted as a good first approximation, the model was soon rounded out with indirect methods for monitoring the solar wind's speed and polarity at high heliographic latitudes. These methods led in turn to a picture of the evolution of the wind's structure through the complete sunspot cycle.

In general terms, the many scientists who helped develop the present picture of the solar wind's three-dimensional structure relied on two related strategies. They used space-based measurements to find means of inferring the solar wind's properties at high heliographic latitudes from ground-based observations. And they used sophisticated modeling to find ways of correlating measured or inferred solar-wind and coronal properties with one another. In forging such connections, they confirmed strong convictions that ground-based techniques still deserve support and that solar research has an underlying unity.

CHAPTER SEVEN

The Inconstant Sun

1964–1990

The solar wind's influence on the earth is feeble compared with that of the sun's radiation. The wind's flux could probably be doubled or halved without much effect on day-to-day life. Comparable changes in the sun's radiative output would speedily roast or freeze us. Changes of only 5 percent would be catastrophic. Our very presence on the planet is evidence that the sun's luminosity is one of the more stable environmental factors. This is not to say, however, that the solar constant is perfectly stable. Today's scientists have grounds for thinking not only that the sun's radiative output is variable but also that such variability is at least partly responsible for major climatic changes.

The question of the constancy of the sun's radiative output has a long history. Since the discovery of sunspots, many scientists have wondered whether spots influenced the sun's luminosity. In the seventeenth and eighteenth centuries, those who considered the question generally presumed— to judge from the popular hypothesis that starspots caused stellar variability—that sunspots diminished the sun's brightness (see chap. 1). In the early nineteenth century, by contrast, William Herschel argued that the larger the number of spots, the greater would be the sun's output of heat (again, see chap. 1). The invention of the first solar radiometers in the 1830s and the discovery of the sunspot cycle in the 1840s set the stage for empirical studies of the issue (see chap. 2). Starting in the 1870s, a few solar physicists dedicated themselves to collecting and analyzing data bearing on the constancy of the sun's luminosity (see chaps. 2 and 3). From time to time they reported measurements that appeared to demonstrate the variability of the solar output, but ultimately, their research was inconclusive.

The most important source of uncertainty in these ground-based measurements was the atmosphere, which absorbed varying amounts of incident radiation depending on the numbers and characteristics of airborne particles, the vertical distribution of its constituent gases, and the degree of turbulence. Another difficulty was that solar radiometers, or pyrheliometers, were less sensitive and stable than most astronomical instruments. In the 1950s and 1960s, therefore, opinion was still divided about the connection between the sun's activity and its luminosity. Some solar physicists, following Charles G. Abbot of the Smithsonian Astrophysical Observatory, whose observing program had amassed a huge amount of radiometric data between 1902 and 1955 (see chap. 3), believed the sun's luminosity varied with the sunspot cycle. Others—evidently a sizable majority—doubted there was any good evidence for such variation (Doel 1990). They were inclined instead to side with those theoretical astrophysicists who, on grounds of simplicity, supposed that both energy generation within the sun and the sun's radiative output have remained virtually constant throughout historic times.

This chapter examines how, since the mid-1960s, scientists have investigated the sun's radiative output. It opens by recounting the developments in precision radiometry and theoretical astrophysics that revived interest in the question. Attention then shifts to three lines of work between 1975 and 1980 that produced tantalizing evidence of the sun's inconstancy—research on prolonged swings in solar activity and their possible implications for the solar-constant problem, fresh probes of Abbot's data for signs of short-term variability in the sun's luminosity, and direct measures of the solar output with radiometers on two meteorological satellites. Next the chapter describes how Richard Willson clinched the case for the inconstancy of the solar constant with his direct measurements of solar irradiance from the *Solar Maximum Mission*. It then traces the efforts of solar physicists to interpret the new irradiance measurements. The conclusion not only summarizes the story but also draws out its implications for our understanding of contemporary solar physics.

A familiar theme that recurs in this chapter is the role of scientists outside the solar physics community in opening up new lines of research on the sun. A fresh theme is the significance of exact measurements for contemporary solar physics. The amplitude of solar-irradiance variations is small—normally at the 0.05 percent level with occasional excursions to the 0.3 percent level. Unambiguous detection of variations in the solar radiative output had to await the orbiting of radiometers with a relative stability of at least 0.01 percent. Once detected and traced, however, the pattern of variability has—despite its small amplitude—provided solar physicists with valuable grist for their interpretive mills.

A Quickening of Interest, 1964–1975

In the early 1970s, solar physicists took a fresh interest in the old problem of the sun's constancy. They did so partly because of public concern about the adequacy and cleanliness of available energy sources (Greenberger 1983; Katz 1984). In June 1971 and April 1973, for instance, Richard M. Nixon sent Congress the first presidential messages to focus on energy issues. In October 1973 the Arab nations sharpened the growing sense of crisis when, angry about American support for Israel, they embargoed oil exports to the United States. As the "energy crisis" unfolded, most politicians and bureaucrats came to regard an increased reliance on solar power as part of the long-range solution. Attention focused here on the practical problems of developing solar power into a viable supplemental source of energy. Still, the federal government stood readier than ever before to nurture scientific research on the sun's radiative output and its influence on the earth's climate.

Solar physicists had, however, more than government patronage to promote their interest in the sun's radiative output. Advances in precision radiometry since the mid-1960s raised the prospect that short-term fluctuations in the solar constant would soon be detected. In addition, challenges to orthodox stellar theory fostered hopes that evidence of intermediate or long-term variability in the sun's luminosity might soon be forthcoming.

Advances in Solar Radiometry

The impetus for the reinvigoration of solar radiometry came from the Jet Propulsion Laboratory in Pasadena (Drummond and Hickey 1968; Willson 1985; Plamondon 1988). About 1964, spacecraft engineers there became concerned that the temperature readings from several *Rangers* and *Mariner 2* did not match the values predicted on the basis of preflight simulation testing. Resolution of this discrepancy was essential for the development of reliable photovoltaic and thermal-control systems. Hence the laboratory initiated investigations to ascertain whether the value used for the solar constant was correct and whether the radiometers used in the simulation chambers were reliable. Sources of error were ultimately found in both directions.

As the investigation unfolded, some of the physicists involved came to believe they would soon be able not merely to correct the solar constant but also to increase the number of significant figures to which it was known. They had two reasons for optimism. They were developing more sensitive, rapid, and portable radiometers, and they had some chance of getting state-of-the-art radiometers out into space to make direct measurements of the sun's radiative output. By 1970, in fact, two men—one a veteran in solar radiometry with a sizable research team, the other a relative newcomer to the specialty who was something of a loner—were confident that a transformation in solar-constant research was in the offing.

The first was Andrew J. Drummond (1917–72), chief scientist at the Eppley Laboratory in Newport, Rhode Island. Born and educated in Scotland, he had worked in meteorological radiometry at the Kew Observatory and in South Africa before coming to the instrument-making firm in 1956 (Thekaekara 1973). His role in the formulation of the International Pyrheliometric Scale for the International Geophysical Year and Eppley's prominence in the radiometer market put the firm in a good position to participate in the Jet Propulsion Laboratory's project to ascertain why spacecraft temperature readings were awry. In 1964, Eppley Laboratory obtained a contract to remeasure the solar constant (Drummond and Hickey 1968).

Drummond's group built a twelve-channel radiometer that could simultaneously measure the total irradiance and the energy flux in several bands along the solar spectrum. The sensor for each channel was a wire-wound thermopile[1] that measured the electrical current engendered by the incident radiation. The total-irradiance channels were open, while the spectral channels were provided with filters that excluded radiation outside the specified bands. Drummond, his colleagues, and their counterparts at the Jet Propulsion Laboratory conducted a series of preliminary tests from inside a CV-990 jet aircraft (fig. 7.1). Then in 1966 and again in 1967, Drummond's team sent their twelve-channel radiometer (fig. 7.2) well above most of the atmosphere in the U.S. Air Force/NASA X-15 rocket aircraft. On the second flight everything went smoothly. During the X-15's thirteen seconds above 77 km, each channel took some forty readings. Soon Drummond and his collaborators were announcing a "new value for the solar constant of radiation" in *Nature* (Drummond et al. 1968, 259) and the "first direct measurements [of the] solar constant" in *Science* (Laue and Drummond 1968, 888). Their new value of 1361 watts/m² was about 2.5 percent lower than the figure most commonly used in the American space program. Having calibrated the radiometer before and after flight by comparing its response with that of standard pyrheliometers, they estimated that their value was accurate to within 1 percent.

This success encouraged Drummond in his hope that an improved version of the Eppley radiometer would qualify for a satellite payload. From the outset, his group had pursued a light, compact, and rugged design in the hope of getting an instrument into space (Drummond et al. 1967). Not long

---

1. Invented in 1831, the thermopile was immediately used by Macedonio Melloni for studying the solar infrared (Schettino 1989).

after the appearance of the articles in *Nature* and *Science*, Drummond and his colleague John R. Hickey (b. 1936) proposed a Planetary Heat Budget Experiment for Nimbus E (Kyle 1987). Their bid was turned down, perhaps because of problems with the design for the earth-viewing channels. Persisting, Drummond and Hickey teamed up with scientists interested in terrestrial radiometry to propose the Earth Radiation Budget Experiment for Nimbus F, a satellite that would be testing several new instruments for climatological research. The Eppley group's part of the proposal was ambitious. They wanted to provide a radiometer that would measure not only the total solar irradiance but also the radiative output within nine narrower spectral bands. They hoped that, besides securing data that would be "practical for both geophysical and spacecraft engineering design purposes" (Hickey 1973, 136), they could advance "knowledge of long-term solar-constant variability" (Drummond 1973, 34). Drummond and Hickey's persistence paid off with NASA approval of the Earth Radiation Budget Experiment as part of the Nimbus-F payload (Naugle 1969, 1971).

Drummond's death in 1972 shifted the prime responsibility for the Nimbus-F radiometer to Hickey, who had acquired a thorough grounding in radiometry during a decade at Eppley. He did everything possible to maximize both the instrument's absolute accuracy—its ability to provide a trustworthy value of the solar constant—and the instrument's precision—its ability to measure changes in the solar output (Hickey and Karoli 1974). Still, wary about promising more than he could deliver, he adopted the conservative design goals of an accuracy of ± 0.75 percent and a precision of ± 0.3 percent for the total-irradiance channel. When the prototype (fig. 7.3) was calibrated in the fall of 1973, Hickey concluded he had been too cautious. That November he announced at the Smithsonian Symposium on Solar Radiation Measurements and Instrumentation that an accuracy of ± 0.2 percent and a precision of ± 0.025 percent had been achieved. He and his collaborators hoped these levels could "be maintained through the spacecraft interface, through production of the flight model instrument and during orbital operation" (Hickey, Hilleary, and Maschhoff 1974, 147). In fact, on account of delays in the launch of Nimbus F, the outcome would not be known until after June 1975.

Even as Drummond, Hickey, and their collaborators were qualifying their radiometer for space, Richard C. Willson (b. 1937) was doing his utmost to achieve the same end. This young physicist joined the Jet Propulsion Laboratory's Instrumentation Section in 1963, eager to make good use of the knowledge of optics he had acquired during graduate studies at the University of Colorado (Willson 1985, 1988a, 1988b). He soon realized, however, that he was too much of a perfectionist to take much pleasure in teamwork.

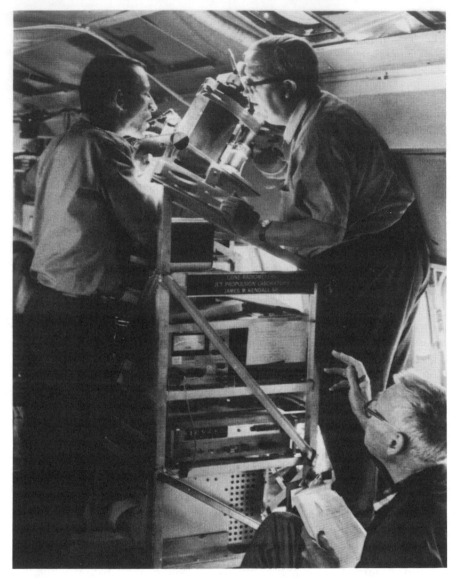

Figure 7.1. Andrew Drummond (upper left) inspecting a radiometer with James Kendall (seated) kibitzing during radiometer tests aboard a CV-990 jet in the mid-1960s. (Source: Hickey and Karoli 1974, cover page.)

He passed through several projects during his first few years in Pasadena and then settled into radiometry, a field where an independent personality could indulge a passion for accuracy and still accomplish something.

Willson's first contact with radiometry was in 1964 (Willson 1988a,

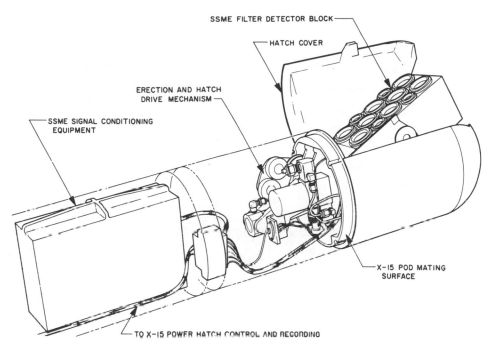

SSME FILTER DETECTOR BLOCK

HATCH COVER

ERECTION AND HATCH
DRIVE MECHANISM

SSME SIGNAL CONDITIONING
EQUIPMENT

X-15 POD MATING
SURFACE

TO X-15 POWER HATCH CONTROL AND RECORDING

Figure 7.2. Schematic of the Eppley radiometer installation in an X-15 wing pod; two of the twelve apertures in the SSME (solar spectrum measurement experiment) were dedicated to total-irradiance measurements. (Source: Preprint of Drummond et al. 1967, 6. By permission of AIAA.)

1988b; Plamondon 1988). That summer he was assigned to Joseph A. Plamondon, a mechanical engineer who was among those working on the problem of the spacecraft temperature readings. Plamondon's focus was on the possibility that the preflight simulation testing had been faulty. To look into the heat-transfer relations used in the simulations, he needed a very small absolute radiometer. Nothing suitable was available, so he set out to develop one. His evaluation of possible designs indicated that cavity sensors had the most promise. It was at this juncture that Plamondon secured Willson's assistance. Together they visited the laboratory of physicist Floyd Haley, who had developed a cavity detector for calibrating measurements of line intensities in the ultraviolet. All three discussed how this specialized instrument might be transformed into a total-irradiance radiometer. For the next couple of months, Willson helped Plamondon carry out a theoretical assessment of cavity designs. Then he returned to the Instrumentation Section to work on ultraviolet spectrophotometers.

Plamondon's next assistant was James M. Kendall, Sr. (1899–1986), a veteran engineering physicist who had just begun a second career in the Jet

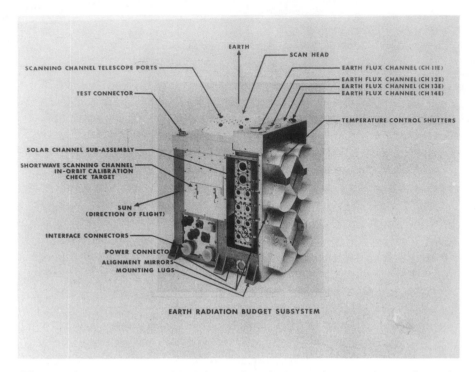

Figure 7.3. Engineering model of the Earth Radiation Budget experiment; the total-irradiance channel is the lowest opening in the left-hand row of the solar subassembly. (Courtesy of John Hickey and the Eppley Laboratory.)

Propulsion Laboratory's Instrumentation Section (Plamondon 1988; Willson 1988a, 1988b). Kendall's work in naval ordnance had given him no experience in radiometry, but he was a gifted hands-on scientist. Starting with Haley's basic design, he soon constructed a total-flux radiometer that, except for being somewhat larger than desired, met Plamondon's specifications. The instrument (fig. 7.4) was in principle simple (Plamondon and Kendall 1965). Its central feature was an insulated black cavity that was maintained at a constant temperature by electronic means. Since the cavity was in thermal equilibrium, the energy entering it must match that exiting. There were two inputs—the unknown irradiance entering the cavity's aperture and the heating power needed to keep the cavity's temperature constant. The sole output was the radiation leaving the cavity's aperture. Hence the incident irradiance could be determined by subtracting the heating power, which could be measured, from the exiting radiation, which could be calculated from the temperature. The cavity radiometer was superior not only to ordinary radiometers such as those sold by the Eppley Laboratory but also to earlier standard instruments. It was not subject to the calibration

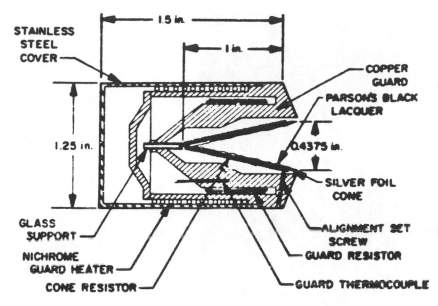

STAINLESS
STEEL
COVER

1.5 in.

1 in.

COPPER
GUARD

PARSONS BLACK
LACQUER

1.25 in.

0.4375 in.

SILVER FOIL
CONE

GLASS
SUPPORT

ALIGNMENT SET
SCREW

NICHROME
GUARD HEATER

GUARD RESISTOR

CONE RESISTOR

GUARD THERMOCOUPLE

Figure 7.4. Sketch of James Kendall's cavity radiometer; note its smallness. (Source: Plamondon and Kendall 1965, 66).

errors that inevitably contaminated the results of ordinary radiometers. And thanks to the cavity's high absorptance and the precise thermal control, the instrument was not subject to some of the uncertainties limiting the accuracy of existing absolute radiometers.

In 1967 Willson was reassigned to radiometric work. His initial task was to assess the accuracy of the Eppley radiometers that were used in the Jet Propulsion Laboratory's space simulators. To ground himself in the field, he attended the Eppley Laboratory's third annual course on radiometry (fig. 7.5). Once back in Pasadena, he conducted his tests using Kendall's absolute cavity radiometer, or ACRAD, as the standard of comparison. By this time, the instrument had "been subjected to rigorous experimental tests and to considerable theoretical scrutiny. The results of these examinations indicate that this radiometer is capable of measuring radiant energy with an uncertainty of less than 1%. In relatively well controlled experimental environments, the actual uncertainty can be less than 0.5% which represents an order of magnitude improvement over the best competing radiometric techniques" (Willson 1967, 2). Finding that there was "considerable uncertainty involved in the calibration and use of the Eppley Mark 1 radiometers," Willson (1967, 12) recommended against "further reliance upon the Eppley devices."

In the course of these comparison tests, Willson came to share Kendall's

Figure 7.5. Richard Willson (standing, far left) and John Hickey (seated, far left) at the Eppley Laboratory's third annual course on "fundamental radiometry for experimental scientists," July 31-August 5, 1967. (Courtesy of John Hickey and the Eppley Laboratory.)

optimism about the cavity radiometer's prospects. But they found it impossible to work together (Willson 1988b). For one thing, there was a difference in style. Kendall approached the job of improving the instrument as a resourceful tinkerer. Willson, by contrast, used a theoretical approach, systematically attacking problems that parametric analysis revealed as major sources of uncertainty. There was also a difference in goals. As Willson (1975a, 321) later observed, while "Kendall's primary interest [was] in laboratory measurements of great accuracy," his own goal was "to produce instrumentation suitable for automatic, remote measurements of solar irradiance in any environment." On account of their differences, Kendall and Willson were "fiercely competitive" (Willson 1988b). Yet as is so often the case with competitors, they were continually learning from one another.

Kendall's first major step beyond this prototype instrument was to transform it into the standard absolute cavity radiometer, or SACRAD (Kendall and Berdahl 1970). He went on to use this instrument, which was completed in 1968, to show that the Stefan-Boltzmann constant relating the intensity of the radiation from a black body to its temperature could be measured to within 0.3 percent of the value given by theory. While conducting this investigation, he continued refining his instrument. The result was the primary absolute cavity radiometer, or PACRAD, which could operate at ambient pressures and temperatures. His objective here was to replace the instruments that had defined the International Pyrheliometric Scale since 1956. This hope was realized in 1970 at the third meeting for the international comparison of pyrheliometers held in Switzerland (Fröhlich et al. 1973). The Eppley Laboratory, which had anticipated this outcome by getting a license to Kendall's patent, soon began producing the Eppley-Kendall radiometer for the commercial market (Hickey 1988).

Meanwhile, Willson was busy transforming Kendall's instruments into radiometers that could be used outside the laboratory (fig. 7.6). He completed his first radiometer—the active cavity radiometer (type) II, or ACR II—about the spring of 1968. Within months, he got two of these radiometers onto a high-altitude balloon that was part of the Jet Propulsion Laboratory's program to standardize solar cells. This was, Willson (1971, 4339) asserted, "the first direct measurement of solar irradiance by standard detectors at an altitude where extinction by atmospheric water vapor and known aerosol distributions [is] negligible." He had to admit, however, that his measure of 1370 watts/m² might be in error by as much as 2 percent because of uncertainties introduced by the quartz windows on his radiometers.

Willson soon went on to develop the active cavity radiometer III. His goal was to transform Kendall's new "primary" radiometer into an instrument capable of remote operation in the atmosphere or in space. Before long, he was reporting success in such important journals as *Nature* (1972) and *Applied Optics* (1973). Parametric analysis of this radiometer's characteristics indicated it could measure the solar constant to within about ±0.2 percent. Comparing its performance with that of Kendall's radiometer yielded the same result. Willson (1973, 810) concluded that the active cavity radiometer had great "potential usefulness in astrophysical, meteorological, and engineering solar radiation measurement programs." Thereafter, Willson (e.g., 1974, 1975a, 1975b) was tireless in extolling the accuracy and versatility of his instrument. By mid-1975, as we see below, his claims on behalf of the instrument's capabilities were often echoed.

Fresh Doubts about the Sun's Constancy

Quite separately from the advances in precision radiometry, fresh doubts about the validity of orthodox stellar theory were leading solar physicists to consider the possibility that the sun's luminosity varied. The doubts surfaced

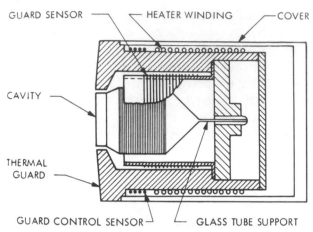

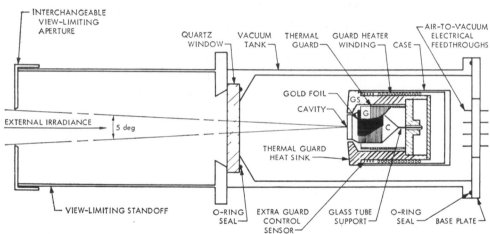

Figure 7.6. Two approaches to radiometry: *Top*, Kendall's standard absolute cavity radi-ometer for laboratory studies; *bottom*, Willson's active cavity radiometer II for field mea-surements; note that though the sensors had the same basic geometry, the instruments differed in most other respects. (Sources: Kendall and Berdahl 1970, 1083; Willson 1971, 4326. © American Geophysical Union.)

in the late 1960s after Raymond Davis, Jr., reported the first results from his solar-neutrino experiment in South Dakota. His subterranean measure-ments indicated that the energy-generating thermonuclear reactions in the sun's core were producing many fewer neutrinos than stellar theory pre-dicted (see chap. 5). By the 1970s, theoretical and nuclear astrophysicists were inclined to regard the gap between Davis's observations and the predic-tions as a significant discrepancy (Pinch 1986; Bahcall 1989). They sought its explanation in two main directions. Perhaps the predictions were based on a faulty understanding of neutrino physics. Or perhaps there was something wrong with the standard model for the sun and other main-sequence stars.

William A. Fowler (b. 1911), Caltech's genial pioneer of nuclear astrophysics, was among the first to explore these possibilities. Of particular interest here was his challenge to the canonical view that the sun's luminosity has slowly increased since the sun settled down on the main-sequence line over 4.5 billion years ago. His idea was that a mixing of the sun's core with overlying material sometime in the last 1 to 10 million years caused a reduction in the sun's central temperature and hence in the neutrino flux. Fowler (1972a) broached this possibility at a solar-neutrino conference in February 1972, then wrote it up for *Nature*. Here he also pointed out that a mixing episode would reduce the sun's luminosity. Perhaps, he suggested, such solar cooling was responsible for the present glacial epoch. Though happy that geology did "not rule out" his scenario, Fowler (1972b, 26) emphasized that it must be regarded as a "desperate explanation for the present low neutrino flux."

Fowler's proposal swiftly attracted the attention of the theoretical astrophysics community. Between June and December 1972, scientists in Cambridge, England, the Los Angeles area, and New York sent follow-up articles to *Nature* (Dilke and Gough 1972; Rood 1972; Ezer and Cameron 1972; Ulrich and Rood 1973). Their shared goals were to identify a mechanism capable of causing sudden mixing and to explore the astrophysical and geophysical consequences of mixing episodes.

The veteran theoretical astrophysicist Alastair G. W. Cameron soon offered a general account of this line of thought. Asserting that something was "badly wrong with our understanding of the way in which the sun operates," Cameron (1973, 505) characterized Fowler's hypothesis as being "among the remedies . . . that show some promise of accounting for the [neutrino] problem." He admitted that the mixing mechanisms proposed to date were controversial. But he encouraged those who wanted to explore whether "large excursions in the solar luminosity may have some interesting terrestrial effects" not to allow themselves to "be intimidated by previous astronomical dogma" (508–9). Cameron thought it possible, indeed, that recurrent mixing episodes were responsible for the ice ages that punctuated the earth's history every 200 or 300 million years. He recognized that astrophysicists and geophysicists would need to revise many of their theories if the sun ordinarily put out much more energy than at present. Biology would also be affected: "Major changes in the solar luminosity may also have had dramatic effects on biological evolution. . . . Certainly the environmental pressures on the evolutionary process are very much greater during a glacial period [than during an interglacial era]. It may have been just such evolutionary pressures that produced mankind and initiated his technological development. From this, one can argue that mankind will find it most difficult to detect solar neutrinos just at the time he first starts looking" (510). He was, to judge from this concluding speculation, more than a little fond of the idea that the sun's luminosity dipped sharply every once in a while.

Over the next couple of years, the mixing hypothesis lost much of its allure (Cameron 1975; Pinch 1986). One reason was that the mechanisms proposed for the initiation of solar mixing were implausible. Another was that the idea that the sun was now in an exceptional phase of its history flew in the face of uniformitarian convictions. Yet another was that the hypothesis was in competition with a growing number of alternative explanations of the neutrino deficit. Despite these difficulties, some theoretical astrophysicists—especially the few with close ties to the solar physics community—kept on investigating the mixing hypothesis and its consequences. This continuing interest was important in the present context because it led solar physicists to consider two possibilities. It induced them to wonder whether the sun's luminosity had fluctuated in the past. It also quickened their interest in the possibility that paleoclimatology might yield evidence of such variability.

Caltech solar physicist Harold Zirin, whose interest in the sun's constancy had been piqued by his colleague Fowler, certainly had these possibilities in mind (Zirin 1988b). In 1974 he raised them when the issue of sun-weather relationships came up at a meeting of the National Science Foundation's panel for atmospheric sciences. Fred D. White, who supervised the foundation's program in this area, persuaded Zirin to organize a symposium on the issue.[2] The outcome was the Workshop on the Solar Constant and the Earth's Atmosphere, which was held at Big Bear Solar Observatory in May 1975 (Zirin 1975). Zirin got Roger K. Ulrich, a solar theoretical astrophysicist at UCLA who had become one of the foremost advocates of the mixing hypothesis, to lead off with a general argument for past variations in solar luminosity (Ulrich 1975). He also got several paleoclimatologists to summarize their specialty's findings. And having apprised himself of recent developments in radiometry at the Jet Propulsion Laboratory, he got Willson to describe the advantages of the active cavity radiometer for making direct measurements of solar-constant variability (Willson 1975a).

The workshop's participants concluded that research on the variability of the sun's luminosity was not only desirable but also feasible:

2. NASA Goddard's sponsorship the year before of a workshop on "Solar Activity and Meteorological Phenomena" (Bandeen and Maran 1975) may well have inspired White's suggestion.

The interdisciplinary workshop on the solar "constant" and the Earth's climate agrees that there is evidence that climate change in the past on scales of a few years to eons could be due to changes in solar irradiance. Such changes would be of great significance to our understanding of the physics of the Sun and future climate change. Accurate measurements of the solar constant with long-term continuity and stability are of the greatest importance; the time to plant a flower that blooms in a hundred years is now. We therefore recommend that a continuous long-term program (essentially a modern continuation of the discontinued Smithsonian program) of solar constant measurement, coordinated from space and ground be instituted. We believe that instruments capable of such measurements presently exist, with absolute accuracy of 0.25% and stability of 0.1% [i.e., Willson's active cavity radiometer]. (Zirin, Moore, and Walter 1976, 380–81)

The stage was set for a concerted attack on the problem of the constancy of the solar constant.

Evidences of Solar-Irradiance Variability, 1975–1980

By mid-1975, to judge from the Big Bear Workshop, many solar and solar-terrestrial physicists had abandoned, or were set to abandon, the orthodox doctrine of the solar constant's constancy. Over the next five years, belief in the inconstancy of the solar radiative output inspired several workshops (most notably White 1977). It was one thing, however, to suspect that the sun's output varied and quite another—as the long history of frustrated attempts indicated—to marshal a compelling case for such variability. Still, by 1980 three approaches suggested at the Big Bear Workshop yielded tantalizing, if not entirely convincing, lines of evidence for solar-irradiance variability

Eddy's Case for the Sun's Inconstancy

The timing of the Big Bear Workshop was ideal for John A. Eddy (b. 1931). An Annapolis graduate, he had served in the Navy before enrolling in the new Ph.D. program in astrogeophysics at the University of Colorado in 1957 (Eddy 1987). During his doctoral studies, he got involved in Gordon Newkirk's project to use high-altitude balloons as platforms for observing the corona (see chap. 5). This undertaking not only provided him with data for his dissertation on the stratospheric scattering of sunlight but also helped him land a position at the High Altitude Observatory. Over the next decade, he enlarged his ambit with papers on Jupiter's atmosphere, the solar corona, the history of solar physics, and American Indian astronomy. However, despite—or perhaps because of—his versatility, he was among the first to be cut from the High Altitude Observatory's staff in 1973 when budget problems forced retrenchment. Eddy kept his head above water by writing a semipopular book about *Skylab's* results for NASA and by teaching part time at the University of Colorado. Meanwhile, with Eugene Parker's encouragement, he embarked on yet another new investigation. This inquiry soon put him in pursuit of an idea that would add an interesting dimension to the solar-constant problem. By the time of the Big Bear Workshop, indeed, Eddy was ready to espouse it with a missionary's zeal.

Eddy's idea was that historical records and sequential carbon-14 data showed that the sun was neither so stable nor so regular as commonly assumed. His test case was the period from 1645 to 1715, when, according to all contemporary documentation, sunspots were quite rare (Eddy 1975a, 1975b). In the nineteenth century William Herschel, Gustav Spörer, and E. Walter Maunder had commented on this prolonged minimum in solar activity and its possible meteorological consequences. But the few twentieth-century solar physicists who were familiar with the issue regarded it as a bogus problem engendered by the insouciance of seventeenth-century astronomers. Eddy's prior historical work led him to question this cavalier dismissal of the early observations. He thought it more prudent to accept the testimony of Hevelius, Cassini, Flamsteed, and their colleagues about the rarity of sunspots during the second half of the seventeenth century.

Eddy's review of evidence that, according to current knowledge of solar activity and its terrestrial effects, could corroborate or contradict the seventeenth-century testimony lent credence to what he dubbed "the Maunder minimum." Between 1645 and 1715, not a single spot achieved sufficient size to be seen by the Chinese astrologers, who by long tradition kept watch on the sun for possible omens. Likewise, during this period very few aurorae made appearances below the polar circle, and not a single one was seen as far south as the British Isles or the Chinese court. Again, not a single eclipse report during this period mentioned the long coronal streamers that ordinarily accompanied intense solar activity. Last, during the Maunder minimum the production of carbon 14 by cosmic rays was—to judge from the isotope's abundance in tree rings—much higher than usual. This was just what one would expect, since it had earlier been observed that periods of low solar activity were accompanied by high cosmic-ray fluxes.

If, as all these proxies for low solar activity suggested, the Maunder minimum really occurred, Eddy (1975a, 365) thought it must have been "the most significant event in the history of solar observation." His guess was that such an event would have exerted an "influence on climate." The conjecture was borne out by a prior historical study showing that European

Figure 7.7. John Eddy in 1977, a leading advocate of the sun's inconstancy. (Courtesy of the National Center of Atmospheric Research/National Science Foundation.)

winters were particularly harsh in the second half of the seventeenth century. Eddy thought, although he only hinted at this position in the abstract of his Big Bear paper (1975b), that the very low level of solar activity during the Maunder minimum was associated with a reduced solar radiative output and hence colder terrestrial temperatures. If so, his data revealed considerable solar-irradiance variability.

During the two years following the Big Bear Workshop, Eddy (fig. 7.7) strengthened his case for the reality of the Maunder minimum. His most striking new argument was based on the sunspot observations of Christoph

Scheiner in 1625–26 and Johannes Hevelius in 1642–44. He and two col-leagues determined the patterns of differential rotation during the two inter-vals (Eddy, Gilman, and Trotter 1976, 1977). The pattern in 1625–26 matched that observed since Richard Carrington discovered the phe-nomenon in 1859 (see chap. 3). But during 1642–44, just at the onset of the Maunder minimum, the sun's equatorial velocity was about one day faster than usual. Eddy thought it probable that this enhancement in differential rotation was associated with the anomalous decline in solar activity.

Meanwhile, by using some of the indirect indicators of the Maunder minimum, Eddy (1976a, 1976b) was finding prolonged minima and max-ima of solar activity before the era of telescopic observations. Records of sunspots visible to the naked eye and of aurorae observed south of the polar circle agreed with levels of carbon 14 in tree rings in suggesting the occur-rence of a major minimum in the sixteenth century, which he named after Spörer, and a major maximum in the twelfth and thirteenth centuries. By relying on carbon-14 data, Eddy was able to trace numerous minima and maxima back to about 3000 B.C. (fig. 7.8). At first he was troubled by the ill-behaved undulations in the envelope of solar activity; but soon he was contending that his findings constituted "one more defeat in our long and losing battle to keep the sun perfect, or, if not perfect, constant, and if inconstant, regular" (Eddy 1976a, 1200).

Although Eddy gave first priority to ascertaining the main features in the solar-activity curve, he remained interested in the implications of his re-search for the sun-weather problem. This interest was sustained by his success in correlating the activity curve with prior climatological studies. Like the Maunder minimum, prolonged minima were associated with se-vere winters and expanding glaciers. And like the medieval grand max-imum, prolonged maxima were associated with mild winters and shrinking

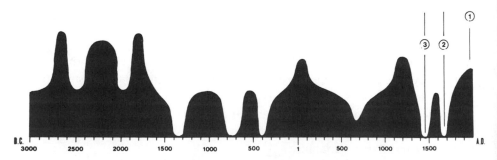

Figure 7.8. Eddy's graph of the long-term envelope of solar activity based on carbon-14 data; (1) indicates the present, (2), the Maunder minimum, and (3), the Spörer minimum. (Source: Eddy 1976b, 966. © American Geophysical Union.)

glaciers. Eddy (1976b, 969) doubted that these major climate shifts were caused by minor "trigger mechanisms" such as fluctuations in solar ultraviolet emission that, by modulating the terrestrial atmosphere's transparency, determined how much of the sun's radiative output reached the earth's surface. Admitting a "distaste" for such complex processes, he proposed that the sun made its influence felt through "the simplest and most straightforward process." His "hunch" was that the sun's radiative output changed "slowly and ponderously," following the same "meandering" path as "the overall envelope of solar activity." On this "notion," the carbon-14 curve stretching back to 3000 B.C. was a record not only of the average sunspot number but also "of the solar constant, with peak-to-peak amplitudes of perhaps 1%."

Eddy's evidence for his proposal was thin. The best he could do was adduce the pyrheliometric data collected by Abbot and his colleagues at the Smithsonian Astrophysical Observatory between 1908 and 1955. Although he found no trace of the eleven-year solar cycle in the Smithsonian data, he thought he discerned a 0.25 percent increase in the solar constant over the entire period. This increase, he suggested, matched the concurrent rise in the envelope of solar activity. Eddy (1976b, 969) realized that the hypothesis was open to a potent objection: "If the solar constant does not follow the wiggles in daily or annual sunspot number, how can it follow the envelope?" He got around this difficulty by pointing out that "the solar constant may not follow the sunspot number at all." Instead, its swings could be the cause, not a consequence, of the corresponding swings in the solar-activity envelope.

Mindful of the anomalous rotation that appeared to be associated with the Maunder minimum, Eddy (1976b, 970) conjectured that changes in the outward flow of heat affected the solar dynamo and thereby the general level of solar activity. If so, the "individual ups and downs of the 11-year cycle . . . tell little of changes in the solar output and predict almost nothing of consequence in terrestrial meteorology. If one sought a solar-weather connection [for short time scales,] he would always be frustrated in what he found, and driven . . . to ever more intricate mechanisms, much as pre-Copernican astronomers were driven into epicycles. And that, I would submit, may be just exactly what has happened in the past century of solar-weather research." Eddy was determined, despite the many fragile links in his chain of reasoning, to challenge orthodox notions about the sun's constancy.

From 1976 through 1979, in fact, Eddy was extraordinarily busy promoting his findings and speculations. He wrote several papers for specialist journals and conferences. He also placed articles in *Science, Scientific Ameri-*

can, and the *Encyclopaedia Britannica Yearbook of Science and the Future* (Eddy 1976a, 1977a, 1978). He even did a program called "The Missing Sunspots" for the Nova television series (Eddy 1977b). His results, and the attention they attracted, enabled Eddy to get his career back on track. In 1978, after a year visiting the Harvard-Smithsonian Center for Astrophysics, he returned to the High Altitude Observatory as a senior scientist.

Foukal's Reappraisal of Abbot's Data

Another of the participants in the Big Bear Workshop of May 1975 was Peter V. Foukal (b. 1945), a solar physicist with the Harvard-Smithsonian Center for Astrophysics. His interest in the solar-constant problem had been aroused by recent endeavors to pin down relationships between the solar cycle and terrestrial weather (Foukal 1987b). Most of the solar physicists who were engaged in such work were investigating one or another esoteric trigger mechanism. Like Eddy, Foukal thought they were being too clever by far. But unlike Eddy, he suspected that the main way solar activity influenced meteorological phenomena was through its direct modulation of solar irradiance. He was invited to the workshop because Harold Zirin, his postdoctoral supervisor from 1969 to 1972, was aware that he had begun to look into this possibility.

At the workshop, Foukal (1975) discussed how active regions might affect the sun's radiative output. He surmised that sunspots diminished the solar flux while faculae—bright areas that often accompanied sunspots— enhanced it. Observations indicated that spots and faculae, though their numbers followed similar curves from solar minimum to maximum, covered different fractions of the solar disk from day to day. He supposed, therefore, that at any given time there would likely be a net imbalance between the deficits caused by the spots and the enhancements caused by the faculae. Foukal estimated that the resulting variability in the solar constant was at the 0.1 percent level, which seemed enough to affect terrestrial climate. But further research would be needed to determine whether the spots or the faculae were predominant. One approach would be to relate improved spot and facular photometric measures with the data that, if all went well, would soon be coming from the Nimbus solar radiometer. Besides recommending this approach, Foukal revealed that his own plan was to search Abbot's data for a twenty-seven-day signal of solar rotation.

Soon after the workshop, Foukal, Pamela E. Mack, and Jorge E. Vernazza set about looking for evidence that active regions modulated the solar constant (Foukal 1987b, 1988). They used—at Willson's urging—data collected in 1969 by cavity radiometers that Plamondon had placed on *Mariner* 6 and 7 to help monitor the thermal balance of these spacecraft. They also used—as Foukal had planned—the pyrheliometric data collected between 1923 and 1952 by Abbot and his assistants at the Smithsonian Astrophysical Observatory. It did not matter that neither data set had high absolute accuracy, since the goal was to detect short-term variability in the sun's luminosity. Instrumental and systematic errors could be eliminated by working with the differences between successive measurements.

To Foukal's surprise, the *Mariner* data yielded no clear evidence that spots or faculae influenced the solar irradiance. The Smithsonian data also failed to yield any sign of a spot signal in the sun's radiative output, but Foukal and his colleagues did find tolerably good evidence that changes in facular area had caused changes of up to 0.1 percent in the Smithsonian measurements. On the basis of the *Mariner* results, they rejected the possibility that this facular signal represented a real variation in the solar output. Instead, resorting to a trigger mechanism, they interpreted their signal as evidence of the influence of facular ultraviolet radiation on atmospheric transmission. Foukal, Mack, and Vernazza (1977, 958) concluded that "the Sun's total luminosity changes remarkably little on a time scale of weeks."

Over the next two years, Foukal and Vernazza extended and deepened their analysis of the Smithsonian data. They went beyond sampling the data to use all of the nearly 11,000 daily measurements. They also employed more sophisticated statistical tests than in the first analysis. Their initial objective seems to have been to get a clearer picture of the facular signal they had detected. But as they proceeded with the study, Foukal and Vernazza (1979, 707) came to believe they were isolating "the first direct evidence of a change in stellar luminosity caused by magnetic field activity."

In advancing this conclusion, Foukal and Vernazza first discussed short-term periodicity in the Smithsonian data. Their autocorrelation analysis of sixty-one overlapping segments of data yielded peaks close to periods of one and two solar rotations. These recurrence peaks constituted good evidence that rotating solar features had modulated the Smithsonian measurements. The amplitude of the effect was at about the 0.07 percent level. Foukal and Vernazza went on to argue that both faculae and sunspots were the features causing the modulations. As in their 1976 analysis, there was strong evidence that faculae had increased the measured flux. This time, thanks to the greater sensitivity achieved by the use of the entire data set, there was also evidence that increases in sunspot area had decreased the measured flux. The possibility remained, however, that faculae and sunspots had influenced the Smithsonian measurements by modulating atmospheric transmission, not the sun's luminosity. Rejecting their earlier hypothesis, Foukal and Vernazza (1979, 707, 713) pointed out the many difficulties with devising "an atmospheric mechanism that would rapidly lower . . . transmission in response to sunspots [and] increase it in response to faculae." They were confident that their study had disclosed "a real variation in total luminosity, caused by the effect of intense magnetic fields on energy transport to the photosphere."

Foukal easily got a hearing for the study's results on the conference circuit (e.g., 1980a) and in *Sky and Telescope* (1980b). Yet few if any solar physicists shared his confidence that "direct" evidence of facular and sunspot modulation of the solar constant had actually been found in the Smithsonian data. One reason was that Foukal and Vernazza's argument against the possibility of atmospheric modulation was not conclusive. Another and more serious reason was that their results were at odds with the findings of Douglas V. Hoyt, a solar-terrestrial physicist at the National Oceanic and Atmospheric Administration's Environmental Research Laboratories in Boulder. Hoyt (1979, 427) was claiming that the Smithsonian program's "solar constant values [were] independent of solar activity." Rather than attempting to ascertain how Foukal and Hoyt had reached their contradictory claims, solar physicists preferred to await reports of the direct measurement under way on *Nimbus 6* and 7 and planned for the Solar Maximum Mission.

Hickey's Radiometers on Nimbus 6 *and* 7

The third approach suggested at the Big Bear Workshop was, of course, to monitor the solar irradiance from the atmosphere. Here the inside track was held by John Hickey. In fact, between July 1975 and early 1980, while Eddy was making his case for slow swings in the sun's radiative output at the 1 percent level and Foukal was developing his case for facular and sunspot modulations at the 0.07 percent level, Hickey and his co-workers had the only solar-irradiance monitors in space. In 1979, after surmounting numerous difficulties, they had the satisfaction of detecting two dips in the solar constant at the 0.3 percent level that could be associated with exceptional solar activity.

In July 1975, some two months after the conference at Big Bear, *Nimbus 6* returned its first solar-irradiance data (Hickey et al. 1976). The measurements were obtained by the Earth Radiation Budget Experiment's solar channel 3, the sensor of which had been calibrated before flight with an Eppley-Kendall primary absolute cavity radiometer. Channel 3 could make a set of measurements once every 107 minutes when, as *Nimbus 6* crossed the antarctic terminator on its sun-synchronous, near-polar, circular orbit, the solar instrument panel pointed straight at the sun. However, readings were not made on every orbit because the spacecraft's data system lacked the capacity to keep up with all the instruments aboard. Instead, the Earth Radiation Budget Experiment was typically on for two or three days and then off for about the same length of time.

Early reports of the *Nimbus* measurements (e.g., Hickey 1975), which were the first solar-irradiance measurements made from earth orbit, generated considerable interest (Willson and Hickey 1977). During the first ten days of observation, the solar constant averaged 1392 watts/m^2 and was stable to within 0.2 percent. The constant's stability was in line with general expectations, since the sunspot cycle was then near minimum. But the reports of the constant's absolute value raised eyebrows. The solar-irradiance channel, which supposedly possessed an accuracy of 0.2 percent, was yielding a value about 1.5 percent higher than the standard value that had just been extracted from measurements made in the late 1960s (Fröhlich and Brusa 1975). The result quickly engendered three main interpretations. The solar constant might have increased since the last sunspot maximum. Or—and these possibilities seemed more likely—either the solar constant's reported value or its new standard value was in error.

That fall solar-radiometry physicists suggested two strategies for getting to the bottom of the matter. Hickey and his colleagues proposed an important modification in the Earth Radiation Budget Experiment planned for the next Nimbus (Kyle 1987). They wanted to provide it with a second solar-irradiance channel by substituting a self-calibrating cavity radiometer for the ultraviolet-irradiance sensor at channel 10. Although NASA approved the modification, this strategy did not promise a fast resolution of the issue, since Nimbus G's launch was three years away. Consequently, Kendall, Willson, and others urged that the accuracy of the radiometer then on *Nimbus 6* be checked by means of a calibration rocket equipped with several absolute radiometers (Willson and Hickey 1977; Duncan et al. 1977). In early 1976, NASA agreed to the calibration rocket.

After frenetic preparations, NASA launched the calibration rocket from White Sands, New Mexico, on June 29, 1976 (Willson and Hickey 1977; Duncan et al. 1977). The *Aerobee*, which reached an altitude of 250 km, carried a duplicate *Nimbus-6* thermopile and a prototype Nimbus-G cavity radiometer provided by Hickey, a primary absolute cavity radiometer provided by Kendall, and two active cavity radiometer IV's provided by Willson. That same day, by prearrangement, the Earth Radiation Budget Experiment on *Nimbus 6* was operating. All six radiometers—the five on the rocket and the one on the orbiting spacecraft—yielded measures of the solar irradiance. The Eppley thermopiles on the rocket and *Nimbus 6* both gave values of 1389 watts/m^2. All the cavity radiometers on the rocket, however, yielded lower values—Hickey's Nimbus-G prototype measured 1369 watts/m^2; Kendall's radiometer, 1364; and both of Willson's radiometers, 1368. Faced with this evidence, Hickey conceded that the *Nimbus-6* measures of the solar constant were, for some unknown reason, too high (Hickey et al. 1976).

Although the calibration rocket did not challenge the ability of the *Nim-*

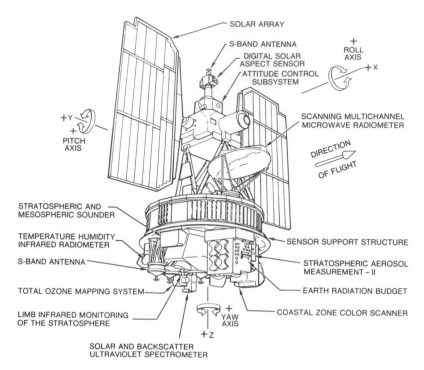

Figure 7.9. Schematic of the *Nimbus 7*, launched October 24, 1978; the Earth Radiation Budget subassembly included the Eppley Laboratory's panel of solar sensors—the vertical array of ten circular apertures; the new cavity radiometer was behind the lowest aperture. (Source: NASA 1978.)

bus-6 radiometer to trace fluctuations in the solar irradiance, Hickey never managed to extract any results of consequence from its measurements. One reason was that various subsystem failures on *Nimbus 6* exacerbated the problems of cleaning up the raw data (Kyle 1987). Another was that, hard pressed for funds, NASA did not provide adequate support for the data analysis, inducing some key computer personnel to drop out of the Earth Radiation Budget Experiment's team. Yet another was that Hickey had turned his attention elsewhere. He and his colleague Roger G. Frieden were busy supervising the development of Eppley's cavity radiometer for Nimbus G (Hickey et al. 1977). They were determined that the new sensor would not suffer the ignominious fate of the *Nimbus-6* radiometer.

In October 1978, Nimbus G was renamed *Nimbus 7* (fig. 7.9) upon its successful launch into an orbit having nearly the same characteristics as that of its predecessor. The spacecraft carried the modified Earth Radiation Budget Experiment with its new cavity sensor (Hickey et al. 1980, 1988). On November 16, after the satellite had outgassed for three weeks, NASA ground controllers brought the experiment into service. As hoped, the sensor in solar channel 10 turned out to be not only faster but also more accurate than *Nimbus 6's* discredited monitor. It could measure the irradiance once a second during the 3-minute period on each 104-minute orbit that the Earth Radiation Budget Experiment faced the sun. Moreover, thanks to a calibration rocket flight on November 16, 1978, there was reason to think its accuracy was better than 0.5 percent.

Hickey and his collaborators came up with two main findings during the first year of *Nimbus-7* solar-irradiance monitoring. The first, which turned out to be spurious, was that the sun's radiative output had increased as the sunspot cycle approached maximum (Hickey, Stowe, et al. 1980). In particular, the cavity radiometer indicated that the solar constant averaged 1376 watts/m² between November 1978 and May 1979 (fig. 7.10)—0.66 percent higher than the average value measured by the four cavity radiometers on the calibration rocket of June 1976. To assure quick publication, the *Nimbus-7* scientists sent a report off to *Science* in October 1979. Notwithstanding their alacrity, Hickey's group was forestalled in announcing the apparent increase in the sun's radiative output. The month before, Willson and two associates had sent *Science* a paper reporting that the active cavity radiometers on the

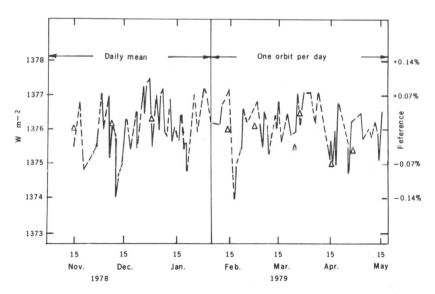

Figure 7.10 Hickey's first published graph of *Nimbus 7's* daily measurements of the solar irradiance; the triangles indicate the solar-constant values found after complete reduction of the data; the right ordinate gives the percentage deviation from the mean value of 1376.0 watts/m². (Source: Hickey, Stowe, et al. 1980, 282. By permission of *Science* © AAAS.)

calibration rockets of June 1976 and November 1978 had detected a 0.4 percent increase in the solar luminosity during this period (Willson, Duncan, and Geist 1980).

The second finding of Hickey's group was more striking, and also reliable. Preliminary data analysis during the fall of 1979 revealed two major decreases in the sun's radiative output. In August the *Nimbus-7* cavity radiometer had recorded a 0.36 percent dip in the solar irradiance. Three months later, it traced a comparable dip. Consulting various solar indicators, Hickey soon ascertained that the dips had coincided with periods of intense solar activity. Early in 1980, he decided to announce the result at the American Geophysical Union's meeting in Toronto that spring (Hickey, Griffith, et al. 1980). This time Hickey beat Willson to the punch. But his abstract was so brief that the way still stood open for Willson to take the initiative. It was not long before he did so.

Willson's Observational Breakthrough, 1975–1980

Even as Eddy, Foukal, and Hickey were seeking, and finding, various evidences of solar-irradiance variability, Willson was preparing to make what would be hailed as the decisive observations. The Big Bear Workshop of May 1975 seems to have been the turning point in his persistent campaign to get an experiment into space. The conferees, as we have seen, acknowledged his active cavity radiometer as the state-of-the-art instrument. They also lobbied the NASA participants about the importance of getting instruments of equal or higher quality into orbit. In particular, they urged Adrienne Timothy, the program officer for solar physics from NASA Headquarters, to add a berth for a solar-irradiance monitor to the payload of the Solar Maximum Mission (Zirin, Moore, and Walter 1976).

That summer, Willson obtained funding from NASA's Weather and Climate Program for a design study of the active cavity radiometer IV (Willson 1975b, 1976). His goal was a radiometer that would be accurate at the 0.1 percent level over long periods in space. To reach this goal, he planned numerous improvements in the instrument. The most important refinement would be to replace diffuse black as the surface coating of the radiometer's conical cavity. On account of this coating's roughness, an indeterminate, if small, amount of radiation was reflected out of the instrument whenever an incident ray interacted with the cavity's wall. The consequence was that the uncertainty in the absorptance could not be pushed below 0.1 percent. By using a coating of specular black, he could be sure that radiation would be trapped in the new radiometer until after the sixth internal reflection. He hoped, accordingly, to attain an absorptance uncertainty of less than 0.0001 percent.

Achieving long-term reliability in space would, Willson realized, take more than fine-tuning of the active cavity radiometer. There was the distinct possibility that its performance would degrade under the steady assault of high-energy radiation and particles. Zirin suggested on the basis of his experience in the U.S. Navy with marine chronometers that redundancy was the solution to this problem (Zirin 1988b). On his advice, Willson decided to make two extra sensors an integral part of any experimental package—one for monitoring the primary detector's performance and the second for resolving any significant disagreements between the primary detector and its monitor.

While Willson was working on the fourth-generation active cavity radiometer, the idea of placing an irradiance monitor on the Solar Maximum Mission was gaining support at NASA Headquarters. An important early proponent was S. Ichtiaque Rasool, the science deputy in the Office of Space Science (Zirin 1988b). Just two months after the Big Bear Workshop, thanks to Rasool's backing, Zirin was able to make a pitch for space research on the solar constant before NASA's influential Physical Sciences Committee (NASA 1975). Another early advocate was Adrienne Timothy. Even with Rasool's high-level support, she had set herself no mean task. The instrument had little bearing on the Solar Maximum Mission's primary objective of expanding knowledge of solar flares. Moreover, it would be a late addition to the payload, which was nearing the final stages of selection. And it would increase the costs of a mission that was already under attack on budgetary grounds. Somehow, she and Harold Glaser, head of the new solar-terrestrial division, managed to surmount all these difficulties (Glaser 1975).

The upshot was that in February 1976, not long after naming the principal investigators for the Solar Maximum Mission's main instruments, NASA officially opened the competition for a solar-irradiance monitor on the spacecraft (Mathews 1976). Timothy and the other agency officials who crafted the announcement invited proposals for both total-irradiance and ultraviolet-irradiance monitors in the hopes that a combined experiment could be launched (Glaser 1976a). They were, in the light of Willson's claims for the active cavity radiometer IV, allowing considerable leeway when they specified design criteria of 0.5 percent accuracy and 0.1 percent precision for the total-irradiance monitor, but they felt they had no choice if they wanted to get a reasonable number of proposals (Glaser 1988).

No matter what the competition, Willson was in an excellent position to get the berth. He had established himself as one of the most exacting workers in precision radiometry. He had also impressed the panel of experts overseeing the *Nimbus-6* calibration rocket with his plans for the fourth-generation active cavity radiometer (Duncan et al. 1977). And assisted by Mac C. Chapman of TRW, who saw an opportunity here for the corporation to get a head start on the next round of earth radiation budget experiments (Willson 1988a), he had put together a strong proposal-drafting team.

In the proposal submitted in May, Willson (1976) audaciously bid to supply a monitor for both total and ultraviolet irradiance. Three active cavity radiometer IVs would be at the heart of the total-irradiance monitor (fig. 7.11), and a fourth would serve as the calibration standard for the ultraviolet detectors. His decision to bid for the entire berth was made possible by Kendall, who would be assisting with the fine-tuning of radiometers, and Zirin, who would be taking the lead in interpreting the ultraviolet measurements. Willson also had the backing of the Jet Propulsion Laboratory and TRW, which would be providing technical and managerial assistance before and during the flight. For all his confidence, however, Willson did not want his proposal to be considered an all-or-nothing proposition. He emphasized his readiness to supply the total-irradiance monitor alone.

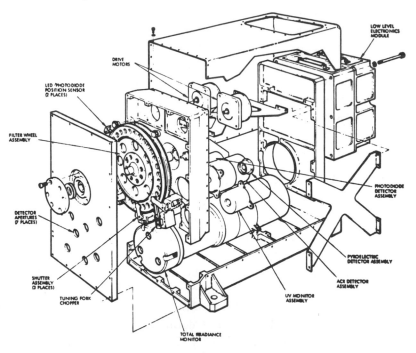

Figure 7.11. Willson's proposed design for a Solar Maximum Mission instrument package that would monitor both total and ultraviolet irradiance. (Source: Willson 1976, technical section, 3-2.)

By July 1976, NASA's reviewers were leaning toward Willson for a $1.6 million total-irradiance monitor and Guenter Brueckner of the Naval Research Laboratory for a $2.5 million ultraviolet-irradiance monitor (Glaser 1976b). However, NASA's Office of Applications, which had agreed to pay for the research on account of its relevance to climatology, no longer had sufficient funds to cover both parts of the solar-monitoring package. It had raided the experiment's allocation to help cover unexpected costs incurred when, on cancellation of the Gamma Ray Explorer, the Solar Maximum Mission was given responsibility for demonstrating the value of spacecraft with standardized subsystems (Hinners 1976). Glaser and Timothy tried for several months to raise the funds needed for both monitors. Unsuccessful, they decided to give Willson's active cavity radiometer irradiance monitor (ACRIM) the nod because of its greater potential for illuminating the sun-weather problem (Glaser 1976c).

In February 1977, shortly after getting the berth on the Solar Maximum Mission, Willson managed to get another irradiance-monitor berth on the shuttle's first scientific mission, Spacelab 1 (Hinners 1977). He devoted himself during the next three years to the irradiance monitors (fig. 7.12) for

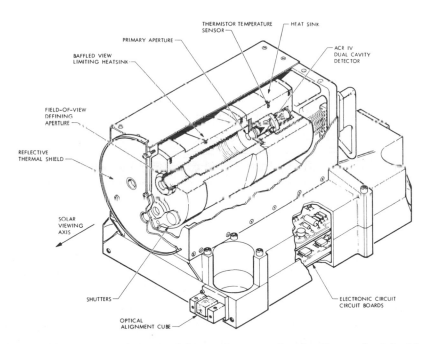

Figure 7.12. Willson's schematic of the irradiance monitor that flew on the *Solar Maximum Mission;* note the many changes since the 1976 design (see fig. 7.11). (Source: Willson 1981, 221.)

these two missions. He proceeded without benefit of Zirin's counsel because the Caltech solar physicist resigned as coinvestigator once it became clear that Willson's contract was too small to support any concurrent ground-based observations (Glaser 1977; Willson 1988a; Zirin 1988b). Willson's presumption was that he would satisfy those in the solar and solar-terrestrial communities who were interested in the sun's radiative output if he gave his instrument the highest possible accuracy, precision, speed, and durability (Willson 1978, 1979).

Although Willson was happy with the way his project was moving ahead, he was not indifferent to the possibility that Hickey might forestall him with the radiometer on *Nimbus 7*. Hence, contrary to his better judgment (Willson 1988b), he was persuaded to release the results obtained by his active cavity radiometers on the 1976 and 1978 calibration rockets. In September 1979, Willson and two coauthors sent *Science* the paper that scooped Hickey in announcing an increase in the solar irradiance in the thirty-month interval between the rocket flights. Anxious to establish priority, they insisted that this was the first "direct measurement of solar luminosity variation" (Willson, Duncan, and Geist 1980, 177).

On February 14, 1980, near the maximum in the sunspot cycle, NASA launched the *Solar Maximum Mission* into an inclined, 96-minute circular orbit. The spacecraft's pointing-control system worked exactly as planned, directing the instrument panel at the sun with an accuracy of 0.5 seconds of arc throughout the 50 to 60 minutes of each orbit outside the earth's shadow. Ground controllers at Goddard Space Flight Center activated Willson's instrument the following day. After a quick checkout, the irradiance monitor began work (Willson and Hudson 1981b; Willson 1982a). It obtained 32 measurements of about 1-second duration every 2-minute shutter cycle, or approximately 750 readings each orbit and 11,500 readings each day. Its absolute accuracy, to judge both from the built-in reference radiometers and from instruments on a calibration rocket sent up in May that year, was better than 0.1 percent and its relative precision better than 0.005 percent. As Willson had hoped, the irradiance monitor on the *Solar Maximum Mission* set new standards of performance.

Willson gave his first report just three months after the *Solar Maximum Mission's* launch. His personal opinion was that this was rushing things (Willson 1988b), but his NASA contacts wanted him to get his results out in a timely fashion. He could hardly refuse, since his data-processing team had efficient routines in place from the outset. The occasion was the American Geophysical Union's meeting in Toronto on May 20, 1980, the same meeting in which Hickey's group announced the detection of major irradiance dips in August and October of the year before. Willson's plot (1980) claimed to track the solar irradiance through several swings having a total amplitude of about 0.1 percent. Those present could see that his results were much more detailed than those of Hickey's group, but they were disappointed that neither Willson nor Hickey could make a compelling case that specific solar phenomena were responsible for the reported irradiance fluctuations (Foukal 1988).

Willson soon had a collaborator to help him trace variations in irradiance back to the sun—Hugh S. Hudson (b. 1939), a solar physicist at the University of California at San Diego. Hudson's quick entry into the collaboration was eased not only by his own foresight but also by the Solar Maximum Mission's Guest Investigator Program. Long interested in solar flares, he had hoped to participate in the mission as a member of the team behind its hard X-ray telescope (Hudson 1988b), but his team's proposal was not accepted. So, encouraged by Timothy and others who were putting together the Guest Investigator Program, he looked over the mission's payload for interpretive opportunities. Hudson (1977) was impressed that, if all went well, Willson's instrument would "provide unique, precise observations of total irradiance over a long period of time, from a stable space platform." He sought and obtained NASA support for a two-pronged project—pursuing "the solar interpretation of variability observed by the SMM" and implementing a ground-based infrared-photometry program "to complement the SMM observations." In the spring of 1980, once he was familiar with the quality of Willson's data, Hudson lost all interest in the second part of his project. He focused instead on identifying the solar sources of the variability in Willson's irradiance records.

In their first collaborative papers, Willson and Hudson (1981a, 1981b) described the instrument and its operation and reported its measurements from February 16 through April 3, 1980. Their choice of this terminal date was no coincidence. When composing these papers in June and July, they did not yet feel ready to reveal that a 0.2 percent dip in the solar irradiance had occurred between April 4 and 9. Willson, who was working with a tape delay of a month or so and a data-reduction delay of half a month, became aware of this dip near the beginning of June (Willson 1988a). Hudson soon managed to associate the dip, which they dubbed the "Big Dipper" (Hudson 1988a), with the meridian passage of a large sunspot group (fig. 7.13). Then, drawing upon subsequent sunspot records, he predicted that a similar dip would turn up in late May when the data set for this period became available. In August, just as Hudson had predicted, Willson found a second major dip in the measurements for May 24–28 (fig. 7.14). NASA and the Jet Propulsion Laboratory issued a joint press release (NASA 1980) that promptly engendered journalistic interest (e.g., Alexander 1980). The fol-

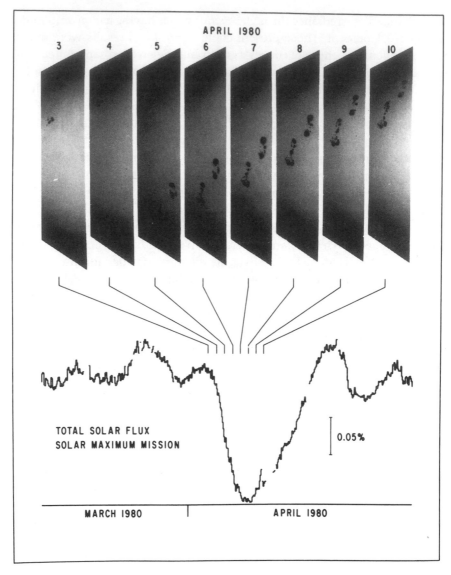

Figure 7.13. The large sunspot group that Hugh Hudson associated with the "Big Dipper" of April 1980. (Courtesy of Hugh S. Hudson.)

lowing month, Willson, Hudson, and their collaborators sent an article (Willson et al. 1981) about the *Solar Maximum Mission's* irradiance monitor and its striking observations to *Science*. Willson was soon off on the conference circuit, giving the first of many reports on the monitor's results at a European conference on the "Physics of Solar Variations" (Willson 1981).

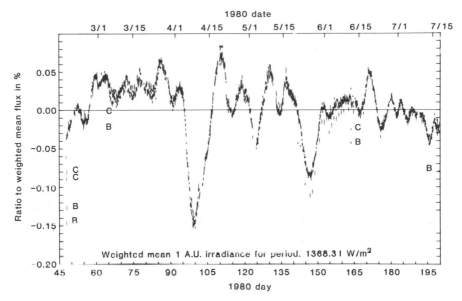

Figure 7.14. Willson's results for February-July 1980; the B and C readings were made by the two built-in reference radiometers; note the large irradiance dips in April and May 1980 as well as the level of detail compared to that in Hickey's results (see fig. 7.10). (Source: Willson et al. 1981, 700. By permission of *Science* © AAAS.)

Willson, Hudson, and their coauthors made three chief points in these papers. They stressed the reliability of the solar-irradiance fluctuations recorded by Willson's monitor. They spotlighted Hudson's success in associating the two dips of April and May 1980 with "large, recently developed or developing spot groups crossing the central solar meridian" (Willson et al. 1981, 701–2). And they emphasized the difficulty of explaining the low-level variability that characterized much of the monitor's record. Besides making these points about the short-term changes in the sun's radiative output, Willson retracted his earlier claim that the solar irradiance had registered a gain of 0.4 percent in the thirty months between the 1976 and 1978 calibration rockets. During the preflight testing of the *Solar Maximum Mission's* monitor, he had begun to suspect that the active cavity radiometer IV gave high readings when operating at low atmospheric pressures (Willson 1988b). If so, the radiometer's loss of vacuum during the 1978 rocket flight could have affected its reading of 1373.4 watts/m². This suspicion was confirmed when he found that the mean value for the first forty-five days of the *Solar Maximum Mission* was 1368.68 watts/m², quite close to the 1976 figure. Willson (1981, 226), sure that his reported "irradiance increase of 0.4% . . . was an artifact," concluded that any revelations about longer trends must await further measurements from the *Solar Maximum Mission's* monitor.

Willson's instrument continued securing superb data for almost 300 days. Then on December 11, 1980, as the result of the progressive deterioration and failure of three fuses, the spacecraft lost its pointing control (Covault 1981a; Willson 1984). The ground controllers managed to stabilize the satellite by putting it into a spin, but they were unable to do better than keep its instruments pointed to within 10° of the sun's center. Willson adjusted to the new observing conditions by leaving the irradiance monitor unshuttered from orbit sunrise to orbit sunset so that measurements would be obtained whenever the sun happened to be fully within the instrument's field of view. The outcome was about a hundred trustworthy readings per day, less than 1 percent of the number obtained during normal operation. The irradiance monitor's heyday, it appeared, was over.

Following up on Willson's Breakthrough, 1980–1990

Before the *Solar Maximum Mission's* loss of pointing control, Willson's monitor had measured the solar irradiance over two million times. On account of their exceptional quality, the measurements quickly attracted favorable attention from solar physicists. Willson's results were, indeed, accorded pride of place at several conferences during the next few years— for example, the Goddard Space Flight Center's "Workshop on Variations of the Solar Constant" in November 1980 (Sofia 1981b), the Sacramento Peak Observatory's "Summer Workshop on the Physics of Sunspots" in July 1981 (Cram and Thomas 1981), and the International Astronomical Union's "Joint Discussion on Solar Luminosity Variations" in August 1982 (Eddy and Foukal 1983).

The solar physicists' interest in Willson's data, and also in Hickey's data from the *Nimbus*-7 radiometer, centered on the problem of interpretation (Newkirk 1983a; Hudson 1988a). It was clear from early on that the solar irradiance varied on different time scales. Attention first went to variations having solar-active-region time scales of a few days to a few months. Interpretive work soon began as well on variations having solar-oscillation time scales of a few minutes. Finally, about 1986, once long-term trends could be discerned in the irradiance data, serious study commenced on changes having solar-cycle time scales of years.

Active Regions as Sources of Irradiance Variation

Before 1980 was out, solar physicists realized they would have to abandon the common practice of treating the solar irradiance as equivalent to the sun's total radiative output or luminosity. This practice was based on the assumption that the sun's radiative output was the same in all directions (Sofia 1981a). The assumption's falsity, however, was evident from the irradiance dips observed by the radiometers on *Nimbus 7* and the *Solar Maximum Mission* when large sunspot groups crossed the solar meridian. How much, if at all, the luminosity varied with the irradiance depended on what happened to the energy that was blocked by the sunspots. The solar physicists who got involved in interpreting Willson's and Hickey's data sets took two main positions on the issue of the blocked energy. Some argued that the energy was diverted to and rapidly emitted by nearby bright faculae. Others maintained that it was stored deep in the sun's turbulent convection zone, then emitted globally over long periods of time. The resulting controversy did much to promote interest in the whole solar-irradiance problem.

Hudson was—partly as the result of strong encouragement from Willson (1988b)—a proponent of local storage and facular reemission. At first he was uncomfortable with this position. He warned, for instance, that Willson was loading their first report to *Science* with "fluff" when writing that "the sunspot-deficit energy is stored or delayed within the convection zone through the effects of the magnetic fields associated with the sunspot groups involved. Subsequent development of related faculae in these regions may provide the mechanism for radiating the stored energy away" (Willson et al. 1981, 702). Willson refused to back off (Hudson et al. 1982; Hudson and Willson 1981; Willson 1982b), however, and Hudson (1983) eventually embraced the position. Hudson's analysis indicated that sunspot deficits accounted for about half of the variance in the daily averages of the *Solar Maximum Mission's* irradiance measurements. He supposed that, once adequate observational data became available, facular excesses would be found to account for much of the remaining variance. In his scenario, this facular reemission of the blocked energy occurred after nearby storage for a week or so. Such storage, he argued, was the most straightforward interpretation of the irradiance dips. He also suggested that it helped explain why faculae had longer lifetimes than spots.

Sabatino Sofia and his collaborators at Goddard Space Flight Center— notably Ludwig Oster, a visitor from Boulder, and Kenneth Schatten, the solar-wind theorist—advanced a stronger version of the facular-emission hypothesis. Like Willson and Hudson, they contended that the energy blocked by sunspots emerged from neighboring faculae (Schatten et al. 1981; Oster, Schatten, and Sofia 1982), but they did more than make a general argument for this position. They used observations of plage—bright areas in the chromosphere overlying photospheric faculae—to construct a daily facular index. Then, employing the first 153 days of Willson's data as a template, they determined the scaling factors for the spot and facular indices that gave the closest fit. This phenomenological model accounted for about 70 percent of the variance in the irradiance record. Gratified, Sofia's group went on to predict the ups and downs of Willson's forthcoming irradiance curve. They soon had the pleasure of announcing that their model, which they began thinking of as the Goddard "standard model," had predicted about 75 percent of the variance in the next 117 days of Willson's data (Sofia, Oster, and Schatten 1982, 82).

In analyzing the model, Sofia's group found some evidence that the radiative deficit from sunspots matched the radiative excess from associated faculae. Like Willson and Hudson, they believed these phenomena must be linked: "The main effect of [solar] activity is then an angular redistribution of the energy, such that the energy reduction which appears normal to the surface in spots is reemitted in the faculae primarily at larger angles to the normal. In a sense, solar activity acts like a lighthouse fanning out a small fraction of the solar energy flow" (Oster, Schatten, and Sofia 1982, 772). This diversion of the blocked energy to "the faculae in the outskirts of activity regions" (773) would take time, so some energy storage would occur on facular lifetimes of two to three months. Strictly speaking, this storage would imply fluctuations on the same time scale in the sun's total radiative output. This effect struck Sofia's group as so minuscule and ephemeral, however, that they made the simplifying assumption that "the solar luminosity is always constant" (Schatten et al. 1982, 51).

Meanwhile, Peter Foukal in Cambridge was developing the deep-storage hypothesis. His initial research after Willson's observational breakthrough focused on the large dips in the irradiance data (Foukal 1981, 1982). Like Hudson, he found that sunspot blocking accounted satisfactorily for the depth and duration of the dips, but he saw no reason to postulate "any detailed energy balance between spots and faculae." It bothered him that those who envisioned a channeling of the blocked energy from sunspots to faculae did not have a "physically convincing mechanism to achieve this detailed balance" (Foukal, Fowler, and Livshits 1983, 863). Foukal supported quite a different view of sunspot blocking. Like the Dutch solar theorist Henk C. Spruit (1977, 1982), he had carried through calculations indicating that the heat blocked by a sunspot was stored first in nearby and then in more distant parts of the turbulent convection zone underlying the photosphere. Once stored there, however, the heat would take millennia to make its way back to the sun's surface. Hence, on account of the efficiency of this long-term storage, sunspots diminished not only the solar irradiance but also the sun's luminosity.

Not surprisingly, John Eddy in Boulder also got involved in modeling Willson's data (Eddy, Hoyt, and White 1982; Eddy 1983). His chief collaborator was Douglas Hoyt, who had gained a close knowledge of earlier solar records while analyzing the Smithsonian data. They first considered the relative contributions of spots and faculae to irradiance variability, concluding that sunspots accounted for much more of the variance than faculae. This conclusion ruled out a detailed balance between spot deficits and facular excesses. Like Foukal, they thought the blocked energy might well be stored in the lower convection zone for very long times. They were attracted, indeed, to "the notion of a frustrated Sun, with stored, and gradually released energy from remembered sunspots" (Eddy, Gilliland, and Hoyt 1982, 690). The dominance of sunspot blocking in the irradiance record created a dilemma for Eddy, who had predicted just the opposite based on his earlier studies of the effects of prolonged minima and maxima. They were hopeful, however, that "continued precision monitoring" would reveal that "other mechanisms [than sunspot blocking were] also at work" (Hoyt and Eddy 1983, 511).

In campaigning for their model, Eddy and Hoyt went after its most successful competitor—the so-called "Goddard standard model." Their basic criticism was that Sofia and his collaborators had scaled spot and facular indices from Willson's data rather than determining the magnitude of the indices directly from solar observations. The result was that the Goddard "sunspot areas consistently exceed the actual measured sunspot areas." As it happened, however, the use of "chromospheric plage areas as a proxy for faculae . . . introduces a large overestimate in these areas that may compensate for the sunspot area excess. Thus two compensating overestimates give the appearance of a successful model. [Although the Goddard] correlation of 0.83 . . . exceeds our value of 0.783 for all 1980, it is not statistically more significant [because it] merely reflects that they use the SMM observations in deriving the values of [their model's] various parameters" (Hoyt and Eddy 1983, 510–11). These were strong words. The solar physicists who were involved in interpreting the data from the radiometers on *Nimbus* 7 and the *Solar Maximum Mission* had come to loggerheads over whether the energy blocked by sunspots was emitted by neighboring faculae or stored deep in the convection zone.

To no one's surprise, the issue gave rise to a vigorous debate in June 1983 at the NASA Workshop on Solar Irradiance Variations on Active Region Time Scales (LaBonte et al. 1984), or WARTS as the workshop at Caltech was fondly called by its organizers (Hudson 1988b). Hudson, Sofia, and Schatten insisted that faculae emitted most of the energy blocked by sunspots, while Foukal and Eddy made the case for long-term storage. Tape recorders caught much of the repartee. Schatten remarked, for example, that "the important point is that once the energy gets to these shallow depths, there is no way to stop it from coming out. It's like trying to dam the Mississippi. You may temporarily store the energy, but there is large horizontal heat transport. So it will get around obstructions. There will not be storage for anything like 11 years, which would be needed to get solar cycle variations" (LaBonte et al. 1984, 145–46). Foukal disagreed with Schatten's use of the Mississippi analogy: "If you build a dam, the water continues to flow downstream, but the elevation of the water propagates upstream. That is what happens with a thermal plug [like a sunspot]. The heat flows outwards, but the thermal signal propagates downwards fast. The heat flow right down to the bottom of the convection is perturbed in such a way as to store energy. The product of the difference in temperature and the specific heat gives you the stored energy" (146). Hudson jumped in on Schatten's side with the opinion that "in active regions energy comes out of spots and goes into faculae" (147). In reply, Foukal dramatized his concern about the lack of a plausible coupling mechanism by observing that "the angular sizes of the Moon and the Sun are the same as seen from the Earth, but you better be careful what you conclude from that" (147).

Tempers cooled in the years following the 1983 workshop. Moreover, a consensus emerged that sunspot deficits were about equal to facular enhancements. Gary A. Chapman, who had been investigating the energy balance of active regions (Chapman 1980), and his colleagues at the California State University at Northridge's San Fernando Observatory played an important role here with a detailed ground-based study of the energy balance of a single active region over its entire lifetime (Chapman, Herzog, and Lawrence 1986; Chapman 1987). They found that facular emission from the region was between 70 percent and 120 percent of the energy blocked by the sunspots. This finding was buttressed by Foukal and Judith Lean (1986). They concluded after a thorough statistical analysis that spot and facular contributions to irradiance variability in 1980 were comparable in magnitude on the evolutionary time scale of active regions.

This agreement did not, however, resolve the underlying issue of the fate of the energy blocked by sunspots. Chapman, an early partisan of detailed balance, insisted that all, or at least "a major portion of the missing sunspot flux is radiated by faculae" (Chapman, Herzog, and Lawrence 1986, 654). Foukal, by contrast, remained doubtful that faculae were "direct conduits for channeling and reradiation of the missing sunspot radiative flux" (Foukal and Lean 1986, 835). Although some attempts had been made to link the phenomena (e.g., Schatten and Mayr 1985; Schatten et al. 1986), he still regarded "a detailed energy balance . . . between spots and faculae" as physically implausible (Foukal 1987a, 804). The issue, it seems, will remain unsettled until solar physicists have a better understanding of the physics of active regions.

Oscillations as Sources of Irradiance Variation

Back in 1980, soon after the first studies of the solar-irradiance variations caused by active regions, Hudson set his graduate student Martin F. Woodard to searching Willson's data for traces of solar oscillations. The possibility must have seemed great that this was a quest for a will-o'-the-wisp. The irradiance fluctuations caused by individual oscillations were in all likelihood well below the sensitivity of Willson's radiometer. Yet, as Hudson had realized in 1977 when he wrote his proposal for the Solar Maximum Mission Guest Investigator Program, there were good reasons to conduct the search. Sophisticated analytical techniques could be used to isolate faint repetitive signals from a data set of sufficient duration. If the search succeeded, the resulting "measurements [would] have broad and basic importance" (Hudson 1977, 2). Hudson was thinking here of the recent demonstration that solar oscillations of about 5 minutes' duration were caused by sound waves resonating in the sun (Deubner 1975). A flurry of follow-up studies had shown how precise knowledge of the oscillations could be used, in analogy to terrestrial seismology, to probe the sun's interior.

Woodard and Hudson began their quest for oscillations with an analysis of the irradiance data collected between March 5 and July 20, 1980. Although Willson had not designed the *Solar Maximum Mission's* monitor or its observing routine with such a study in mind, the measurements turned out to have sufficient precision and temporal resolution for the task. Woodard and Hudson looked for irradiance oscillations having periods of about 5 and 160 minutes, a choice suggested by the claims of earlier helioseismologists. In the 5-minute region, they managed to detect more than a dozen power peaks with relative amplitudes of a few parts per million (fig. 7.15). They were confident that the peaks were irradiance oscillations because a Birmingham group had just reported finding peaks with nearly identical periods by means of sustained observations of the sun's surface velocity. In the 160-minute region, Woodard and Hudson were unsuccessful. This outcome, however, was not disheartening, since other groups had also failed to confirm claims for these slower oscillations.

In September 1981, Woodard and Hudson announced their initial results at an International Astronomical Union meeting on solar and stellar oscillations held at the Crimean Astrophysical Observatory. At this juncture, their focus was on Willson's "beautiful" data set and their procedures for analyzing it (Woodard and Hudson 1983a, 73). There was something quite satisfying in the combination of technical and analytical virtuosity that had made their detection of the 5-minute oscillations possible. Soon, however, Woodard and Hudson were not only analyzing the irradiance data but also exploring the implications of their findings. In the process, they placed two articles

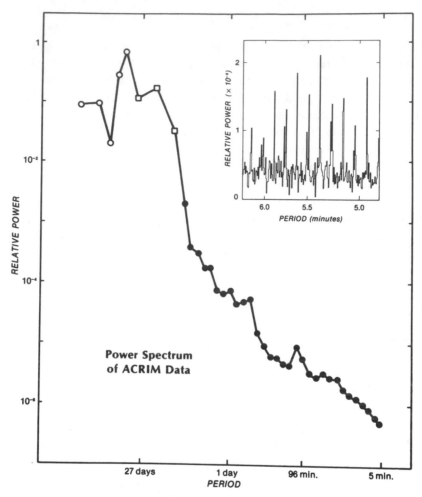

Figure 7.15. Woodard and Hudson's power spectrum for all of Willson's high-quality irradiance data obtained during 1980; the inset is at the scale needed for resolving details in the 5-minute band. (Source: Willson, Hudson, and Woodard 1984, 502.)

in *Nature* (Woodard and Hudson 1983b; Woodard 1984a), and Woodard earned his doctorate (1984b).

Thanks to the rapid advance of helioseismology, Woodard and Hudson knew a good deal from the outset about the irradiance oscillations they had found in the 5-minute band of the power spectrum. Their particular oscillations were detectable by Willson's monitor because they were produced by some of the simplest and strongest sound-wave resonances among the millions occurring in the sun. They were of three types—zero-degree pulsations in which the entire sun periodically brightened and dimmed, first-degree pulsations in which one solar hemisphere was brightening as the other was dimming, and second-degree pulsations in which opposite quarter sectors were brightening as their neighbors were dimming. Within each type there were several orders, each caused by a specific overtone of the degree's fundamental standing wave. Woodard and Hudson knew, for instance, that the strong 5.4-minute oscillation in their power spectrum (see fig. 7.15) was the twenty-first overtone of the first-degree pulsation.

The most interesting point Woodard and Hudson made in interpreting their results had to do with the sun's internal rotation. The orthodox view was that the rate of rotation did not change appreciably along the solar radius. However, some maverick theorists—variously motivated by a desire to overthrow general relativity or explain the paucity of solar neutrinos— thought the sun's core rotated much faster than its surface. If they were correct, the oscillation peaks of the first- and second-degree waves, which passed close to the sun's center, should split into three peaks of slightly different period. This prediction was, it seemed, confirmed by the Birmingham group's prior study of low-degree solar oscillations. Woodard and Hudson ended up, however, regarding this evidence for nonuniform rotation as weak. Their power spectrum had better resolution than did that of the Birmingham group because it was based on a much longer set of observations. Even so, they could not detect the "fine structure" that would be engendered by "rapid internal rotation of the Sun" (Woodard and Hudson 1983b, 589). Woodard soon refined the analysis. As before, he concluded that the *Solar Maximum Mission* measurements "removed [the reported] conflict between general relativity and solar oscillation data" (Woodard 1984a, 532).

By this time Woodard had squeezed Willson's high-quality data set— secured before the *Solar Maximum Mission's* loss of fine-pointing control in December 1980—for all it was worth. Fortunately for him, NASA had decided that a good way to demonstrate the versatility of the manned shuttle would be to restore the ailing spacecraft to normal working order. Despite some tense moments, the repair mission in April 1984 was a great success (see chap. 5). The quality of the ensuing data from Willson's monitor was comparable to that of the data collected during the mission's first ten months (Willson et al. 1986). The way was open, therefore, for a comparison of the 1980 and 1984 power spectra.

Woodard carried through the comparison while on a postdoctoral fellowship at the Harvard-Smithsonian Center for Astrophysics. His collaborator there was Robert W. Noyes, who in the early 1960s was a member of the Caltech group that had conducted the first studies of 5-minute oscillations. Woodard and Noyes's objective was to ascertain whether the solar cycle affected the oscillation pattern. Just as a drum's resonances would change if its dimensions were altered, so should the sun's oscillatory periods change if, for example, the solar radius varied in the course of the sunspot cycle. In comparing the power spectra for 1980 (near solar maximum) and 1984 (approaching solar minimum), Woodard and Noyes (1985) did find that eight of the nine strongest oscillation peaks in the two data sets were at slightly longer periods. They tentatively interpreted this result as evidence that the sun's radius had increased by about 100 km during the solar cycle's declining phase. Two years later, Woodard (1987) reported additional evidence for a shift in the sun's oscillatory periods between the sunspot cycle's maximum and minimum. However, as a new member of Harold Zirin and Kenneth G. Libbrecht's helioseismology group at Caltech's Big Bear Solar Observatory, he cautiously refrained from speculating about the solar-cycle changes underlying this shift. More ambitious studies of irradiance oscillations will probably not be forthcoming until radiometer sensitivity has been increased by a couple of orders of magnitude. This technique could then play an important role not only in helioseismology but also in stellar seismology (Hudson et al. 1986; Libbrecht 1988).

Network Faculae as Sources of Solar-Cycle Irradiance Variation

While solar physicists were investigating active regions and oscillations as sources of solar-irradiance variation, the *Nimbus-7* and *Solar Maximum Mission* radiometers were monitoring the sun. Hickey, Willson, and their co-workers hoped to determine how the total irradiance varied through the solar cycle. This was an ambitious goal. If, as seemed likely, such variability was small, success depended not only on an instrument's sensitivity but also on its long-term stability. Instrumental drift was a very real danger, since the performance of key components was liable to degrade with the passage of time in the space environment.

In late 1985, Willson, Hudson, and two collaborators (Willson et al. 1985, 1) reported the "observation of a long-term downward trend in total solar irradiance" at the American Geophysical Union meeting in San Francisco. Daily averages from the *Solar Maximum Mission's* irradiance monitor indicated a decline of 0.019 percent a year during the first five years of the mission. Willson's confidence in the trend's reality rested on the fact that similar results were obtained from the calibrating radiometers built into his monitor, from radiometers carried aloft by sounding rockets and high-altitude balloons, and from the radiometer on *Nimbus 7*. Although certain about the trend's existence, Willson and his coauthors cautiously refused to say anything about its cause other than that it would probably be found to be "related to the solar cycle" (Willson et al. 1985, 6). By 1988, Willson was

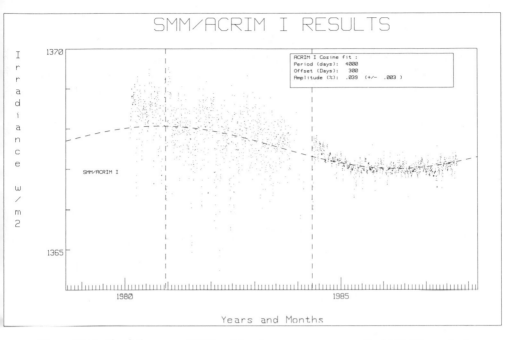

Figure 7.16. The daily mean of Willson's irradiance measurements for 1980-87; much of the scatter of the data between the vertical dashed lines is attributable to the low sampling rate during the period when the *Solar Maximum Mission* spacecraft lacked fine pointing control; the dashed curve indicates the solar-cycle modulation. (Courtesy of Richard C. Willson.)

sure that the decline observed between 1980 and 1985 was a solar-cycle effect. The daily means from the *Solar Maximum Mission's* monitor had leveled out and then begun to climb again as sunspot activity passed through its minimum and resumed its ascent. Willson and Hudson (1988) believed the time had come to publish the irradiance data with a curve superimposed on it to highlight solar-cycle modulation (fig. 7.16). Shortly after the appearance of their report, teams associated with the radiometers on *Nimbus 7* and the *Earth Radiation Budget Satellite* announced similar results (Kerr 1988; Hickey et al. 1988).

In the meantime, a few solar physicists were already trying to explain why the irradiance varied in phase with the solar cycle. Foukal and Lean took the lead, giving papers at meetings of the American Geophysical Union (Kerr 1987) and the American Astronomical Society's Solar Physics Division and submitting articles to the *Astrophysical Journal* (Foukal and Lean 1988) and *Science* (Lean and Foukal 1988). Their analysis focused on the irradiance measurements collected from 1981 through 1984, a period in which the annual trend lines for the data collected by both Hickey and Willson had similar slopes. This agreement, which was not present in the two data sets for 1980, struck them as strong evidence that during 1981–84 the observed long-term "variations were caused by a solar, rather than an instrumental, effect" (Foukal and Lean 1988, 349). Implicit here, as will become evident, was a suspicion that the trend lines for 1980 were not solar in origin.

In analyzing the 1981–84 irradiance measurements, Foukal and Lean first removed all the variation that could be attributed to sunspot blocking. Then they examined how the adjusted irradiance curve correlated with two different indices of facular radiative output. One index—the area of chromospheric plage—indicated the output of faculae associated with active regions. The other was a global index that reflected the radiative output not only from the active-region faculae but also from the bright facular grains associated with the magnetic network crisscrossing the sun. All the major ups and downs in the adjusted curve were matched by both facular indices, but the downward trend was matched only by the global index. This finding, Foukal and Lean believed, implied that the waxing and waning of the global radiative output from the network faculae played the dominant role in solar-cycle modulation of solar irradiance.

Once having accounted for the observed irradiance curve for 1981–84, Foukal and Lean computed its shape since the beginning of solar cycle 21 in 1975. Their calculated curve represented the net effect of output deficits from sunspot blocking and output enhancements from active-region and network faculae. It gave a graphic portrayal of solar-cycle modulation of the irradiance. At the same time, it called into question the downward trends in the irradiance data that Hickey had reported for late 1978 through 1980 and Willson for 1980. The computed curve suggested instead that the general trend of the solar irradiance was upward until 1981. Foukal and Lean (1988, 356) thought it likely that both Hickey's and Willson's "radiometers declined in responsivity after their respective launches and then stabilized," but they readily conceded that their model would need to be compared with the radiometric observations over a longer period to reveal the source of the disagreement.

Foukal and Lean's interpretation of solar-cycle modulation of the sun's radiative output enjoyed a favorable reception (e.g., Kerr 1987, 1988; Willson and Hudson 1988; Livingston, Wallace, and White 1988; Fröhlich and Pap 1989). Schatten (1988) has suggested that polar faculae may join with network faculae in causing the irradiance to vary in phase with the solar cycle. But in an article applying their model to the entire period from 1874 to 1988, Foukal and Lean (1990) have argued that Schatten's model overestimates the influence of the polar faculae. No matter what the fate of Foukal and Lean's interpretation, solar physicists generally think that space observations have established that the sun's luminosity depends on the level of solar activity. If so, the sun's radiative output must have been lower during the Maunder minimum than it has been at any time since. Perhaps, as Eddy argued in the mid-1970s, that prolonged minimum did indeed give rise to the Little Ice Age.

Conclusion

Willson's measurements transformed research on the sun's radiative output. During the 1970s, solar physicists had grown increasingly interested in the possibility that the solar irradiance varied with solar activity. But neither they nor specialists in precision radiometry were able to respond to this interest with more than tantalizing evidences of variability. Hindsight indicates that the tools brought to bear on the issue lacked the precision to do anything further. In 1980, thanks to NASA's decision to include the active cavity radiometer irradiance monitor on the *Solar Maximum Mission,* Willson got an instrument with the requisite sensitivity into orbit. His subsequent graphs of the irradiance's fluctuations were accepted as conclusive evidence of the inconstancy of the solar constant.

Our story illustrates the continuing importance of outsiders to the advance of solar physics. The scientists who reopened the issue of the constancy of the solar constant between the mid-1960s and mid-1970s rarely attended solar conferences or contributed to *Solar Physics.* In particular, those who developed the instruments with the sensitivity required to measure solar-irradiance variability—Drummond, Hickey, Kendall, Willson, and their collaborators—were specialists in precision radiometry. And those who led the way in questioning the steadiness of the sun's radiative output—Fowler and Cameron—were specialists in nuclear and in theoretical astrophysics. Not only did outsiders reopen the issue of the sun's constancy, but one—Willson—also played the central role in settling it. His success in getting an irradiance monitor of unprecedented accuracy, precision, speed, and durability into space enabled him to put solar-irradiance variability beyond a shadow of a doubt. Even as it resolved this old issue, however, his monitor posed a host of new questions about the sources of the variability it was measuring.

Our story also illustrates how the solar physics community, despite its inherent self-referencing tendency, continues to enlarge its research agenda in response to the initiatives of outsiders. The entire community is too large and diverse for all or even a majority of its members to play active parts in responding to such opportunities. Rather, a few members whose research interests or institutional affiliations give them a special sympathy for or awareness of the outsiders' endeavors will generally take the lead. In the present case, Zirin and Eddy were especially important in the years preceding Willson's observational breakthrough. With the Big Bear Workshop of May 1975, Zirin not only directed the solar physics community's attention to the problem of the solar constant's constancy but also marshaled support for an irradiance-monitor berth on the *Solar Maximum Mission*. Thereafter, Eddy drew upon old observational records and radiocarbon data to give substance to speculations that the sun might well be much less stable and regular than customarily assumed. The success of Zirin, Eddy, and other solar physicists who took an early interest in the irradiance problem was manifest in the warm welcome accorded Willson's results. Since 1980, indeed, several solar physicists have gotten involved in interpreting evidence of solar-irradiance variability.

Finally, our story sheds light on the importance of precision in modern solar research. The solar constant is actually rather well named. Willson's measurements indicate that it remained between 1364.3 and 1369.4 watts/m^2 from February 1980 through September 1987. In everyday life, we would regard this degree of stability as constancy. Solar physics, however, is not the same as everyday life. Since Galileo, the field has advanced to the point that the small fluctuations of the solar irradiance can be made to yield interesting insights. Solar physicists have focused their attention, therefore, not on the solar constant's high degree of stability but rather on its variability. In particular, they have sought to correlate Willson's and Hickey's irradiance records with concurrent ground-based observations of various solar features. They have concluded that sunspots and associated faculae account for a large fraction of the variance over the time scales characterizing active-region evolution. They have also begun to think that bright facular grains along the active network are responsible for the variance over solar-cycle time scales. If further research bears out these ideas, solar physicists will have not only a fuller understanding of the sun but also a good start in explaining terrestrial ice ages.

The successes of solar-irradiance research since 1980 have engendered interest in maintaining a sustained, continuous record of the sun's radiative output. For the time being, however, monitoring capabilities are very limited. The *Solar Maximum Mission* recently reentered the atmosphere, which, as the result of the heating by the current solar maximum, had expanded out toward its decaying orbit (Nichols 1989). This spacecraft's premature demise left direct irradiance monitoring, for the present, to *Nimbus* 7 and three other meteorological satellites (Hickey et al. 1988). Their instruments and observing routines cannot match those used by Willson on the *Solar Maximum*

Mission. Still, they may suffice to continue the solar-irradiance record until Willson's active cavity radiometer irradiance monitor II goes up on the Upper Atmosphere Research Satellite, which is now scheduled for flight in August 1991. Plans are less definite for the period following this spacecraft's thirty-month mission, but the value of irradiance monitoring for both solar physics and terrestrial climatology will probably be well enough established by then to ensure that NASA continues the effort.

CHAPTER EIGHT

Conclusion

Scientific knowledge of the sun has progressed, not merely changed, in the nearly four centuries since Galileo and his contemporaries first directed telescopes at the heavens. At least that has been the perspective of this book. Recall that Galileo, one of the best natural philosophers of his day, could say little more about the sun than that it was an immense sphere standing at the center of the universe, that it rotated with a period of approximately twenty-eight days, and that the generation and decay of spots in its equatorial zone proved its mutability. Since his day, solar science has gotten ever more specific and robust.

Successive conceptions of the sun have been increasingly rich in that they have embodied more and more detailed representations of its properties and workings. This increase in specificity has taken two forms. First, a growing number of characteristics have been used in scientific descriptions of the sun. Since Galileo, solar descriptions have included data and inferences not only about the sun's shape, size, rotation, and spots but also—in rough chronological order—about its faculae, mass and density, motion through space, infrared and ultraviolet radiation, dark-line spectrum, radiative output, activity cycle, prominences and chromosphere, corona, age, flares, chemical composition, bright-line spectrum, magnetic field, corpuscular emanations, radio transmissions, X-ray emissions, neutrino output, coronal holes, active network, and whole-body oscillations. Second, the growth in the number of solar characteristics has been accompanied by a rise in the precision of the parametric values given for each attribute. Since Newton's estimate of the sun's mass, for example, the values assigned to this parameter have closed in on the current value of $332,946.0\pm0.3$ earth masses along a damped swinging path (Prologue). Likewise, since William Herschel's use of wheat prices as an indicator for the sun's radiative output, the measures of this parameter at the earth's mean distance first closed in on a value of 1370 ± 20 watts/m^2 and then in 1980 began unambiguously tracing its variability (chap. 7).

Besides their growing richness, successive conceptions of the sun have embodied ever sturdier interpretations of the sun's structure and behavior.

The solar theories of, say, the early 1600s, the early 1800s, the early 1900s, the late 1930s, and the present have been hardier than their antecedents in two ways. They have linked more and more of the sun's characteristics to one another. They have also withstood increasingly severe observational and consistency tests.

These interpretive trends have, for example, characterized successive theories of the sun's radiative output. During the seventeenth and eighteenth centuries, most scientists interested in the sun supposed that sunspot blocking diminished its luminosity. In 1801 William Herschel broke with this position, invoking a correlation between sunspots and grain prices to support his view that spots enhanced the sun's output (chap. 1). Some four decades later, Mayer and Waterston opened up a second line of theorizing about solar radiation by suggesting that the only sources capable of powering the sun for more than a few centuries were infalling meteors and gravitational contraction. Soon thereafter, Helmholtz propelled Waterston's contraction hypothesis into the scientific mainstream and, in doing so, linked the sun's mass, size, age, and luminosity (chap. 2). The contraction theory remained dominant until the early 1900s, when various physical and astronomical investigations discredited its paltry time scale. During the next two decades, Eddington, Atkinson, and especially Bethe led the way in developing the theory that chains of thermonuclear reactions deep in the solar interior powered our sun as well as other ordinary stars. In linking the sun's mass, size, chemical composition, age, and luminosity to one another, the thermonuclear theory satisfied much more stringent conditions than the contraction theory it superseded (chap. 3).

Since 1940, Bethe's theory has been modified to accommodate the results of further research on nuclear reactions and the sun's internal conditions. However, Davis's failure to find the solar-neutrino flux predicted by the modified theory has put the orthodox solar model under something of a cloud (chap. 5). Although the difficulty may originate in neutrino physics, the set of interlocking theories currently used to account for the sun's radiative output could be flawed, perhaps seriously (chap. 7). Whatever the case, the explanation of energy generation that prevails after the next rounds of research on solar-neutrino production (using new detectors) and solar internal structure (using oscillation data) will surely be even sturdier than the present one. Meanwhile, various inquiries have recently enlivened theorizing on the old question of the relation between solar activity and luminosity. In particular, Hickey's and especially Willson's measurements of solar irradiance from earth orbit have enabled theorists to make considerable progress in linking spots, faculae, and the active network to fluctuations in the sun's output of energy (chap. 7). This research, it seems reasonable to suppose, will feed into the development of more robust models of the sun's convective zone.

The pace and direction of solar science's advance have depended most immediately on the number, objectives, backgrounds, and interconnections of those investigating the sun and on the tools and techniques they have been able to bring to bear in their inquiries. Between 1610 and 1810, the field's progress was slow and, except for the insight that the sun was a star, almost entirely within the descriptive realm (chap. 1). The astronomers and natural philosophers who sought to enlarge knowledge of the sun in this era were few. Furthermore, their solar inquiries were typically brief excursions carried out in the course of either advocating their general approaches to nature (e.g., Galileo, Descartes, and Newton) or pursuing broad questions (e.g., the transit-of-Venus observers' interest in the dimensions of the solar system and William Herschel's concern with the natural history of the heavens). Finally, their tools and techniques—ordinary astronomical telescopes at low-altitude sites in wet climates, occasional campaigns of coordinated observing, and Newtonian mechanics—were suitable only for improving descriptions of the sun's most visible features (especially its spots) and mechanical properties (its size, mass, and motions).

In the century between 1810 and 1910, solar science progressed at a more rapid pace than before (chap. 2). The record was particularly impressive on the descriptive side, but this era also witnessed the beginning of quantitative modeling on the theoretical side. For a good while, the nineteenth-century contributors to solar science resembled their forerunners in that their numbers were low and their primary interests elsewhere. Nonetheless, the development of fresh observing programs for eclipses (especially Baily) and sunspots (Schwabe) and the use of new physical theories to create solar thermodynamics (notably Waterston and Helmholtz) and solar spectroscopy (Kirchhoff and Bunsen) opened the way for a midcentury flowering of solar research. During the heady 1860s, some dozen or so European and American astronomers led the way in constituting a self-conscious, if informal, "solar physics" community. Thereafter they and their recruits served as the field's chief promoters and custodians, doing all they could to increase solar observational capabilities, to systematize and extend prior work, and to assimilate promising contributions from newcomers and outsiders. Hale capped off this century by devising the spectroheliograph, by establishing the *Astrophysical Journal* and the International Union for Cooperation in Solar Research, by stressing the symbiotic relationship of solar and stellar physics, by building a solar observatory with a state-of-the-art spectroscopic laboratory atop Mount Wilson, and by inaugurating the study of solar magnetism.

Between 1910 and 1940 there was a further quickening in the pace of solar science as substantial headway was made in both the descriptive and interpretive arenas (chap. 3). The most steadfast contributors to this progress were the solar physicists, whose numbers and observational capabilities grew throughout the period. Their focus was on refining and augmenting descriptive knowledge of the sun's radiative output, activity cycle, and absorption and emission spectra. The most dazzling contributors to this progress were scientists whose primary ambits, at least when they made their first contributions, were outside the solar physics community. Thus a planetary astrophysicist (Lyot) and an amateur astronomer (McMath) played lead roles in improving descriptions of the corona and fast-moving solar phenomena by developing the coronagraph and solar cinematography. Again, four theoretical astrophysicists (Payne, Russell, Eddington, and Strömgren) led the way in establishing hydrogen's abundance in the sun, and an experimental physicist (Edlén) led in identifying the ions producing the coronal lines. And three theoretical astrophysicists (Eddington, Milne, and Cowling), an experimental physicist (Atkinson), and a theoretical physicist (Bethe) were at the forefront in theorizing about the physical conditions and energy-generating reactions in the sun's interior. Evidently, outsiders who hit upon ideas for innovative instruments or enjoyed ready access to the latest physical research could learn enough about solar research to do pioneering work more easily than the community's insiders could take timely advantage of pertinent developments in neighboring fields.

Solar science's continuing advance in the past half-century has involved an increasingly effective interplay among the field's growing number of observational and theoretical specialities. Since 1940, the kind of solar phenomena observed have been substantially extended with the introduction of radio telescopes, magnetographs, neutrino detectors, and spacecraft-borne instruments for studying radiations and particles blocked by the earth's atmosphere (chaps. 4–5). Meanwhile, the resolution of the data obtained with these and earlier instruments has been steadily enhanced by upgrading optical and electronic components and, in the case of space observations, by increasing spacecraft range, pointing, reliability, and durability. And the speed and subtlety of both data processing and model testing have risen along with the power of electronic computers and the scope and depth of relevant physical and astrophysical theories.

This development of research capabilities has been accompanied by a major transformation in the solar physics community (chaps. 4–5). Since 1940, the world population of solar physicists holding secure positions and publishing with some frequency has risen from about fifty to about three hundred. This numerical gain, though not insignificant, tells but a small part of the story of the field's transformation. On the eve of World War II, the typical solar physicist had a colleague or two and a small number of assistants. His postwar counterpart has been increasingly likely to be one of several solar physicists heading up a sizable team of scientists, engineers, technicians, and secretaries. Such teams, moreover, have been working ever more closely both with solar teams at other sites and with teams having complementary objectives in an ever wider array of scientific, industrial, space, and military institutions. Finally, the increasing scale and complexity of the solar research enterprise have given rise to subdisciplinary forums—a journal, associations, symposia, and workshops—that, by nurturing interactivity, have contributed to the field's pace and coherence.

The case studies on the emergence of research specialties dealing with the solar wind and solar-irradiance variability (chaps. 6–7) illustrate the roles of new observational capabilities and increasing intrafield interactions in the post-*Sputnik* advancement of scientific knowledge of the sun. Between 1959 and 1962, space-science teams at the Soviet Academy's Radio Engineering Institute (led by Gringauz), the Massachusetts Institute of Technology (led by Bridge and Rossi), and the Jet Propulsion Laboratory (led by Neugebauer and Snyder) moved the solar wind from the realm of conjecture into the realm of observational science by demonstrating the wind's existence with the aid of plasma traps and spectrometers on *Lunik 2*, *Explorer 10*, and *Mariner 2*. Likewise, some two decades later, space-science teams at Eppley Laboratory (led by Hickey) and especially the Jet Propulsion Laboratory (led by Willson) did the same for solar-irradiance variability with the aid of their radiometers on *Nimbus 7* and the *Solar Maximum Mission*. In each instance, once the phenomenon's existence was established, an increasingly diverse array of observers and theorists have sought to relate it to other solar phenomena. In the case of the solar wind, direct measurements of its flux and energy have been related to observations of interplanetary and solar fields (e.g., Ness and Wilcox), coronal holes (e.g., Krieger's group), and high-latitude wind speeds (e.g., Coles and Rickett's group). Using this growing observational base, a succession of theorists (notably Rosenberg, Schulz, Hundhausen, and Wilcox's group) have developed a convincing picture of the solar wind's three-dimensional structure and its evolution throughout the solar cycle. Similarly in the case of the sun's radiative output, direct measurements of its variability have been related to solar phenomena of short, intermediate, and longer duration (oscillations, sunspots and faculae, and the activity cycle). Moreover, though much probably remains to be done, a growing number of solar physicists (most notably Hudson and Woodard, Sofia's group, and Foukal and Lean) have interpreted the observational relations with increasing success.

Although research on the solar wind and on solar-irradiance variability has made great strides since 1960 and 1980, substantial further progress may well depend upon NASA's success in revitalizing its space-science program in the next few years. On account of the *Challenger* disaster, NASA delayed the launch of *Ulysses*, which will conduct the first in situ observations of the solar wind and related phenomena at high heliographic latitudes, until October 1990. The agency has not yet been able to launch the Upper Atmosphere Research Satellite, which will continue the high-quality solar-irradiance record begun by the *Solar Maximum Mission*. With any luck, these missions' results will play an important role in subsequent research on the sun's wind and radiative output.

The prolonged delay of the solar-polar and upper-atmosphere missions dramatizes something that has been true since Galileo's day—that solar science has never functioned as a closed, autonomous system. Social-cultural trends have always influenced which scientists have chosen to study the sun and what tools they have brought to the task. One important trend has been the intensification of popular interest in the sun. The roots of this interest go back to the early human cultures that acknowledged the importance of the sun in day-to-day life by inventing solar deities or religions. The modern legacy of ancient solar worship has been a widespread fascination with the sun and its secrets. This fascination has surely accounted, in part at least, for the frequency with which scientists of the first rank have temporarily turned aside from their primary pursuits to consider the sun—for example, Galileo, Descartes, Newton, and Laplace in the seventeenth and eighteenth centuries (chap. 1), Humboldt, Helmholtz, Thomson, and Kirchhoff in the nineteenth century (chap. 2), and Eddington, Chandrasekhar, Bethe, and Alfvén in the twentieth century (chaps. 3, 6). Since the mid-nineteenth century, this popular curiosity has been complemented and amplified by a growing interest in the sun's precise influence on the climate, on terrestrial magnetism, on radio propagation, and most recently on spacecraft subsystems, including nuclear-test monitors (chaps. 2–5). This pragmatic interest has surely accounted, again at least in part, for a certain predisposition of government officials to meet the rising costs of solar research—for example, at the Kew and Greenwich observatories in the nineteenth century (chap. 2), at the Smithsonian Astrophysical Observatory in the first half of this century (chap. 3), at the High Altitude and Sacramento Peak observatories after World War II (chap. 4), and in the American and Soviet space programs since *Sputnik* (chap. 5).

Another important trend has been the general progress of science and technology since the Renaissance. Time and again, instruments, techniques, and theories originating in other contexts have been integrated into solar science's armamentarium—especially the telescope and Newtonian mechanics in the seventeenth century (chap. 1); photography, thermodynamics, and spectral analysis in the nineteenth century (chap. 2); stellar-structure theory, quantum mechanics, and nuclear physics in the second through fourth decades of this century (chap. 3); and radio, rockets and spacecraft, electronics and computing, and plasma physics since 1940 (chaps. 4–7). Thus, notwithstanding solar science's many contributions over the years to physics, astrophysics, geophysics, applied optics, and space technology, the field has been on balance a beneficiary rather than a donor science. In a sense, that is, progress in describing and interpreting the sun has been virtually a derivative of the overall success of modern science and technology in interrogating and mastering nature.

To recapitulate the argument to this point, solar science has made unrelenting progress for almost four centuries. Successive conceptions of the sun have embodied ever more specific representations of its properties and ever more robust theories of its structure and workings. In the first instance, solar science's advance as a body of knowledge has depended on the success of the scientists choosing to investigate the sun in improving the field's tools and techniques. In turn, this improvement in collective research capabilities has been influenced by such trends as the pragmatic intensification of Western culture's abiding fascination with the sun and the general advance of science and technology. Ultimately, to add a final step to the argument, the ability of successive generations of scientists to enlarge knowledge of the sun has been an indirect consequence of the emergence of new values and beliefs, new means of production, new forms of government, and new levels of warfare.

This last point could conceivably be summarized with the claim that solar science's advance since the early seventeenth century has been, in the final analysis, a by-product of humanity's progress in the modern age. This claim, however, strikes me as a very optimistic reading of the situation. In the light of this century's wars and environmental catastrophes, it seems clear that many of the developments that have undergirded the recent progress of solar science have been closely intertwined with developments that may lead to humanity's demise. The world's most prosperous and powerful nations still have much work to do if they are to solve the difficult problems of safeguarding humanity from all present and future genocidal weapons and technologies. If they succeed, our progeny will, I am confident, know much more about the sun than do we. If these nations fail, our descendants— should there be any—will be fortunate if they can worship the sun as did the ancient Egyptians.

BIBLIOGRAPHICAL ESSAY

In writing this history of solar science since Galileo, I have used—as the text citations and Bibliography make clear—an immense number and variety of materials. Here I provide those who wish to read further with a guide to the historical studies I have found most pertinent, dependable, and insightful. Nearly all the works mentioned have appeared in the past two decades. One reason for this recency is that the history of science, also a beneficiary of post-*Sputnik* patronage, has grown rapidly since 1960. Another is that historians have been devoting more attention to nineteenth- and twentieth-century science. Yet another is that partly as the result of professionalization and competition, recent historical studies of science are generally more reliable and discerning than earlier works.

My chapter on the seventeenth and eighteenth centuries offers the first overview of the origins of modern solar science between Galileo and Herschel. In part, it makes use of thematic studies that have worked out one part or another of the story—Van Helden (1977) on the invention of the telescope; North (1974) and Shea (1972) on the earliest telescopic observations of sunspots; Dick (1982) and Crowe (1986) on the concept of the plurality of worlds, including its premise that the stars are suns; Van Helden (1985) and Woolf (1959) on endeavors to find the dimensions of the solar system, including the sun; and Hendry (1983) and Hoskin (1980) on early investigations of the sun's direction of motion. It also draws upon biographical studies by Drake (1978), Westfall (1985), and Biagioli (1990) of Galileo, by Westfall (1980) of Newton, and by Schaffer (1980a, 1980b, 1981) and Turner (1977) of Herschel.

In dealing with the period from 1810 to 1910, I have been able to rely on a rich body of relevant historical literature. Clerke (1902) includes some classic chapters on the rise of solar physics in her monumental history of astronomy during the nineteenth century. Meadows (1970, 1984a, 1984b) and Herrmann (1985) also provide useful orientation. Besides these general treatments, my narrative depends on many thematic works—Ranyard (1879) on the rise of eclipse observing; Meadows and Kennedy (1981) and Schröder (1984) on the correlation of the solar and geomagnetic cycles; James (1983, 1985) on the beginnings of solar spectroscopy; Kidwell (1981), James (1982), Smith and Wise (1989), Powell (1988), Burchfield (1975), and DeVorkin (1984) on the inception and development of the contraction theory of solar-energy generation; Bartholomew (1976) on British solar physics in the 1860s; Lankford (1981) and Hufbauer (1986) on the role of amateurs in the rise of astrophysics; Warner (1986) on refinements in the techniques for ruling diffraction gratings; and DeVorkin (1981) on the early history of the Solar Union. The chapter also relies on several biographies—especially Meadows (1972) of Lockyer; Warner (1971) of Rutherfurd; Herrmann (1982) of Zöllner; Beardsley (1981) and Eddy (1990) of Langley; Osterbrock (1984) of Keeler; and Wright (1966) and Wright, Warnow, and Weiner (1972) of Hale.

In investigating solar science since the early twentieth century, I soon realized that many important developments lack full histories—for example, Lyot's invention and exploitation of the coronagraph, the background and reception of Bethe's solution to the stellar-energy problem, and Edlén's identification of the coronal emission lines, just to mention three subjects I intend to study further in the future. Still, a good deal of relevant work has already been done by historians of science and scientists with a historical bent, and more is under way.

The chapter on solar research between 1910 and 1940 makes use of a number of thematic studies—Kevles (1971) and Schroeder-Gudehus (1978) on the background of the International Astronomical Union; Jones (1965), Hoyt (1979), and DeVorkin (1990a) on Abbot's attempts to monitor the solar constant; DeVorkin (1975, 1984), Hufbauer (1981), and Kenat (1987) on the evolution of Eddington's theory of stellar structure down to 1924; and DeVorkin and Kenat (1983a, 1983b) on the development of a compelling case for hydrogen's abundance in the sun's atmosphere. It also relies on biographies by DeVorkin (1989c) of Russell, by Chandrasekhar (1980) of Milne, and by Kidwell (1987) of Payne.

The attention I give to the military context for solar observing during and after World War II reflects, in large part, the influence of Kevles (1979), DeVorkin (1980, 1985, 1990b), Roland (1985), Forman (1987), Kidwell (1990), and the papers in De Maria, Grilli, and Sebastiani (1989). The chapter's treatment of efforts to observe the sun's short- and long-wavelength radiations exploits a variety of studies—Newell (1980), Friedman (1981), Hirsh (1983), Van Allen (1983), and particularly DeVorkin (1986a, 1987, 1989b, 1990c) and Hevly (1987) on the establishment of rockets as platforms for observing solar ultraviolet and X-ray emissions; and Hey (1973), Edge and Mulkay (1976), Wild (1987), and particularly Sullivan (1984, 1990) on the rise of radio solar astronomy. It also makes use of Newell (1980) and Needell (1989) as well as the official histories of Spencer-Jones (1959) and Nicolet (1958) in describing the preparations for the International Geophysical Year.

My general orientation to science in the space age derives—in ways too complicated for me to sort out—from careful reading of Newell (1980), an early version of Tatarewicz (1990), McDougall (1985), Stares (1985), and Smith (1989), in that order. The account of the advance of solar observational capabilities since 1957 utilizes only a limited number of historical studies—Goldberg (1981) on the rise of space solar physics; Compton and Benson (1983) on the preparations for and conduct of the *Skylab* mission; Eddy (1979) on *Skylab's* place in the history of solar science; Howard (1985) on solar observing at Mount Wilson Observatory; and Pinch (1986) on the rise of solar-neutrino research and controversy. The treatment of the solar physics community since *Sputnik* is a revised and abbreviated version of Hufbauer (1989).

Most of the historical works that I found for the case studies dealt with background and context. The account of solar-wind research uses Schröder (1984) the evolution of auroral physics, Simpson (1985) and Needell (1987) the postwar study of solar modulations of cosmic rays, Hirsh (1983) the background of Rossi's move into space science, and Koppes (1982) the Jet Propulsion Laboratory's transformation after *Sputnik*. That on solar-irradiance variation draws on Doel (1990) on Abbot's solar-constant program and its reception, and Pinch (1986) on the solar-neutrino controversy.

The trend is clear. The closer historians get to the present, the greater are their opportunities to break fresh ground. Opening up recent subjects to historical inquiry is, to be sure, no easy task. But as my detailed references attest, research materials are abundant and many scientists are forthcoming.

BIBLIOGRAPHY

Manuscript and Microfilm Collection Abbreviations

CWS = Conway W. Snyder Papers, Canyon Country, California

JMW = John M. Wilcox Papers, Center for Space Science and Astrophysics, Stanford University, Stanford, California

LG = Leo Goldberg Papers, Kitt Peak National Observatory, Tucson, Arizona

NHO = NASA History Office, Washington, D.C.

NHQ = NASA Headquarters, Washington, D.C.

RLN = Ray L. Newburn Papers, Space Science and Exploration Division, National Air and Space Museum, Washington, D.C.

SHMA = Sources for the History of Modern Astrophysics, American Institute of Physics, New York, New York

WOR = Walter Orr Roberts Papers, Library, University of Colorado, Boulder, Colorado

Journal Abbreviations

Annals IGY = Annals of the International Geophysical Year

ARAA = Annual Review of Astronomy and Astrophysics

Bull. AAS — Bulletin of the American Astronomical Society

Comptes Rendus = Comptes Rendus de l'Académie des Sciences

JHA = Journal for the History of Astronomy

JOSO: AR — Joint Organization for Solar Observations: Annual Report

MNRAS = Monthly Notices of the Royal Astronomical Society

Phil. Trans. = Philosophical Transactions of the Royal Society

QJRAS = Quarterly Journal of the Royal Astronomical Society

Trans. IAU = Transactions of the International Astronomical Union

Aaboe, A. 1974. Scientific astronomy in antiquity. *Phil. Trans.* A276:21–42.

AAS Division on Solar Physics. 1972. *Bull. AAS* 4:376.

Abbot, C. G. 1906. Samuel Pierpont Langley. *Astronomische Nachrichten* 171:91–96.

———. 1929a. *The sun and the welfare of man.* Washington D.C.: Smithsonian Institution.

———. 1929b. Variations of solar radiations. In International Research Council, *Second report of the commission appointed to further the study of solar and terrestrial relationships,* 14–15. Paris: Chiron.

———. 1939. Recent work of the Smithsonian Astrophysical Observatory. In International Council of Scientific Unions, *Cinquième rapport de la Commission pour l'étude des Relations entre les Phénomènes Solaires et Terrestres*, 11–12. Florence: Barbéra.

Abstracts of papers presented at the meeting of the Division on Solar Physics, held 17–19 November 1970 at Huntsville, Alabama. 1971. *Bull. AAS* 3:259–66.

Aiton, E. J. 1972. *The vortex theory of planetary motions*. London: Macdonald.

Akasofu, S. -I., B. Fogle, and B. Haurwitz, eds. 1968. *Sydney Chapman: Eighty from his friends*. Fairbanks and Boulder: University of Alaska, University of Colorado, and University Corporation for Atmospheric Research.

Aleksandrov, S. G., and R. E. Federov. 1961. *Sovetskie sputniki i kosmicheskie korabli*. Moscow: Akademie nauk SSSR.

Alexander, G. 1980. Dips in sun's energy output detected. *Los Angeles Times*, August 7, 1:3, 26.

Alfvén, H. 1957. On the theory of comet tails. *Tellus* 9:92–96.

———. 1975. Electric current structure of the magnetosphere. In *Physics of the hot plasma in the magnetosphere*, ed. B. Hultqvist and L. Stenflo, 1–22. New York: Plenum.

———. 1977. Electric currents in cosmic plasmas. *Reviews of Geophysics and Space Physics* 15:271–84.

Allen, C. W. 1955. *Astrophysical quantities*. 1st ed. London: Athlone.

———. 1973. *Astrophysical quantities*. 3d ed. London: Athlone.

Alvensleben, A. von. 1975. Report of the eighth meeting of the provisional board of JOSO in Florence, February 24–28. *JOSO: AR*, 1974, 27–33.

American Institute of Physics. 1964. IMP-1 experiment supports water-sprinkler hypothesis (October 14). JMW.

Appleton, E. V. 1938. Remarks on Dr. Chapman's note on radio fade-outs and the associated magnetic disturbances. *Terrestrial Magnetism* 43:487.

Arago, F. 1836. Observations de l'eclipse de soleil du 15 mai 1836. *Comptes Rendus* 2:503–4.

Aristotle. ca. 340 B.C./1960. *On the heavens*. Trans. W. K. C. Guthrie. Cambridge: Harvard University Press.

Ash, M. E., I. I. Shapiro, and W. B. Smith. 1967. Astronomical constants and planetary ephemerides deduced from radar and optical observations. *Astronomical Journal* 72:338–50.

Astronomy in Moscow to-day. 1944. *Observatory* 65:143–44.

Atkinson, R. d'E. 1931. Atomic synthesis and stellar energy. *Astrophysical Journal* 73:250–95, 308–47.

Atkinson, R. d'E., and F. G. Houtermans. 1929. Zur Frage der Aufbaumöglichkeit der Elemente in Sterne. *Zeitschrift für Physik* 54:656–65.

Babcock, H. D. 1959. The sun's polar magnetic field. *Astrophysical Journal* 130:364–65.

Babcock, H. D., and H. W. Babcock. 1951. The ruling of diffraction gratings at the Mount Wilson Observatory. *Journal of the Optical Society of America* 41:776–86.

Babcock, H. W. 1953. The solar magnetograph. *Astrophysical Journal* 118:387–96.

————. 1961. The topology of the sun's magnetic field and the twenty-two-year cycle. *Astrophysical Journal* 133:572–87.

————. 1977. Interview with S. Weart (July 25). Niels Bohr Library, American Institute of Physics, New York, N.Y.

————. 1986. Diffraction gratings at the Mount Wilson Observatory. *Vistas in Astronomy* 29:153–74.

Babcock, H. W., and H. D. Babcock. 1952. Mapping the magnetic fields of the sun. *Publications of the Astronomical Society of the Pacific* 64:282–87.

————. 1955a. The sun's magnetic field, 1952–1954. *Astrophysical Journal* 121:349–66.

————. 1955b. The sun's magnetic field and corpuscular emission. *Nature* 175:296.

Bahcall, J. N. 1969. Neutrinos from the sun. *Scientific American* 221, 1:28–37.

————. 1989. *Neutrino astrophysics*. Cambridge: Cambridge University Press.

Bahcall, J. N., and R. Davis, Jr. 1982. An account of the development of the solar neutrino problem. In *Essays in nuclear astrophysics presented to William A. Fowler on the occasion of his seventieth birthday*, ed. C. A. Barnes, D. D. Clayton, and D. Schramm, 243–85. Cambridge: Cambridge University Press.

Bahcall, J. N., R. Davis, Jr., and L. Wolfenstein. 1988. Solar neutrinos: A field in transition. *Nature* 334:487–93.

Baily, F. 1836. On a remarkable phenomenon that occurs in total and annular eclipses of the sun. *MNRAS* 4:15–19.

————. 1842. Some remarks on the total eclipse of the sun, on July 8th, 1842. *MNRAS* 5:208–14.

Bandeen, W. R., and S. P. Maran, eds. 1975. *Possible relationships between solar activity and meteorological phenomena: A symposium held at Goddard Space Flight Center, Greenbelt, Maryland, November 7–8, 1973*. NASA SP-366. Washington D.C.: NASA.

Bartels, J. 1932. Terrestrial-magnetic activity and its relations to solar phenomena. *Terrestrial Magnetism and Atmospheric Electricity* 37:1–52.

Bartholomew, C. F. 1976. The discovery of the solar granulation. *QJRAS* 17:263–89.

Baumgartner, F. J. 1987. Sunspots or sun's planets: Jean Tarde and the sunspot controversy of the early seventeenth century. *JHA* 18:44–54.

Beardsley, W. R. 1981. Samuel Pierpont Langley: Early conflict between teaching and research at the Western University of Pennsylvania. *Western Pennsylvania Historical Magazine* 64:345–72.

Beggs, J. M. 1982. Letter to Congressman L. Winn, Jr., April 16. SMM File, NHO.

Behn, R. C. 1968. Ionospheric electron density studies. In Wukelic 1968, 81–213.

Behr, A., and H. Siedentopf. 1953. Untersuchungen über Zodiakallicht und Gegenschein nach lichtelektrischen Messungen auf dem Jungfraujoch. *Zeitschrift für Astrophysik* 32:19–50.

Benjamin, F. S., Jr., and G. J. Toomer. 1971. *Campanus of Novara and medieval planetary theory: Theorica planetarum*. Madison: University of Wisconsin Press.

Berkner, L. V., ed. 1958. Manual on rockets and satellites. *Annals IGY* 6:1–508.

Berkner, L. V., and H. Odishaw. 1961. A note on the space science board. In *Science in space*, ed. L. V. Berkner and H. Odishaw, 429–36. New York: McGraw-Hill.

Bernheimer, W. E. 1929. Strahlung und Temperatur der Sonne. In *Handbuch der Astrophysik*, ed. G. Eberhard, A. Kohlschütter, and H. Ludendorff, 5:1–56. Berlin: Springer.

———. 1936. Strahlung und Temperatur der Sonne. In *Handbuch der Astrophysik*, ed. G. Eberhard, A. Kohlschütter, and H. Ludendorff, 7:333–49. Berlin: Springer.

Bernstein, J. 1980. *Hans Bethe: Prophet of energy.* New York: Basic.

Berry, R. 1988. Renaissance on Mount Wilson. *Astronomy* 16, 4:18–23.

Berthold, G. 1894. *Der Magister Johann Fabricius und die Sonnenflecken, nebst einem Excurse über David Fabricius.* Leipzig: Veit.

Bessel, F. W. 1848. *Populäre Vorlesungen über wissenschaftliche Gegenstände.* Ed. H. C. Schumacher. Hamburg: Perthes-Besser und Maucke.

Bester, A. 1966. *The life and death of a satellite.* Boston: Little, Brown.

Bethe, H. A. 1938a. Energy production in stars. *Cornell Daily Sun*, May 2, 7.

———. 1938b. Letter to C. L. Critchfield (May 10). Courtesy of Charles Critchfield, Los Alamos, N. Mex.

———. 1939. Energy production in stars. *Physical Review* 55:434–56.

———. 1968. Energy production in stars. *Science* 161:541–47.

Biagioli, M. 1990. Galileo's system of patronage. *History of Science* 28:1–62.

Biermann, L. F. 1951. Kometenschweife und solare Korpuskularstrahlung. *Zeitschrift für Astrophysik* 29:274–86.

———. 1952. Über den Schweif des Kometen Halley im Jahre 1910. *Zeitschrift für Naturforschung* 7A:127–36.

———. 1953. Physical processes in comet tails and their relation to solar activity. In *La physique des comètes: Communications présentées au quatrième colloque international d'astrophysique tenu à Liège les 19, 20 et 21 septembre 1952*, 251–62. Louvain: Ceuterick.

———. 1957. Solar corpuscular radiation and the interplanetary gas. *Observatory* 77:109–10.

———. 1985. On the history of the solar wind concept. In *Historical events and people in geosciences: Selected papers from the symposia of the Interdivisional Commission on History of IAGA during the IUGG General Assembly, held in Hamburg, 1983*, ed. W. Schröder, 39–47. Frankfurt am Main: Lang.

Biography of Dr. John C. Lindsay. [1962]. NHO.

Bohlin, J. D. 1976. Division of solar physics. *Bull AAS* 8:589.

———. 1977. Division of solar physics. *Bull. AAS* 9:698.

———. 1978. Division of solar physics. *Bull. AAS* 8:723–24.

Bok, B. J. 1934. The stability of moving clusters. *Harvard College Observatory Circular* 384:1–41.

Boss, L., W. W. Campbell, and G. E. Hale. 1903. *Report of Committee on Southern and Solar Observatories.* Washington D.C.: Carnegie Institution of Washington.

Bowen, E. G. 1984. The origins of radio astronomy in Australia. In Sullivan 1984, 85–111.

Bowen, I. S. 1952. Mount Wilson and Palomar observatories. *Carnegie Institution of Washington Year Book* 51:3–33.

————. 1956. Mount Wilson and Palomar observatories. *Carnegie Institution of Washington Year Book* 55:31–63.

Bowen, I. S., and B. Edlén. 1939. Forbidden lines of Fe VIII in the spectrum of Nova RR Pictoris (1925). *Nature* 143:374.

Bracher, K. 1981. Getting to the 1860 solar eclipse. *Sky and Telescope* 61:120–22.

Brandt, J. C. 1970. *Introduction to the solar wind.* San Francisco: Freeman.

Bridge, H. S. 1963. Plasmas in space. *Physics Today* 16, 3:31–37.

————. 1987. Letter to K. Hufbauer (August 10). NHO.

Bridge, H. S., C. Dilworth, A. J. Lazarus, E. F. Lyon, B. Rossi, and F. Scherb. 1962. Direct observations of the interplanetary plasma. *Journal of the Physical Society of Japan* 17, supp. A-II:553–59.

Bridge, H. S., C. Dilworth, B. Rossi, and F. Scherb. 1960. An instrument for the investigation of interplanetary plasma. *Journal of Geophysical Research* 65:3053–55.

Bringing the space-spin down to earth. 1957. *Newsweek* 50 (October 21): 35.

Brown, R. A., and R. Giacconi. 1987. New directions for space astronomy. *Science* 238:617–19.

Brush, S. G. 1979. Looking up: The rise of astronomy in America. *American Studies* 20, 2:41–67.

————. 1987. The nebular hypothesis and the evolutionary worldview. *History of Science* 25.246–78.

Bruzek, A. 1975. K. O. Kiepenheuer (1910–1975). *Solar Physics* 43:3–7.

Bunsen, R. W. 1859. Letter to H. E. Roscoe (November 15). Quoted in Kirchhoff 1972, 11–12.

Burchfield, J. D. 1975. *Lord Kelvin and the age of the earth.* New York: Science History.

Bushnell, D. 1962. [History of Sacramento Peak Observatory]. Manuscript, Sacramento Peak Observatory, Sunspot, N. Mex.

Cameron, A. G. W. 1973. Major variations in solar luminosity? *Reviews of Geophysics and Space Physics* 11:505–10.

————. 1975. Solar models in relation to terrestrial climatic variations. In Bandeen and Maran 1975, 143–47.

Carrington, R. C. 1858. On the distribution of the solar spots in latitude since the beginning of the year 1854. *MNRAS* 19:1–3.

————. 1859a. On certain phenomena in the motions of solar spots. *MNRAS* 19:81–84.

————. 1859b. Description of a singular appearance seen in the sun on September 1, 1859. *MNRAS* 20:13–15.

Catalogue of [solar] data in the world data centers. 1963. *Annals IGY* 36:243–69.

Catholic University of America. 1963. Program for the Symposium of Plasma Space Science (June 11–14). JMW.

Chaikin, A. 1982. Rescuing "Solar Max." *Sky and Telescope* 63:236.

————. 1984. Solar Max: Back from the edge. *Sky and Telescope* 67:494–97.

Chamberlain, J. W. 1960. Interplanetary gas: II. Expansion of a model solar corona. *Astrophysical Journal* 131:47–56.

Chandrasekhar, S. 1962. Letter to M. G. Inghram (March 12). Subrahmanyan Chandrasekhar Papers, Regenstein Library, University of Chicago, Chicago, Ill.
——. 1980. Edward Arthur Milne and the development of modern astrophysics. *QJRAS* 21:93–107.
Chandrasekhar, S., G. Gamow, and M. A. Tuve. 1938. The problem of stellar energy. *Nature* 141:982.
Chapman, G. A. 1980. Variation in the solar constant due to solar active regions. *Astrophysical Journal* 242:L45–L48.
——. 1987. Variations of solar irradiance due to magnetic activity. *ARAA* 25:633–67.
Chapman, G. A., A. D. Herzog, and J. K. Lawrence. 1986. Time-integrated energy budget of a solar activity complex. *Nature* 319:654–55.
Chapman, S. 1957. Notes on the solar corona and the terrestrial atmosphere. *Smithsonian Contributions to Astrophysics* 2, 1:1–11.
——. 1959. Interplanetary space and the earth's outermost atmosphere. *Proceedings of the Royal Society* A253:462–81.
Chapman, S., and J. Bartels. 1940. *Geomagnetism.* 2 vols. Oxford: Clarendon.
Chapman, S., and V. C. A. Ferraro. 1929. The electrical state of solar streams of corpuscles. *MNRAS* 89:470–79.
Charlesworth, J. H. 1977. Jewish astrology in the Talmud, Pseudepigrapha, the Dead Sea Scrolls, and early Palestinian synagogues. *Harvard Theological Review* 70:183–200.
Chevalier, A. 1952. Notice nécrologique sur M. Bernard Lyot, astronome titulaire de l'Observatoire de Paris. *Comptes Rendus* 234:1501–5.
Christiansen, W. N. 1984. The first decade of solar radio astronomy in Australia. In *The early years of radio astronomy: Reflections on fifty years after Jansky's discovery,* ed. W. T. Sullivan III, 112–31. Cambridge: Cambridge University Press.
Christiansen, W. N., D. S. Mathewson, and J. L. Pawsey. 1957. Radio pictures of the sun. *Nature* 180:944–46.
Clemence, G. M. 1964. The system of astronomical constants. *ARAA* 3:93–112.
Clerke, A. M. 1902. *A popular history of astronomy during the nineteenth century.* 4th ed. London: Black.
Clowse, B. B. 1981. *Brainpower for the cold war: The Sputnik crisis and National Defense Education Act of 1958.* Westport, Conn.: Greenwood Press.
Cochrane, R. C. 1966. *Measures for progress: A history of the National Bureau of Standards.* Washington D.C.: National Bureau of Standards.
Coffey, H. E. 1986. Problems in gathering and archiving ground-based solar synoptic data: Talk to the National Academy Panel on Long Term Observations (February 24). NHO.
Cohen, I. B. 1971. *Introduction to Isaac Newton's "Principia."* Cambridge: Harvard University Press.
Coleman, P. J., Jr. 1986. Telephone interview with K. Hufbauer (October 15). NHO.
Coles, W. A., B. J. Rickett, V. H. Rumsey, J. J. Kaufman, D. G. Turley, S. Ananthakrishnan, J. W. Armstrong, J. K. Harmon, S. L. Scott, and D. G. Sime. 1980. Solar cycle changes in the polar solar wind. *Nature* 286:239–41.

Compton, W. D., and C. D. Benson. 1983. *Living and working in space: A history of Skylab.* NASA SP-4208. Washington D.C.: NASA.

Copernicus, N. 1543. *De revolutionibus orbium coelestium.* . . . Nuremberg: Petri.

———. 1543/1976. *On the revolutions of the heavenly spheres.* Trans. and ed. A. M. Duncan. New York: Barnes and Noble.

Corliss, W. R. 1967. *Scientific satellites.* NASA SP-133. Washington D.C.: NASA.

———. 1971. *NASA sounding rockets, 1958–1968.* NASA SP-4401. Washington D.C.: NASA.

Coutrez, R. 1953. Table of solar optical instruments. In Kuiper 1953, 728–33.

Covault, C. 1979. Solar Max Mission will increase sun data twofold. *Aviation Week and Space Technology* 111 (December 17): 52–56.

———. 1981a. Fuse failure curbs Solar Max vehicle. *Aviation Week and Space Technology* 114 (January 5): 14–15.

———. 1981b. Solar Maximum repair mission studied. *Aviation Week and Space Technology* 114 (April 27): 138–42.

———. 1984. Orbiter crew restores Solar Max. *Aviation Week and Space Technology* 120 (April 16): 18–20.

Cowling, T. G. 1935. The stability of gaseous stars, II. *MNRAS* 96:42–60.

———. 1971. The solar wind. *QJRAS* 12:447–52.

Cram, L. E. 1981. Sacramento Peak Observatory. *Solar Physics* 69:411–18.

Cram, L. E., and J. H. Thomas, eds. 1981. *The physics of sunspots, Sunspot, New Mexico, 14–17 July 1981.* Sunspot, N. Mex.: Sacramento Peak Observatory.

Crelinsten, J. M. 1982. The reception of Einstein's general theory of relativity among American astronomers, 1910–1930. Ph.D. diss., Université de Montréal.

Crowe, M. J. 1986. *The extraterrestrial life debate, 1750–1900: The idea of a plurality of worlds from Kant to Lowell.* Cambridge: Cambridge University Press.

Dauzère, J. C. 1931. Observatoire du Pic du Midi. In Ministère de l'instruction publique et des beaux-arts, *Enquêtes et documents relatifs a l'enseignment supérieur.* Vol. 125. *Rapports sur les observatoires astronomiques de province et les observatoires et instituts de physique du globe: Année 1929,* 77–84. Paris: Imprimerie Nationale.

Davis, L., Jr. 1955. Letter to J. A. Simpson (July 15). John A. Simpson Papers, Laboratory of Astrophysics and Space Research, University of Chicago, Chicago, Ill.

———. 1972. The interplanetary magnetic field. In Sonett, Coleman, and Wilcox 1972, 93–103.

Davis, R. E., and W. A. Ostaff. 1966. Memorandum for L. T. Hogarth (September 21). NHO.

D'Azambuja, L. 1934. Éruptions chromosphériques observées au spectrohélioscope. International Astronomical Union, *Bulletin for Character Figures of Solar Phenomena* 26:9.

———. 1939. La coopération internationale pour l'observation continue du soleil: Son développement et ses premiers résultats. In International Council of Scientific Unions, *Cinquième rapport de la commission pour l'étude des Relations entre les Phénomènes Solaires et Terrestres,* 45–50. Florence: Barbéra.

———. 1952. L'oeuvre de Bernard Lyot. *Astronomie* 66:265–77.

D'Azambuja, L., and H. W. Newton. 1938. Commission II (chromospheric phenomena). *Trans. IAU* 6:370–72.

De Jager, C. 1970. Introduction. *JOSO: AR 1970*, 3.

———. 1973a. Western European cooperation in space astronomy. In *Proceedings of the first European astronomical meeting, Athens, September 4–9, 1972*, vol. 1, *Solar activity and related interplanetary phenomena*, ed. J. Xanthakis, 156–64. Berlin: Springer.

———. 1973b. The formal organization of JOSO, in relation to general solar research in Europe. *JOSO: AR 1973*, 13–18.

———. 1978. K. O. Kiepenheuer, 1910–1975. In *Small scale motions on the sun: Proceedings of a colloquium held on the occasion of the change of the name of the former Fraunhofer-Institut into "Kiepenheuer-Institut für Sonnenphysik," Freiburg, 1–3 November 1978*, 1–11. Freiburg: Kiepenheuer-Institut.

———. 1988. Letter to K. Hufbauer (October 11). NHO.

De Jager, C., and P. Maltby. 1970. Report of activities of JOSO in 1970 and preceding years. *JOSO: AR 1970*, 32–36.

De Jager, C., and Z. Svestka. 1966. Letter to potential contributors (June). LG.

———. 1967. Editorial. *Solar Physics* 1:3–4.

De la Rue, W. 1859. Report on the present state of celestial photography in England. *Report of the British Association for the Advancement of Science* 29:130–53.

———. 1862. On the total solar eclipse of July 18th, 1860, observed at Rivabellosa, near Miranda de Ebro, in Spain. *Phil. Trans.* 152:333–416.

Dellinger, J. H. 1935. A new radio transmission phenomenon. *Physical Review* 48:705.

———. 1937. Sudden disturbances of the ionosphere. *Journal of Research of the National Bureau of Standards* 19:111–41.

Dellinger, J. H., and N. Smith, 1948. Developments in radio sky-wave propagation research and applications during the war. *Proceedings of the Institute of Radio Engineers* 36:258–66.

De Maria, M., M. Grilli, and F. Sebastiani, eds. 1989. *The restructuring of physical sciences in Europe and the United States, 1945–1960*. London: World Scientific.

Descartes, R. 1632/1967. *Le monde, ou Le traité de lumière*. Facsimile of the first posthumous edition. In *Oeuvres de Descartes*, vol. 11, ed. C. Adams and P. Tannery, 2d ed. Paris: Vrin.

———. 1644. *Principia philosophiae*. Amsterdam: Elzevir.

———. 1644/1982. *Principia philosophiae*. Facsimile. In *Oeuvres de Descartes*, vol. 8, pt. 1, ed. C. Adams and P. Tannery, 3d ed. Paris: Vrin.

Dessler, A. J. 1967. Solar wind and interplanetary magnetic field. *Reviews of Geophysics* 5:1–41.

Deubner, F. -L. 1975. Observations of low wavenumber nonradial eigenmodes of the sun. *Astronomy and Astrophysics* 44:371–75.

DeVorkin, D. H. 1975. Michelson and the problem of stellar diameters. *JHA* 6:1–18.

———. 1980. The maintenance of a scientific institution: Otto Struve, the Yerkes Observatory, and its Optical Bureau during the Second World War. *Minerva* 18:595–623.

———. 1981. Community and spectral classification in astrophysics: The acceptance of E. C. Pickering's system in 1910. *Isis* 72:29–49.

———. 1984. Stellar evolution and the origin of the Hertzsprung-Russell diagram. In Gingerich 1984, 90–108.

———. 1985. Electronics in astronomy: Early applications of the photoelectric cell and photomultiplier for studies of point-source celestial phenomena. *Proceedings of the IEEE* 73:1205–20.

———. 1986a. Richard Tousey and his beady-eyed V-2s. *Air and Space* 1 (June/July): 86–94.

———. 1987. Organizing for space research: The V-2 rocket panel. *Historical Studies in the Physical and Biological Sciences* 18:1–24.

———. 1989a. *Race to the stratosphere: Manned scientific ballooning in America.* New York: Springer.

———. 1989b. Along for the ride: The response of American astronomers to the possibility of space research, 1945–1950. In De Maria, Grilli, and Sebastiani 1989, 55–74.

———. 1989c. Henry Norris Russell. *Scientific American* 260, 5:126–33.

———. 1990a. Defending a dream: Charles Greeley Abbot's years at the Smithsonian. *JHA* 21:121–136.

———. 1990b. Back to the future: The response of astronomers to the prospect of government funding for research in the decade following World War II. In *Science and the federal patron: Post–World War II government support of American science,* ed. D. van Keuran and N. Reingold. Forthcoming.

———. 1990c. Science with a vengeance: The military origins of the space sciences in the V-2 era. Courtesy of David DeVorkin, National Air and Space Museum, Washington, D.C.

DeVorkin, D. H., and R. C. Kenat, Jr. 1983a. Quantum physics and the stars: 1. The establishment of a stellar temperature scale. *JHA* 14:102–32.

———. 1983b. Quantum physics and the stars: 2. Henry Norris Russell and the abundances of the elements in the atmospheres of the sun and the stars. *JHA* 14:180–222.

Dick, S. J. 1982. *Plurality of worlds: The origins of the extraterrestrial life debate from Democritus to Kant.* Cambridge: Cambridge University Press.

Dicks, D. R. 1960. *The geographical fragments of Hipparchus.* London: Athlone.

———. 1970. *Early Greek astronomy to Aristotle.* Ithaca, N.Y.: Cornell University Press.

Dickson, D. 1986. Europe assesses its options. *Science* 231:665.

———. 1987. Space: It is expensive in the major leagues. *Science* 237:1110–11.

Dieminger, W. 1948. Ionosphere. In *FIAT Review of German Science, 1939–1946: Geophysik,* ed. J. Bartels, 93–163. Wiesbaden: Klemm.

———. 1974. Early ionospheric research in Germany. *Journal of Atmospheric and Terrestrial Physics* 36:2085–93.

———. 1986. Letter to K. Hufbauer (March 19). NHO.

Dilke, F. W. W., and D. O. Gough. 1972. The solar spoon. *Nature* 240:262–64, 293–94.

Doel, R. 1990. Redefining a mission: The Smithsonian Astrophysical Observatory on the move. *JHA* 21:137–53.

Dollfus, A. 1983. Bernard Lyot, l'invention du coronographe et l'étude de la couronne: Un cinquantenaire. *Astronomie* 97:107–29, 315–29.

Donahue, W. H. 1972. *The dissolution of the celestial spheres, 1595–1650.* Facsimile of dissertation. New York: Arno, 1981.

Dornberger, W. R. 1958. *V 2.* Trans. J. Cleugh and G. Halliday. New York: Viking.

Drake, S. 1978. *Galileo at work: His scientific biography.* Chicago: University of Chicago Press.

Drummond, A. J. 1973. A review of determinations of the solar constant from measurements within and above the terrestrial atmosphere. In Drummond and Thekaekara 1973, 1–37.

Drummond, A. J., and J. R. Hickey. 1968. The Eppley-JPL solar constant measurement program. *Solar Energy* 12:217–32.

Drummond, A. J., J. R. Hickey, W. J. Scholes, and E. G. Laue. 1967. Multichannel radiometer measurement of solar irradiance. *Journal of Spacecraft and Rockets* 4:1200–1206.

———. 1968. New value for the solar constant of radiation. *Nature* 218:259–61.

Drummond, A. J., and M. P. Thekaekara, eds. 1973. *The extraterrestrial solar spectrum.* Mount Prospect, Ill: Institute of Environmental Sciences.

Duncan, C. H., R. G. Harrison, J. R. Hickey, J. M. Kendall Sr., M. P. Thekaekara, and R. C. Willson. 1977. Rocket calibration of the Nimbus 6 solar constant measurements. *Applied Optics* 16:2690–97.

Duncombe, R. L., W. Fricke, P. K. Seidelmann, and G. A. Wilkins. 1976. Joint report of the working groups of IAU Commission 4 on precession, planetary ephemerides, units and time-scales. *Trans. IAU* 16B:56–63.

Dunn, R. B. 1964. An evacuated tower telescope. *Applied Optics* 3:1353–57.

———. 1969. Sacramento Peak's new solar telescope. *Sky and Telescope* 38:368–75.

———. 1985. High resolution solar telescopes. *Solar Physics* 100:1–20.

Dupree, A. K., J. M. Beckers, L. W. Fredrick, J. W. Harvey, J. L. Linsky, L. E. Peterson, and A. B. C. Walker, Jr. 1976. Report of the ad hoc committee on interaction between solar physics and astrophysics (June 18). Courtesy of A. K. Dupree. NHO.

Durand, E., J. J. Oberly, and R. Tousey. 1949. Analysis of the first rocket ultraviolet solar spectra. *Astrophysical Journal* 109:1–16.

Dyson, F. W. 1912. Address . . . on presenting the gold medal of the society to Mr. A. R. Hinks. *MNRAS* 72:352–65.

Eddington, A. S. 1898. A total eclipse of the sun. Manuscript of a paper given at the Brynmelyn Literary Society (July 3). Library of Trinity College, Cambridge, England.

———. 1916. On the radiative equilibrium of the stars. *MNRAS* 77:16–35.

———. 1917. Further notes on the radiative equilibrium of the stars. *MNRAS* 77:596–612.

———. 1919. The sources of stellar energy. *Observatory* 42:371–76.

———. 1920. The internal constitution of the stars. *Nature* 106:14–20.

————. 1922. Applications of the theory of the stellar absorption-coefficient. *MNRAS* 83:98–109.

————. 1924. On the relation between the masses and luminosities of the stars. *MNRAS* 84:308–32.

————. 1926. *The internal constitution of the stars.* Cambridge: Cambridge University Press.

————. 1930a. Effect of boundary conditions on equilibrium of a star. *MNRAS* 90:279–84.

————. 1930b. *The rotation of the galaxy.* Oxford: Clarendon Press.

————. 1932. The hydrogen content of the stars. *MNRAS* 92:471–81.

————. 1935. *New pathways in science.* New York: Macmillan.

Eddy, J. A. 1975a. The case of the missing sunspots. *Bull. AAS* 7:365.

————. 1975b. The last 500 years of the sun. In Zirin 1975, 98–107.

————. 1976a. The Maunder minimum. *Science* 192:1189–1202.

————. 1976b. The sun since the Bronze Age. In *Physics of solar planetary environments: Proceedings of the International Symposium on Solar-Terrestrial Physics, June 7–18, 1976, Boulder, Colorado,* ed. D. J. Williams, 2:958–72. Washington D.C.: American Geophysical Society.

————. 1977a. The case of the missing sunspots. *Scientific American* 236, 5:80–88, 92.

————. 1977b. Those mysterious sunspots. *TV Guide* 25, 5:30–32.

————. 1978. A changing sun and climate. In *1979 Encyclopaedia Britannica yearbook of science and the future,* 145–59. Chicago: Encyclopaedia Britannica.

————. 1979. *A new sun: The solar results from Skylab.* NASA SP-402. Washington D.C.: NASA.

————. 1983. An historical review of solar variability, weather, and climate. In McCormac 1983, 1–15.

————. 1987. Telephone interview with K. Hufbauer (December 22). NHO.

————. 1990. Founding the Astrophysical Observatory: The Langley years. *JHA* 21:111–120.

Eddy, J. A., and P. V. Foukal. 1983. A summary of the joint discussion at Patras on solar luminosity variations. *Highlights of Astronomy* 6:79–94.

Eddy, J. A., R. L. Gilliland, and D. V. Hoyt. 1982. Changes in the solar constant and climatic effects. *Nature* 300:689–93.

Eddy, J. A., P. A. Gilman, and D. E. Trotter. 1976. Solar rotation during the Maunder minimum. *Solar Physics* 46:3–14.

————. 1977. Anomalous solar rotation in the early seventeenth century. *Science* 198:824–29.

Eddy, J. A., D. V. Hoyt, and O. R. White. 1982. Reconstructed values of the solar constant from 1874 to the present. In *A collection of extended abstracts presented at the Symposium on the Solar Constant and the Spectral Distribution of Solar Irradiance: IAMAP third scientific assembly, 17–28 August 1981, Hamburg, Federal Republic of Germany,* ed. J. London and C. Fröhlich, 29–34. Boulder, Colo.: National Center for Atmospheric Research.

Eddy, J. A., and R. M. MacQueen. 1986. Gordon Allen Newkirk, Jr. *Solar Physics* 104:257–58.

Eddy, J. A., F. R. Stephenson, and K. K. C. Yau. 1989. On pre-telescopic sunspot records. *QJRAS* 30:65–73.

Edge, D. O., and M. J. Mulkay. 1976. *Astronomy transformed: The emergence of radio astronomy in Britain.* New York: Wiley.

Edlén, B. 1941a. Identification des bandes brillantes dans le stade nébulaire des spectres de novae. In *Les novae et les naines blanches: Colloque international d'astrophysique tenu au Collège de France du 17 au 23 juillet 1939,* ed. A. J. Shaler, 55–68. Paris: Hermann.

———. 1941b. An attempt to identify the emission lines in spectrum of the solar corona. *Arkiv för matematik, astronomi och fysik* 28B, 1:1–4.

———. 1945. The identification of the coronal lines. *MNRAS* 105:323–33.

———. 1986. Letters to K. Hufbauer (March 3, May 23). NHO.

Encke, J. F. 1831. Ueber die nächste Wiederkehr des Cometen von Pons. *Astronomische Nachrichten* 8:317–47.

———. 1842. Wiederkehr des *Pons*schen Cometen 1842. *Astronomische Nachrichten* 19:185–91.

Engvold, O. 1985a. The large European solar telescope. In *High resolution in solar physics: Proceedings of a specialized session of the eighth IAU European regional astronomy meeting, Toulouse, September 17–21, 1984,* ed. R. Muller, 15–28. Berlin: Springer.

———. 1985b. Solar telescopes. *Trans IAU* 19A:43–44.

Eppley Laboratory, Inc. 1967. Eppley lecture series: Precision thermal radiometry. Course no. 3: Fundamental radiometry for experimental scientists. Brochure.

Evans, J. W. 1956. The Sacramento Peak Observatory. *Sky and Telescope* 15:436–41.

———. 1977. Sacramento Peak Observatory. *Bull AAS* 9:243–44.

———. 1986. Letter to K. Hufbauer (April 9). NHO.

Ezell, L. N. 1988a. *NASA Historical Data Book.* Vol. 2. *Programs and projects, 1958–1968.* NASA SP-4012. Washington D.C.: NASA.

———. 1988b. *NASA historical data book.* Vol. 3. *Programs and projects, 1969–1978.* NASA SP-4012. Washington D.C.: NASA.

Ezer, D., and A. G. W. Cameron. 1972. Effects of sudden mixing in the solar core on solar neutrinos and ice ages. *Nature: Physical Science* 240:180–82.

Faye, H. 1865. Sur la constitution physique du soleil. *Comptes Rendus* 60:89–96, 138–50.

Finney, J. W. 1962. Mariner 2 data disclose a constant "solar wind." *New York Times,* October 11, 1, 15.

Firor, J. W., Jr. 1965. Letter to L. Goldberg with "An annual solar astronomy meeting: A proposal to the American Astronomical Society" (July 16). LG.

———. 1968. Letter to G. C. McVittie (March 4). LG.

Fokker, A. D. 1973. CESRA. In *Proceedings of the first European astronomical meeting, Athens, September 4–9, 1972,* vol. 1, *Solar activity and related interplanetary phenomena,* ed. J. Xanthakis, 178–79. Berlin: Springer.

———. 1976. Cooperation in solar physics. *JOSO: AR 1976*, 88–90.

Fontenelle, B. de. 1686–1742/1966. *Entretiens sur la pluralité des mondes: Edition critique.* Ed. A. Calame. Paris: Didier.

———. 1688. *A discovery of new worlds.* Trans. A Behn. London: Canning.

Forbes, E. G. 1971. Carrington, Richard Christopher. In *Dictionary of Scientific Biography*, ed. C. C. Gillispie et al., 3:92–94. New York: Scribner's, 1970.

Forbes, J. D. 1836. Lumière du bord et du centre du soleil. *Comptes Rendus* 2:576.

Forman, P. 1987. Behind quantum electronics: National security as basis for physical research in the United States, 1940–1960. *Historical Studies in the Physical and Biological Sciences* 18:149–229.

Foukal, P. V. 1975. The contribution of active regions to solar variation in the visible and near infrared. In Zirin 1975, 109–10.

———. 1980a. Solar luminosity variation on short time scales: Observational evidence and basic mechanisms. In *The ancient sun: Fossil record in earth, moon, and meteorites, proceedings of the conference in Boulder, Colorado, October 16–19, 1979*, ed. R. O. Pepin, J. A. Eddy, and R. B. Merrill, 29–44. New York: Pergamon.

———. 1980b. Does the sun's luminosity vary? *Sky and Telescope* 59:111–14.

———. 1981. Sunspots and changes in the global output of the sun. In Cram and Thomas 1981, 391–423.

———. 1982. Variations in solar luminosity. In *Second Cambridge workshop on cool stars, stellar systems, and the sun*, ed. M. S. Giampapa and L. Golub, 17–29. Cambridge, Mass.: Smithsonian Astrophysical Observatory.

———. 1987a. Physical interpretation of variations in total solar irradiance. *Journal of Geophysical Research* 92D:801–7.

———. 1987b. Letter to K. Hufbauer (August 5). NHO.

———. 1988. Letters to K. Hufbauer (February 10, August 2). NHO.

Foukal, P. V., L. A. Fowler, and M. Livshits. 1983. A thermal model of sunspot influence on solar luminosity. *Astrophysical Journal* 267:863–71.

Foukal, P. V., and J. Lean. 1986. The influence of faculae on total solar irradiance and luminosity. *Astrophysical Journal* 302:826–35.

———. 1988. Magnetic modulation of solar luminosity by photospheric activity. *Astrophysical Journal* 328:347–57.

———. 1990. An empirical model of total solar irradiance variation between 1874 and 1988. *Science* 247:556–58.

Foukal, P. V., P. E. Mack, and J. E. Vernazza. 1977. The effect of sunspots and faculae on the solar constant. *Astrophysical Journal* 215:952–59.

Foukal, P. V., and J. E. Vernazza. 1979. The effect of magnetic fields on solar luminosity. *Astrophysical Journal* 234:707–15.

Fowler, W. A. 1972a. Comments. In *Proceedings, Solar Neutrino Conference, University of California, Irvine, and the Western White House, San Clemente, California, 25–26 February 1972*, ed. F. Reines and V. Trimble, A-10–13, B-24, C-7–8. Irvine: Physical Sciences Library, University of California.

———. 1972b. What cooks with solar neutrinos? *Nature* 238:24–26.

Fraunhofer, J. 1817. Bestimmung des Brechungs- und Farbenzerstreuungs-

Vermögens verschiedener Glasarten, in Bezug auf die Vervollkommnung achromatischer Fernröhre. *Denkschriften der königlichen Akademie der Wissenschaften zu München: Classe der Mathematik und Naturwissenschaften* 5:193–226.

Friedman, H. 1958. Astrophysical measurements from rockets. *Trans. IAU* 707–8.

———. 1960. Recent experiments from rockets and satellites. *Astronomical Journal* 65:264–71.

———. 1981. Rocket astronomy—an overview. In *Space science comes of age: Perspectives in the history of the space sciences*, ed. P. A. Hanle and V. Del Chamberlain, 31–44. Washington D.C.: Smithsonian Institution.

Fröhlich, C., and R. W. Brusa. 1975. Measurement of the solar constant: A critical review. In Zirin 1975, 111–26.

Fröhlich, C., J. Geist, J. M. Kendall, Sr., and R. M. Marchgraber. 1973. The third international comparisons of pyrheliometers and a comparison of radiometric scales. *Solar Energy* 14:157–66.

Fröhlich, C., and J. Pap. 1989. Multi-spectral analysis of total solar irradiance variations. *Astronomy and Astrophysics* 220:272–80.

Frutkin, A. W. 1965. *International cooperation in space*. Englewood Cliffs, N.J.: Prentice-Hall.

Galilei, G. 1610/1989. Sidereus nuncius, or The sidereal messenger. Trans. and ed. A. van Helden. Chicago: University of Chicago Press.

———. 1613. *Istoria e dimostrazioni intorno alle macchie solari e loro accidenti comprese in lettere scritte all'illvstrissimo signor Marco Velsersi*. Rome: Mascardi.

———. 1613/1957. History and demonstrations concerning sunspots and their phenomena. In *Discoveries and opinions of Galileo*, trans. and ed. S. Drake, 87–144. Garden City, N.Y.: Doubleday.

Gamow, G. 1940. *The birth and death of the sun: Stellar evolution and subatomic energy*. New York: Viking.

Gascoigne, S. C. B. 1984. History of Australian astronomy: Astrophysics at Mount Stromlo—the Woolley era. *Proceedings of the Astronomical Society of Australia* 5:597–605.

Gassendi, P. 1654. *Tychonis Brahei, Equitis Dani, astronomorum coryphaei vita . . . accessit Nicolai Copernici, Georgii Peurbachii, et Iohannis Regiomontani astronomorum celebrium vita*. Paris: Dupuis.

Gassiot, J. P. 1854. Report of the Kew Committee. *Report . . . of the British Association for the Advancement of Science* 24:xxvii–xxxvi.

———. 1858. Report of the Kew Committee. *Report . . . of the British Association for the Advancement of Science* 28:xxxiii–xxxvi.

Gernsheim, H. 1982. *The origins of photography*. 3d ed. New York: Thames and Hudson.

Gingerich, O., ed. 1984. *Astrophysics and twentieth-century astronomy to 1950*, Vol. 4 of *The general history of astronomy*, ed. M. A. Hoskin. Cambridge: Cambridge University Press.

Glaser, H. 1969. Orbiting solar observatories (OSO) workshop (August 28). OSO File, NHO.

————. 1975. Log (September 29, December 11, 18). NHO.

————. 1976a. Letter to N. W. Hinners with draft of the announcement of opportunity for the "Total and spectral irradiance experiment for flight on the Solar Maximum Mission" (January 14). NHO.

————. 1976b. Log (July 6). NHO.

————. 1976c. Letter to N. W. Hinners (November 26). NHO.

————. 1977. Letter to N. W. Hinners (April 29). NHO.

————. 1988. Telephone interview with K. Hufbauer (January 28). NHO.

Glasstone, S. 1965. *Sourcebook on the space sciences.* Princeton, N.J.: Van Nostrand.

Goldberg, L. 1945. Letter to D. H. Menzel (September 28). Quoted in DeVorkin 1989b, 59.

————. 1948. Letter to L. Spitzer (October 5). Quoted in DeVorkin 1990c.

————. 1965a. Letter to J. W. Firor, Jr. (May 11). LG.

————. 1965b. Letter to J. W. Firor, Jr. (September 30). LG.

————. 1966. Letter to N. U. Mayall (November 22). LG.

————. 1967. Letter to A. K. Pierce (April 17). LG.

————. 1968. Letter to H. W. Babcock (July 22). LG.

————. 1977. Donald Howard Menzel. *Sky and Telescope* 53:244–51.

————. 1981. Solar physics. In *Space science comes of age: Perspectives in the history of the space sciences,* ed. A. Hanle and V. D. Chamberlain, 15–30. Washington D.C.: Smithsonian Institution.

————. 1986a. Letter to K. Hufbauer (June 4). NHO.

————. 1986b. Chronology: Solar satellite projects. Manuscript, NHO.

Grant, E. 1990. Celestial incorruptibility in medieval cosmology, 1200–1687. *Boston Studies in the Philosophy of Science.* In press.

Grec, G., E. Fossat, and M. A. Pomerantz. 1983. Full-disk observations of solar oscillations from the geographic south pole. *Solar Physics* 82:55–66.

Greenberger, M. 1983. *Caught unawares: The energy decade in retrospect.* Cambridge, Mass.: Ballinger.

Grigg, W. 1961. Explorer X sustains solar wind theory. *Washington Star,* April 19, 9.

Gringauz, K. I. 1961. Some results of experiments in interplanetary space by means of charged particle traps on Soviet space probes. In *Space research II: Proceedings of the Second International Space Science Symposium, Florence, April 10–14, 1961,* ed. H. C. van de Hulst et al., 539–53. Amsterdam: North-Holland.

————. 1987. Letter to K. Hufbauer (September 17). NHO.

Gringauz, K. I., V. V. Bezrukikh, V. D. Ozerov, and R. E. Rybchinskii. 1960. A study of the interplanetary ionized gas, high-energy electrons, and corpuscular radiation from the sun by means of the three-electrode trap for charged particles on the second Soviet cosmic rocket. *Soviet Physics: Doklady* 5:361–64.

Gringauz, K. I., and M. K. Zelikman. 1957. Measurement of the positive ion concentration along the orbit of artificial earth satellites. *Advances in Physical Sciences* 63:321–39.

Grosser, M. 1962. *The discovery of Neptune.* Cambridge: Harvard University Press.

Grotrian, W. 1933. Die Beobachtung der Sonnenkorona ausserhalb totaler Sonnenfinsternisse. *Sterne* 13:73–80.

————. 1934. Über die physikalische Natur der Sonnenkorona. *Sterne* 14:145–57.

————. 1937. Letter to B. Edlén (February 13). Bengt Edlén Papers, Lund, Sweden.

————. 1939a. Zur Frage der Deutung der Linien in Spektrum der Sonnenkorona. *Naturwissenschaften* 27:214.

————. 1939b. Sonne und Ionosphäre. *Naturwissenschaften* 27:555–63, 569–77.

Guendel, H. H. 1968. Solar and cosmic electromagnetic and charged-particle radiations. In Wukelic 1968, 215–302.

Hale, G. E. 1891. Letter to H. M. Goodman (May 15). Quoted in Wright 1966, 77.

————. 1893. Letter to H. M. Goodman (March 5). Quoted in Wright 1966, 102.

————. 1904. Co-operation in solar research. *Astrophysical Journal* 20:306–12.

————. 1908a. The tower telescope of the Mount Wilson Solar Observatory. *Astrophysical Journal* 27:204–12.

————. 1908b. Solar vortices. *Astrophysical Journal* 28:100–116.

————. 1908c. On the probable existence of a magnetic field in sun-spots. *Astrophysical Journal* 28:315–43.

————. 1915. The direction of rotation of sun-spot vortices. *Proceedings of the National Academy of Sciences* 1:382–84.

————. 1922. Commission de l'atmosphère solaire. *Trans. IAU* 1:31–34.

————. 1924a. Sun-spots as magnets and the periodic reversal of their polarity. *Nature* 113:105–12.

————. 1924b. The law of sun-spot polarity. *Proceedings of the National Academy of Sciences* 10:53–55.

————. 1929. The spectrohelioscope and its work: Part 1. History, instruments, adjustments, and methods of observation. *Astrophysical Journal* 70:265–311.

————. 1931a. The spectrohelioscope and its work: Part 3. Solar eruptions and their apparent terrestrial effects. *Astrophysical Journal* 73:379–412.

————. 1931b. Cooperative research with the spectrohelioscope. In International Research Council, *Third report of the commission appointed to further the study of solar and terrestrial relationships,* 78–79. London: Lund, Humphries.

Hale, G. E., F. Ellerman, S. B. Nicholson, and A. H. Joy. 1919. The magnetic polarity of sun-spots. *Astrophysical Journal* 49:153–78.

Hale, G. E., and S. B. Nicholson. 1938. *Magnetic observations of sunspots, 1917–1924.* Publication 498. Washington D.C.: Carnegie Institution.

Hall, R. C. 1977. *Lunar impact: A history of Project Ranger.* NASA SP-4210. Washington D.C.: NASA.

Hallgren, E. L. 1974. *The University Corporation for Atmospheric Research and the National Center for Atmospheric Research, 1960–1970: An institutional history.* Boulder, Colo.: [National Center for Atmospheric Research].

Hamel, J. 1983. Karl Friedrich Zöllners Tätigkeit als Hochschullehrer an der Universität Leipzig: Ein Beitrag zur Geschichte der Institutionaliesierung der Astrophysik. *NTM—Schriftenreihe für Geschichte der Naturwissenschaft, Technik, und Medizin* 20:35–43.

Hamon, A. 1931. Communications verbales. *Astronomie* 45:72–74.

Hansen, P. A. 1863. Calculation of the sun's parallax from the lunar theory. *MNRAS* 24:8–12.

Harvey, J. W., M. A. Pomerantz, and T. L. Duvall, Jr. 1982. Astronomy on ice. *Sky and Telescope* 64:520–23.

Hassenstein, W. 1941. Das Astrophysikalische Observatorium Potsdam in den Jahren 1875–1935. *Mitteilungen des Astrophysikalischen Observatoriums Potsdam* 1:1–56.

Heines, N. J. 1944. Sun-spot observations needed. *Popular Astronomy* 52:519.

Helmholtz, H. 1854/1962. On the interaction of natural forces. Trans. J. Tyndall. In Helmholtz's *Popular scientific lectures*, ed. M. Kline, 58–92. New York: Dover.

Hendry, J. 1983. Mayer, Herschel and Prevost on the solar motion. *Annals of Science* 39:61–75.

Hermann, D. B. 1973. Zur Frühentwicklung der Astrophysik in Deutschland und in den USA. *NTM—Schriftenreihe für Geschichte der Naturwissenschaft, Technik, und Medizin* 10:38–44.

———. 1975. Zur Vorgeschichte des Astrophysikalischen Observatorium Potsdam (1865 bis 1874). *Astronomische Nachrichten* 296:245–59.

———. 1976. Vogel, Hermann Carl. In *Dictionary of scientific biography*, ed. C. C. Gillispie et al., 14:54–57. New York: Scribner's, 1970.

———. 1982. *Karl Friedrich Zöllner*. Leipzig: Teubner.

———. 1985. *The history of astronomy from Herschel to Hertzsprung*. 3d ed. Trans. K. Krisciunas. Cambridge: Cambridge University Press.

Herschel, J. F. W. 1833. *Treatise on astronomy*. London: Longman.

Herschel, W. 1780. Short account of some experiments upon light that have been made by Zanottus, with a few remarks upon them. In Herschel 1912, xcv–xcvii.

———. 1781. On the periodical star in Collo Ceti. In Herschel 1912, civ–cv.

———. 1782. Letter to N. Maskelyne (April 28). Quoted in Lubbock 1933, 110–11.

———. 1783. On the proper motion of the sun and solar system; with an account of several changes that have happened among the fixed stars since the time of Mr. Flamsteed. *Phil. Trans.* 73:247–83.

———. 1785. Some account of the life and writings of William Herschel, Esq. *European Magazine and London Review, containing the Literature, History, Politics, Arts, Manners and Amusements of the Age* 7:1–3.

———. 1791. On nebulous stars, properly so called. *Phil. Trans.* 81:71–88.

———. 1795. On the nature and construction of the sun and fixed stars. *Phil. Trans.* 85:46–72.

———. 1796. On the method of observing the changes that happen to the fixed stars; with some remarks on the stability of the light of our sun; to which is added, a catalogue of comparative brightness, for ascertaining the permanency of lustre of stars. *Phil. Trans.* 86:166–226.

———. 1800a. Investigation of the powers of the prismatic colours to heat and illuminate objects, with remarks, that prove the different refrangibility of radiant heat; to which is added, an inquiry into the method of viewing the sun advantageously, with telescopes of large apertures and high magnifying powers. *Phil. Trans.* 90:255–83.

———. 1800b. Experiments on the refrangibility of the invisible rays of the sun. *Phil. Trans.* 90:284–92.

———. 1801. Observations tending to investigate the nature of the sun, in order to find the causes or symptoms of its variable emission of light and heat; with remarks on the use that may possibly be drawn from solar observations. *Phil. Trans.* 91:265–318.

———. 1803. Observations of the transit of Mercury over the disk of the sun; to which is added an investigation of the causes which often prevent the proper action of mirrors. *Phil. Trans.* 93:214–32.

———. 1805. On the direction and velocity of the motion of the sun, and solar system. *Phil. Trans.* 95:233–56.

———. 1806. On the quantity and velocity of the solar motion. *Phil. Trans.* 96:205–37.

———. 1811. Astronomical observations relating to the construction of the heavens, arranged for the purpose of a critical examination, the result of which appears to throw some new light upon the organization of the celestial bodies. *Phil. Trans.* 101:269–336.

———. 1912. *The scientific papers of Sir William Herschel.* Ed. J. L. E. Dreyer. 2 vols. London: Royal Society and Royal Astronomical Society.

Hetherington, N. S. 1975a. Adriaan van Maanen's measurements of solar spectra for a general magnetic field. *QJRAS* 16:235–44.

———. 1975b. Winning the initiative: NASA and the U.S. space science program. *NASA Prologue,* Summer, 99–108.

Hevly, B. W. 1987. Basic research within a military context: The Naval Research Laboratory and the foundations of extreme ultraviolet and X-ray astronomy, 1923–1960. Ph.D. diss., Johns Hopkins University.

Hey, J. S. 1973. *The evolution of radio astronomy.* New York: Science History Publications.

Hibbs, A. R. 1958. Inter-office memo to distribution regarding preliminary experimental design—NASA missions (December 5). RLN.

Hickey, J. R. 1973. A satellite experiment to establish the principal extraterrestrial solar energetic fluxes and their variance. In Drummond and Thekaekara 1973, 135–60.

———. 1975. Solar radiation measuring instruments, terrestrial and extra-terrestrial. *Proceedings of the Society of Photo-optical Instrumentation Engineers* 68 (August): 53–61.

———. 1988. Letter to K. Hufbauer (February 11). NHO.

Hickey, J. R., B. M. Alton, H. L. Kyle, and D. Hoyt. 1988. Total solar irradiance measurements by ERB/Nimbus-7: A review of nine years. *Space Science Reviews* 48:321–42.

Hickey, J. R., R. G. Frieden, F. J. Griffin, S. A. Cone, R. H. Maschhoff, and J. Gnaidy. 1977. The self-calibrating sensor of the eclectic satellite pyrheliometer (ESP) program. In *Proceedings of the 1977 annual meeting: American Section of the International Solar Energy Society, June 6–19, 1977, Orlando, Florida,* ed. C. Beach and E. Fordyce, 1:15-1–4. Cape Canaveral, Fla.: American Section of the International Solar Energy Society.

Hickey, J. R., F. J. Griffin, D. T. Hilleary, and H. B. Howell. 1976. Extraterrestrial solar irradiance measurements from the Nimbus 6 satellite. In *Sharing the sun: Solar technology in the seventies—Joint conference, American Section, International Solar Energy Society and Solar Energy Society of Canada, Inc., August 15–20, 1976, Winnipeg,* ed. K. W. Boeer, 329–37. Cape Canaveral Fla.: American Section of the International Solar Energy Society.

Hickey, J. R., F. J. Griffin, H. Jacobowitz, L. L. Stowe, P. Pellegrino, and R. H. Maschhoff. 1980. Comments on solar constant measurements from Nimbus 6 and 7. *Eos* 61:355.

Hickey, J. R., D. T. Hilleary, and R. H. Maschhoff. 1974. Solar radiation measurements capability of the Earth Radiation Budget Experiment of the Nimbus F satellite. In Hickey and Klein 1974, 127–50.

Hickey, J. R., and A. R. Karoli. 1974. Radiometric calibrations for the earth radiation budget experiment. *Applied Optics* 13:523–33.

Hickey, J. R., and W. H. Klein, eds. 1974. *Proceedings of Symposium on Solar Radiation Measurements and Instrumentation, November 13–15, 1973.* Washington D.C.: Radiation Biology Laboratory, Smithsonian Institution.

Hickey, J. R., L. L. Stowe, H. Jacobowitz, P. Pellegrino, R. H. Maschhoff, F. House, and T. H. Vonder Haar. 1980. Initial solar irradiance determinations from Nimbus 7 cavity radiometer measurements. *Science* 208:281–83.

Hinners, N. W. 1976. Letter to J. F. Clark (January 15). NHO.

———. 1977. Letter to R. C. Willson (February 14). NHO.

Hirsh, R. F. 1979. The riddle of the gaseous nebulae. *Isis* 70:197–212.

———. 1983. *Glimpsing an invisible universe: The emergence of X-ray astronomy.* Cambridge: Cambridge University Press.

Hoeksema, J. T. 1984. Structure and evolution of the large scale solar and heliospheric magnetic fields. Ph.D. diss., Stanford University.

Hoeksema, J. T., J. M. Wilcox, and P. H. Scherrer. 1982. Structure of the heliospheric current sheet in the early portion of sunspot cycle 21. *Journal of Geophysical Research* 87A:10331–38.

Holzer, T. E. 1989. Interaction between the solar wind and the interstellar medium. *ARAA* 27:199–234.

Hoskin, M. A. 1963. *William Herschel and the construction of the heavens.* New York: Norton.

———. 1980. Herschel's determination of the solar apex. *JHA* 11:153–63.

House, L. L. 1976. Division of Solar Physics: 1975. *Bull. AAS* 8:401.

Howard, R. A., and M. J. Koomen. 1974. Observation of sectored structure in the outer solar corona: Correlation with interplanetary magnetic field. *Solar Physics* 37:469–75.

Howard, R. F. 1974. Studies of solar magnetic fields. *Solar Physics* 38:283–99.

———. 1976. The Mount Wilson solar magnetograph: Scanning and data system. *Solar Physics* 48:411–16.

———. 1985. Eight decades of solar research at Mount Wilson. *Solar Physics* 100:171–87.

Hoyt, D. V. 1979. The Smithsonian Astrophysical Observatory solar constant program. *Reviews of Geophysics and Space Physics* 17:427–58.

Hoyt, D. V., and J. A. Eddy. 1983. Solar irradiance modulation by active regions from 1969 through 1981. *Geophysical Research Letters* 10:509–12.

Hudson, H. S. 1977. Solar analysis of the Solar-Constant Monitoring Package. Proposal for the SMM Guest Investigator Program. NHO.

———. 1983. Variations of the solar radiation input. In McCormac 1983, 31–41.

———. 1988a. Observed variability of the solar luminosity. *ARAA* 26:473–507.

———. 1988b. Letter to K. Hufbauer (February 10). NHO.

Hudson, H. S., T. M. Brown, J. Christensen-Dalsgaard, A. N. Cox, P. Demarque, J. W. Harvey, J. T. McGraw, and R. W. Noyes. 1986. A concept study for an asteroseismology explorer: proposal submitted to NASA (July 31). Center for Astrophysics and Space Sciences, University of California at San Diego.

Hudson, H. S., S. Silva, M. F. Woodard, and R. C. Willson. 1982. The effects of sunspots on solar irradiance. *Solar Physics* 76:211–19.

Hudson, H. S., and R. C. Willson. 1981. Sunspots and solar variability. In Cram and Thomas 1981, 434–45.

Hufbauer, K. 1981. Astronomers take up the stellar-energy problem, 1917–1920. *Historical Studies in the Physical Sciences* 11:277–303.

———. 1982. *The formation of the German chemical community, 1720–1795.* Berkeley: University of California Press.

———. 1985. Dopo la vittoria: Le valutazioni degli addetti ai lavori sulla fisica teorica (1928–1939). In *La ristrutturazione delle scienze tra le due Guerre Mondiali,* ed. G. Battimelli, M. De Maria, and A. Rossi, 2:81–85. Rome: Goliardica.

———. 1986. Amateurs and the rise of astrophysics, 1840–1910. *Berichte zur Wissenschaftsgeschichte* 9:183–90.

———. 1989. Solar physics' evolution into a subdiscipline (1945–1975). In *New trends in the history of science: Proceedings of a conference held at the University of Utrecht,* ed. R. P. W. Visser, H. J. M. Bos, L. C. Palm. and H. A. M. Snelders, 73–91. Amsterdam: Rodopi.

———. 1990. Spacecraft with instruments that successfully observed solar electromagnetic radiations, 1960–1985. Manuscript, NHO.

Huggins, W. 1866a. On the physical and chemical constitution of the fixed stars and nebulae. *Proceedings of the Royal Institution* 4:441–49.

———. 1866b. *On the results of spectrum analysis applied to the heavenly bodies.* London: Ladd.

Humboldt, A. von. 1851. *Kosmos: Entwurf einer physischen Weltbeschreibung.* Vol. 3, no. 2. Stuttgart: Cotta.

Hundhausen, A. J. 1972. *Coronal expansion and solar wind.* Berlin: Springer.

———. 1977. An interplanetary view of coronal holes. In Zirker 1977, 223–329.

International Research Council. 1920. *Constitutive assembly held at Brussels, July 18th to July 28th, 1919: Reports of proceedings.* Ed. A. Schuster. London: Harrison.

———. 1925. *Third assembly held at Brussels, July 7th to July 9th, 1925: Reports of proceedings.* Ed. A. Schuster. London: Harrison.

Jacob, M. C. 1988. *The cultural meaning of the scientific revolution.* New York: Knopf.

Jacobsen, S., and E. N. Parker. 1973. Eugene Parker on the solar wind, magnetic fields, and earth weather. *Bulletin of the Atomic Scientists* 21, 5:25–30.

Jaeggli, A. 1968. *Die Berufung des Astronomen Joh. Rudolf Wolf nach Zürich.* Zurich: Eidgenössische Technische Hochschule.

James, F. A. J. L. 1981. The early development of spectroscopy and astrophysics. Ph.D. diss., University of London.

———. 1982. Thermodynamics and sources of solar heat, 1846–1862. *British Journal for the History of Science* 15:155–81.

———. 1983. The establishment of spectro-chemical analysis as a practical method of qualitative analysis. *Ambix* 30:30–53.

———. 1985. The discovery of line spectra. *Ambix* 32:53–70.

Janssen, P. J. 1896. Création de l'observatoire. *Annales de l'Observatoire d'Astronomie Physique de Paris* 1:49–55.

Jeans, J. H. 1917. The radiation of the stars. *Nature* 99:365.

Jet Propulsion Laboratory. 1959. Scientific missions for space exploration: Preliminary information generated by the Mission Group of the NASA Study Committee (January 5). RLN.

———. 1963a. *The Mariner R project: Progress report, September 1, 1961–August 31, 1962.* JPL Technical Report 32-353. Pasadena: Jet Propulsion Laboratory.

———. 1963b. *Mariner mission to Venus.* New York: McGraw-Hill.

———. 1965. *Mariner-Venus 1962: Final project report.* NASA SP-59. Washington D.C.: NASA.

Johnson, F. S., J. D. Purcell, R. Tousey, and N. Wilson. 1954. The ultraviolet spectrum of the sun. In *Rocket exploration of the upper atmosphere*, ed. R. L. F. Boyd and M. J. Seaton, 279–88. London: Pergamon.

Johnson, M. J. 1857. Address . . . on presenting the medal of the society to M. Schwabe. *MNRAS* 17:126–32.

Jones, B. Z. 1965. *Lighthouse of the skies: The Smithsonian Astrophysical Observatory—background and history, 1846–1955.* Washington D.C.: Smithsonian Institution.

Jordan, D. S. 1984. Science with the solar optical telescope (SOT). In *The hydromagnetics of the sun: Proceedings of the fourth European meeting on solar physics, Noordwijkerhout, the Netherlands, 1–3 October 1984*, 165–75. Paris: European Space Agency.

Josias, C. S., M. M. Neugebauer, and C. W. Snyder. 1959. Recommendations on solar plasma experiments for early Vega flights (November). CWG.

Jungnickel, C., and R. McCormmach. 1986. *Intellectual mastery of nature: Theoretical physics form Ohm to Einstein.* vol. 1. *The torch of mathematics, 1800–1870.* Chicago: University of Chicago Press.

Kargon, R. H. 1986. Henry Rowland and the physics discipline in America. *Vistas in Astronomy* 29:131–36.

Karth, J. E. 1965. NASA authorization for fiscal year 1966. *Congressional Record: Appendix*, June 17, A3178.

Katz, J. E. 1984. *Congress and national energy policy.* New Brunswick, N.J.: Transaction.

Kawaler, S., and J. Veverka. 1981. The habitable sun: One of William Herschel's stranger ideas. *Journal of the Royal Astronomical Society of Canada* 75:46–55.

Kenat, R. C., Jr. 1987. Physical interpretation: Eddington, idealization and the origin of stellar structure theory. Ph.D. diss., University of Maryland.

Kendall, J. M., Sr., and C. M. Berdahl. 1970. Two blackbody radiometers of high accuracy. *Applied Optics* 9:1082–91.

Kerr, R. A. 1987. Monitoring earth and sun by satellite. *Science* 236:1624–25.

———. 1988. Relax, the sun is brightening again. *Science* 240:1734.

———. 1989. NASA racing the sun to save a satellite. *Science* 244:1443.

Kerwin, M. D. 1982. Letter to P. Fetters (May 19). SMM File, NHO.

Kevles, D. J. 1971. "Into hostile political camps": The reorganization of international science in World War I. *Isis* 62:47–60.

———. 1979. *The physicists*. New York: Vintage.

Kidwell, P. A. 1979. Solar radiation and heat from Kepler to Helmholtz (1600–1860). Ph.D. diss., Yale University.

———. 1981. Prelude to solar energy: Pouillet, Herschel, Forbes and the solar constant. *Annals of Science* 38:457–76.

———. 1987. Cecilia Payne-Gaposchkin: Astronomy in the family. In *Uneasy careers and intimate lives: Women in science, 1789–1979*, ed. P. G. Abir-Am and D. Outram, 216–38, 336–42. New Brunswick, N.J.: Rutgers University Press.

———. 1990. Harvard astronomers and World War II: Disruption and opportunity. In *Science at Harvard University: Historical perspectives*, ed. C. A. Elliott and M. W. Rossiter. Bethlehem, Pa.: Lehigh University Press. Forthcoming.

Kienle, H. 1955. Walter Grotrian. *Mitteilungen der Astronomischen Gesellschaft*, 5–9.

Kienle, H., and H. Siedentopf. 1929. Beobachtung der Sonnenkorona ausserhalb der Finsternis. *Astronomische Nachrichten* 235:9–10.

Kiepenheuer, K. O. 1945. Die Entwicklung des Fraunhofer-Instituts und seine wissenschaftlichen Arbeiten 1939–1945. Manuscript (June 12), Kiepenheuer-Institut, Freiburg im Breisgau, Federal Republic of Germany. Copy on deposit in NHO.

———. 1948. Solar-Terrestrische Erscheinungen. In *FIAT review of German science, 1939–1946: Astronomy, astrophysics and cosmogony*, ed. P. ten Bruggencate, 230–84. Wiesbaden: Klemm.

———. 1953. Solar activity. In Kuiper 1953, 322–465.

———. 1966. Freiburg: Fraunhofer Institut. *Mitteilungen der Astronomischen Gesellschaft*, 31–36.

———. 1970. The aims and prospect of the Joint Organization for Solar Observations. *JOSO: AR 1970*, 8–20.

———. 1974. European site survey for a solar observatory. *Sky and Telescope* 48:84–87.

———. 1975. The JOSO Observatory coming into sight. *JOSO: AR 1974*, 4–9.

Killian, J. R., Jr. 1977. *Sputnik, scientists, and Eisenhower: A memoir of the first special assistant to the president for science and technology*. Cambridge: MIT Press.

Kirchhoff, G. R. 1859–62/1972. *Untersuchungen über das Sonnenspectrum und die Spektren der chemischen Elemente und weitere ergänzende Arbeiten aus den Jahren 1859–1862*. Ed. H. Kangro. Osnabrück: Zeller.

————. 1860. Letter to O. Kirchhoff (May 11). Quoted in Kirchhoff 1859–62/1972, 8.

Kloeppel, J. E. 1983. *Realm of the long eyes: A brief history of Kitt Peak National Observatory.* San Diego: Univelt.

Koppes, C. R. 1982. *JPL and the American space program: A history of the Jet Propulsion Laboratory.* New Haven: Yale University Press.

Koyré, A. 1957. *From the closed world to the infinite universe.* Baltimore: Johns Hopkins Press.

Krafft, F. 1981. Astrophysik contra Astronomie: Das Zurückdrängen einer alten Disziplin durch die Begründung einer neuen. *Berichte zur Wissenschaftsgeschichte* 4:89–110.

————. 1986. Innovationsschube durch Aussenseiter: Das Beispiel des Amateur-Astronomen William Herschel. *Berichte zur Wissenschaftsgeschichte* 9:201–25.

Kreplin, R. W., T. A. Chubb, and H. Friedman. 1962. X-ray and Lyman-alpha emission from the sun as measured from the NRL SR-1 satellite. *Journal of Geophysical Research* 67:2231–53.

Krieger, A. S., A. F. Timothy, and E. C. Roelof. 1973. A coronal hole and its identification as the source of a high velocity solar wind stream. *Solar Physics* 29:505–25.

Krieger, A. S., A. F. Timothy, G. S. Vaiana, A. J. Lazarus, and J. D. Sullivan. 1974. X-ray observations of coronal holes and their relation to high velocity solar wind streams. In *Solar wind three*, ed. C. T. Russell, 132–39. Los Angeles: Institute of Geophysics and Planetary Physics, UCLA.

Kröber, R., ed. 1984. Geschichte der Abt. F: Vorgeschichte und Übersicht Ionosphärenforschung Funkberatung Sonnenforschung. *Rechliner Briefe*, no. 58. Copy on deposit in NHO.

Kuhn, T.S. 1962. *The structure of scientific revolutions.* Chicago: University of Chicago Press.

Kuiper, G. P. 1946. German astronomy during the war. *Popular Astronomy* 54:263–87.

————, ed. 1953. *The sun.* Chicago: University of Chicago Press.

Kundu, M. R. 1982. Probing the radio sun. *Sky and Telescope* 64:6–10.

Kundu, M. R., E. J. Schmahl, and A. P. Rao. 1981. VLA observation of solar regions at 6 cm wavelength. *Astronomy and Astrophysics* 94:72–79.

Kyle, H. L. 1987. A brief history of the Nimbus Earth Radiation Budget (ERB) Experiment. NASA/Goddard Space Flight Center: *Nimbus-7 ERB/Cloud Newsletter* 2, 1:12–16.

LaBonte, B. J., G. A. Chapman, H. S. Hudson, and R. C. Willson, eds. 1984. *Solar irradiance variations on active region time scales: Proceedings of a workshop held at the California Institute of Technology, Pasadena, California, June 20–21, 1983.* NASA Conference Publication 2310. Washington D.C.: NASA.

Lalande, J. -J. L. de. 1762. Mémoire sur les passages de Vénus devant le disque du soleil, en 1761 et 1769. *Histoire de l'Académie Royale des Sciences. Année M.DCCLVII, avec les mémoires mathématique et de physique, pour la même année,* 232–50.

————. 1774. *Abrégé d'astronomie.* Paris: Desaint.

Langley, S. P. 1883. The selective absorption of solar energy. *American Journal of Science*, ser. 3, 25:169–96.

Lankford, J. 1981. Amateurs and astrophysics: A neglected aspect in the development of a scientific specialty. *Social Studies of Science* 11:275–303.

———. 1984. The impact of photography on astronomy. In Gingerich 1984, 16–39.

Laplace, P. S. 1796. *Exposition du système du monde*. Vol. 2. Paris: Cercle-Social.

———. 1813. *Exposition du système du monde*. Paris: Courcier.

Lasby, C. G. 1971. *Project paperclip: German scientists and the cold war*. New York: Atheneum.

Laudan, L. 1977. *Progress and its problems: Toward a theory of scientific growth*. Berkeley: University of California Press.

Laue, E. G., and A. J. Drummond, 1968. Solar constant: First direct measurements. *Science* 161:888–91.

Lean, J., and P. Foukal. 1988. A model of solar luminosity modulation by magnetic activity between 1954 and 1984. *Science* 240:906–8.

Levy, E. H. 1975. Origin of the twenty year wave in the diurnal variation, 1215–1220. In Conference papers: *Fourteenth International Cosmic Ray Conference, . . . August 15–29, 1975*, vol. 4. Munich: Max-Planck-Institut für Extraterrestrische Physik.

Libbrecht, K. G. 1988. Solar and stellar seismology. *Space Science Reviews* 47:275–301.

Liebowitz, R. P. 1985. Chronology: From the Cambridge Field Station to the Air Force Geophysics Laboratory. 1945–1985. Special Report 252. Cambridge, Mass.: Air Force Geophysics Laboratory.

Livingston, W. C., J. W. Harvey, A. K. Pierce, D. Schrage, B. Gillespie, J. Simmons, and C. Slaughter. 1976. Kitt Peak 60-cm vacuum telescope. *Applied Optics* 15:33–39.

Livingston, W. C., L. Wallace, and O. R. White. 1988. Spectrum line intensity as a surrogate for solar irradiance variations. *Science* 240:1765–67.

Lockyer, J. N. 1874. *Contributions to solar physics*. London: Macmillan.

Logsdon, J. M. 1986. The space shuttle program: A policy failure? *Science* 232:1099–1105.

Lovell, A. C. B. 1964. Joseph Lade Pawsey, 1908–1962. *Biographical Memoirs of Fellows of the Royal Society* 10:228–43.

Lovell, D. J. 1968. Herschel's dilemma in the interpretation of thermal radiation. *Isis* 59:46–60.

Lubbock, C. A. 1933. *The Herschel chronicle: The life-story of William Herschel and his sister Caroline Herschel*. Cambridge: Cambridge University Press.

Lüst, R. 1967. Solar corpuscular radiation. *Trans. IAU* 13A:1004–13.

Lyot, B. 1929. Polarisation de la planète Mercure. *Comptes Rendus* 189:425–26.

———. 1930a. La polarisation de Mercure comparée à celle de la lune: Résultats obtenus au Pic-du-Midi en 1930. *Comptes Rendus* 191:703–5.

———. 1930b. La couronne solaire étudiée en dehors des éclipses, with comments by E. Esclangon and H. Deslandres. *Comptes Rendus* 191:834–39.

———. 1931. L'étude de la couronne solaire en dehors des éclipses. *Astronomie* 45:248–53.

———. 1932. Etude de la couronne solaire en dehors des éclipses. *Zeitschrift für Astrophysik* 5:73–95.

———. 1936a. Observations des protubérances solaires faites au Pic du Midi en 1935. *Comptes Rendus* 202:392–94.

———. 1936b. La couronne solaire en 1935. *Comptes Rendus* 202:1259–61.

———. 1936c. Le spectre de la couronne solaire en 1936, longueurs d'ondes et intensités des raies d'émission. *Comptes Rendus* 203:1327–29.

———. 1937. Quelques observations de la couronne solaire et des protubérances en 1935. *Astronomie* 51:203–18.

———. 1939. A study of the solar corona and prominences without eclipses. *MNRAS* 99:580–94.

———. 1944. Le filtre monochromatique polarisant et ses applications en physique solaire. *Annales d'Astrophysique* 7:31–49.

———. 1945. Planetary and solar observations on the Pic du Midi in 1941, 1942, and 1943. *Astrophysical Journal* 101:255–59.

McCormac, B. M., ed. 1983. *Weather and climate responses to solar variations.* Boulder: Colorado Associated University Press.

McDougall, W. A. 1985. . . . *The heavens and the earth: A political history of the space age.* New York: Basic.

McGucken, W. 1969. *Nineteenth-century spectroscopy: Development of the understanding of spectra, 1802–1897.* Baltimore: Johns Hopkins Press.

Mackin, R. J., Jr., and M. M. Neugebauer, eds. 1966. *The solar wind: Proceedings of a conference held at the California Institute of Technology, Pasadena, California, U.S.A., April 1–4, 1964, and sponsored by the Jet Propulsion Laboratory.* Oxford: Pergamon.

McMath, R. R. 1937. The tower telescope of the McMath-Hulbert Observatory. *Publications of the Observatory of the University of Michigan* 7:1–56.

McMath, R. R., and R. M. Petrie. 1933. The spectroheliokinematograph. *Publications of the Observatory of the University of Michigan* 5:103–17.

McMath, R. R., and E. Pettit. 1937. Prominences of the active and sun-spot types compared. *Astrophysical Journal* 85:279–303.

McMath, R. R., and A. K. Pierce. 1960. The large solar telescope at Kitt Peak. *Sky and Telescope* 20:64–67, 132–35.

McMullin, E. 1987. Bruno and Copernicus. *Isis* 78:55–74.

Mandel'shtam, S. L. 1967. Studies of shortwave solar radiation in the U.S.S.R. *Applied Optics* 6:1834–44.

Maran, S. P., and B. E. Woodgate. 1984. A second chance for Solar Max. *Sky and Telescope* 67:498–500.

Massey, H. S. W., and M. O. Robins. 1986. *History of British space science.* Cambridge: Cambridge University Press.

Mathews, C. W. 1976. Announcement of opportunity: Solar constant measurement experiment for flight on the Solar Maximum Mission (February 9). NHO.

Maunder, E. W. 1904a. Note on the distribution of sun-spots in heliographic latitude. *MNRAS* 64:747–61.

———. 1904b. Magnetic disturbances, 1882–1903, as recorded at the Royal Observatory, Greenwich, and their association with sun-spots. *MNRAS* 65:2–34.

Meadows, A. J. 1970. *Early solar physics*. Oxford: Pergamon.

―――. 1972. *Science and controversy: A biography of Sir Norman Lockyer*. Cambridge: MIT Press.

―――. 1975. *Greenwich Observatory: The royal observatory at Greenwich and Herstmonceux*. Vol. 2. *Recent history (1836–1975)*. London: Taylor and Francis.

―――. 1984a. The origins of astrophysics. In Gingerich 1984, 3–15.

―――. 1984b. The new astronomy. In Gingerich 1984, 59–72.

Meadows, A. J., and J. E. Kennedy. 1981. The origin of solar-terrestrial studies. *Vistas in Astronomy* 25:419–26.

Megerian, G. K. 1946. Letter to J. G. Bain with minutes of the V-2 panel's meeting on February 17 (March 7). NHO.

Mellor, D. P. 1958. *The role of science and industry*. Vol. 5 of *Australia in the War of 1939–1945*, series 4. Canberra: Australian War Memorial.

Menzel, D. H. 1929. Letter to H. N. Russell (March 24). SHMA Russell, reel 30.

―――. 1940. Harvard coronagraph in Colorado. *Sky* 4 (June): 3–4.

―――. 1941. What is the solar corona? *Telescope* 8 (May–June): 64–67.

Merrill, P. W. 1922. Interferometer observations of double stars. *Astrophysical Journal* 56:40–53.

Meyer, P., and J. A. Simpson. 1955. Changes in the low-energy particle cutoff and primary spectrum of cosmic radiation. *Physical Review* 99:1517–23.

Miller, H. S. 1970. *Dollars for research: Science and its patrons in nineteenth-century America*. Seattle: University of Washington Press.

Miller, J. D. 1970. Henry Augustus Rowland and his electromagnetic researches. Ph.D. diss., Oregon State University.

Milne, E. A. 1924. Statistical equilibrium in relation to the photoelectric effect and its application to the determination of absorption coefficients. *Philosophical Magazine* 47:209–41.

―――. 1929. The masses, luminosities, and effective temperatures of the stars. *MNRAS* 90:17–54.

―――. 1945. The president's address on the award of the gold medal to Professor Bengt Edlén. *MNRAS* 105:138–45.

Mitchell, S. A. 1936. Eclipses of the sun. In *Handbuch der Astrophysik*, ed. G. Eberhard, A. Kohlschütter, and H. Ludendorff, 7:382–409. Berlin: Springer.

Mögel, E. H. 1931. Beziehungen zwischen Empfangsstörungen bei Kurzwellen und der magnetischen Tätigkeit der Erde. *Forschungen und Fortschritte* 7:22–23.

Mohler, O. C., and H. Dodson-Prince. 1978. Robert Raynolds McMath, May 11, 1891–January 2, 1962. *Biographical Memoirs of the National Academy of Sciences* 49:185–202.

More, H. 1647. Democritus Platonissans, or An essay upon the infinity of worlds out of Platonick principles. In *Philosophical poems*, 2d ed., 187–218. Cambridge: Daniel.

Murphy, J. F. 1982. Memo to J. M. Beggs, H. Mark, and Culbertson reporting congressional support for the Solar Maximum Mission pending DOD's fulfillment of its agreement to reimburse part of the mission (July 21). SMM file, NHO.

NASA. 1961a. Press conference Explorer X (April 18). NHO.
———. 1961b. Summary minutes of the Particles and Fields Subcommittee of the Space Sciences Steering Committee (May 25–26). NHQ.
———. 1961c. Ranger spacecraft (July 24). NHO.
———. 1961d. Summary minutes of the Particles and Fields Subcommittee of the Space Sciences Steering Committee (October 4–6). NHQ.
———. 1962a. Summary minutes of the Particles and Fields Subcommittee of the Space Sciences Steering Committee (January 16). NHQ.
———. 1962b. OSO fact sheet. News Release 62–32 (February 22). NHO.
———. 1962c. Summary minutes: Solar Physics Subcommittee of the Space Sciences Steering Committee (September 13–14). NHQ.
———. 1962d. Mariner II scientific experiments (October 10). NHO.
———. 1965a. *Orbiting Solar Observatory satellite OSO I: A project summary.* NASA SP-57. Washington D.C.: NASA.
———. 1965b. Summary minutes: Solar Physics Subcommittee of the Space Science Steering Committee (April 29–30). NHQ.
———. 1975. Minutes of the NASA Space Program Advisory Council's Physical Sciences Committee (July 16–17). NHQ.
———. 1976. Structure of the sun's magnetic field found. Ames Research Center public relations release 76–85 (December 6). NHO.
———. 1978. Nimbus-G press kit (September 8). NIIO.
———. 1980. NASA satellite detects changes in energy output from sun. NASA News Release 80–124 (August 6). SMM File, NHO.
———. ca. 1982. *Major NASA satellite missions and key participants, 1958–1981.* Greenbelt, Md.: NASA.
Naugle, J. E. 1969. Letter to W. L. Smith (December 23). NHO.
———. 1971. Letter to W. L. Smith (April 30). NIIO.
Needell, A. A. 1987. Preparing for the space age: University-based research, 1946–1957. *Historical Studies in the Physical and Biological Sciences* 18:89–109.
———. 1989. Lloyd V. Berkner on organizing American science for social purposes. In De Maria, Grilli, and Sebastiani 1989, 85–95.
Ness, N. F. 1963. Letters to J. M. Wilcox (June 18, 25). JMW.
———. 1964. Letter to E. W. Greenstadt (September 10). JMW.
———. 1965. The earth's magnetic tail. *Journal of Geophysical Research* 70:2989–3005.
———. 1966. Interplanetary magnetic-field measurements by the IMP-1 satellite. In Mackin and Neugebauer 1966, 83–107.
———. 1986. Telephone interview with K. Hufbauer (September 19). NHO.
Ness, N. F., C. S. Scearce, and J. B. Seek. 1964. Initial results of the Imp 1 magnetic field experiment. *Journal of Geophysical Research* 69:3531–69.
Ness, N. F., and J. M. Wilcox. 1964. Solar origin of the interplanetary magnetic field. *Physical Review Letters* 13:461–64.
———. 1965. Sector structure of the quiet interplanetary magnetic field. *Science* 148:1592–94.

——. 1966. Extension of the photospheric magnetic field into interplanetary space. *Astrophysical Journal* 143:23–31.

Neugebauer, M. M. 1958. Inter-office memo to Mission Survey Group regarding preliminary design of an experiment to measure the solar corpuscular radiation (December 11). RLN.

——. 1959a. Inter-office memo to R. V. Meghreblian regarding Ames vs. JPL solar corpuscular radiation instrumentation (June 15). CWS.

——. 1959b. Inter-office memos to A. R. Hibbs regarding Rossi's experiment and a comparison of the Rossi and the JPL corpuscular radiation experiments (August 25). CWS.

——. 1982. Mariner 2 and the discovery of the solar wind. *Planetary Report* 2, 6:11.

Neugebauer, M. M., and C. W. Snyder. 1962. Solar plasma experiment. *Science* 138:1095–96.

Neupert, W. M., and V. Pizzo. 1974. Solar coronal holes as sources of recurrent geomagnetic disturbances. *Journal of Geophysical Research* 79:3701–9.

Newburn, R. L., Jr., and M. M. Neugebauer. 1958. Preliminary consideration of a limited class of problems suitable for study by interplanetary probes and/or satellites (November 13). RLN.

Newcomb, S. 1895. *The elements of the four inner planets and the fundamental constants of astronomy.* Washington D.C.: Government Printing Office.

Newell, H. E., Jr. 1953. *High altitude rocket research.* New York: Academic.

——. 1958. The U.S. rocket-satellite program for the International Geophysical Year. *Annals IGY* 2A:267–71.

——, ed. 1959a. *Sounding rockets.* New York: McGraw-Hill.

——. 1959b. Conference report—PSAC Space Science Panel (December 18). NHO.

——. 1980. *Beyond the atmosphere: Early years in space science.* NASA SP-4211. Washington D.C.: NASA.

Newkirk, G. A., Jr. 1973. Letter to *Solar Physics* editorial board [April 12–13]. Director's Office, High Altitude Observatory, Boulder, Colo.

——, ed. 1979. Commission 10: Solar activity. *Trans. IAU* 17A:11–48.

——. 1983a. Variations in solar luminosity. *ARAA* 21:429–67.

——. 1983b. Interview with D. H. DeVorkin (June 1). Space astronomy oral history project, National Air and Space Museum, Smithsonian Institution, Washington, D.C.

——. 1984. Curriculum vitae (June 15) and autobiographical statement (October 26). NHO.

Newkirk, G. A., Jr., and J. D. Bohlin. 1965. Coronascope II: Observation of the white light corona from a stratospheric balloon. *Annales d'Astrophysique* 28:234–38.

Newkirk, R. W., I. D. Ertel, and C. G. Brooks. 1977. *Skylab: A chronology.* NASA SP-4011. Washington D.C.: NASA.

Newlan, I. 1963. *First to Venus: The story of Mariner II.* New York: McGraw-Hill.

Newton, I. 1687–1726/1972. *Isaac Newton's Philosophiae naturalis principia mathematica: The third edition (1726) with variant readings.* Ed. A. Koyré and I. B. Cohen. Vol. 2. Cambridge: Harvard University Press.

Nichols, R. W. 1971. Mission-oriented R&D: Senator Mansfield's questions sharpen congressional uncertainties about federal R&D patterns. *Science* 172:29–37.

——. 1989. Solar Max: 1980–89. *Sky and Telescope* 78:600–601.

Nicholson, S. B. 1938. Solar and terrestrial relationships. *Carnegie Institution of Washington Publications* 501:103–14.

Nicolet, M., ed. 1958. The International Geophysical Year meetings. *Annals IGY* 2A:1–395.

North, J. 1974. Thomas Harriot and the first telescopic observations of sunspots. In *Thomas Harriot: Renaissance scientist*, ed. J. W. Shirley, 129–65. Oxford: Clarendon Press.

Noyes, R. W. 1965. Letter to L. Goldberg (July 26). LG.

——. 1982. *The sun, our star.* Cambridge: Harvard University Press.

Numbers, R. L. 1977. *Creation by natural law: Laplace's nebular hypothesis in American thought.* Seattle: University of Washington Press.

O'Day, M. D. 1954. Upper air research by use of rockets in the U.S. Air Force. In *Rocket exploration of the upper atmosphere*, cd. R. L. F. Boyd and M. J. Seaton, 1–10. London: Pergamon.

Odishaw, H. 1958. International Geophysical Year. *Science* 128:1599–1609.

Öhman, K. Y. 1938a. A new monochromator. *Nature* 141:157–58.

——. 1938b. A new monochromator. *Nature* 141:291.

——. 1938c. Über den Quarz-Säule-Monochromator. In *Festskrift tillägnad Östen Bergstrand*, 138–49. Uppsala and Stockholm: Almqvist och Wiksells.

——. 1938d. Internationella Astronomiska Unions Kongress i Stockholm, 3–10 Augusti 1938. *Populaer Astronomisk Tidskrift* 19:141–48.

——, ed. 1958. IGY Instruction manual: Solar activity. *Annals IGY* 5:249–301.

Olmsted, J. W. 1942. The scientific expedition of Jean Richer to Cayenne (1672–1673). *Isis* 34:117–28.

Oster, L., K. H. Schatten, and S. Sofia. 1982. Solar irradiance variations due to active regions. *Astrophysical Journal* 256:768–73.

Osterbrock, D. E. 1984. *James E. Keeler, pioneering American astrophysicist.* Cambridge: Cambridge University Press.

——. 1986. Failure and success: Two early experiments with concave gratings in stellar spectroscopy. *JHA* 17:119–29.

Osterbrock, D. E., J. R. Gustafson, and W. J. S. Unruh. 1988. *Eye on the sky: Lick Observatory's first century.* Berkeley: University of California Press.

Parker, E. N. 1956. Modulation of primary cosmic-ray intensity. *Physical Review* 103:1518–32.

——. 1958a. Dynamical instability in an anisotropic ionized gas of low density. *Physical Review* 109:1874–76.

——. 1958b. Plasma instability in the interplanetary magnetic field. In *The plasma in a magnetic field: A symposium on magnetohydrodynamics*, ed. R. K. M. Landshoff, 77–84. Stanford: Stanford University Press.

——. 1958c. Dynamics of the interplanetary gas and magnetic field. *Astrophysical Journal* 128:664–76.

———. 1958d. Cosmic-ray modulation by solar wind. *Physical Review* 110:1445–49.

———. 1958e. Interaction of the solar wind with the geomagnetic field. *Physics of Fluids* 1:171–87.

———. 1958f. Suprathermal particle generation in the solar corona. *Astrophysical Journal* 128:677–85.

———. 1959. Extension of the solar corona into interplanetary space. *Journal of Geophysical Research* 64:1673–81.

———. 1962. Interplanetary dynamics and cosmic ray modulation. *Journal of the Physical Society of Japan* 17, supp. A-II:563–67.

———. 1963. *Interplanetary dynamical processes*. New York: Wiley.

———. 1965. Letter to N. F. Ness (March 26). JMW.

———. 1978. Solar physics in broad perspective. In *The new solar physics*, ed. J. A. Eddy, 1–8. Boulder: Praeger.

———. 1979. A broad look at solar physics: Adapted from the solar physics study of August 1975. In *Solar system plasma physics*, ed. E. N. Parker, C. G. Kennel, and L. J. Lanzerotti, 1:3–49. Amsterdam: North-Holland.

———. 1985a. The future of solar physics. *Solar Physics* 100:599–619.

———. 1985b. Telephone interview with K. Hufbauer (April 8). NHO.

Pawsey, J. L. 1946. Observation of million degree thermal radiation from the sun at a wavelength of 1.5 metres. *Nature* 158:633–34.

———. 1953. Radio astronomy in Australia. *Journal of the Royal Astronomical Society of Canada* 47:137–52.

———. 1961. Australian radio astronomy: How it developed in this country. *Australian Scientist* 1:181–86.

Pawsey, J. L., R. V. Payne-Scott, and L. L. McCready. 1946. Radio-frequency energy from the sun. *Nature* 157:158–59.

Payne, C. H. 1925. *Stellar atmospheres: A contribution to the observational study of high temperature in the reversing layers of stars*. Cambridge: Harvard College Observatory.

Pecker, J. -C., and R. N. Thomas. 1976. Solar astrophysics: Ghettosis from, or symbiosis with, stellar and galactic astrophysics? *Space Science Reviews* 29:217–43.

Pedersen, O., and M. Pihl. 1974. *Early physics and astronomy: A historical introduction*. New York: American Elsevier.

Petrie, R. M. 1962. Builder of solar observatories. *Sky and Telescope* 23:187–91.

Pinch, T. J. 1980. Theoreticians and the production of experimental anomaly: The case of solar neutrinos. In *The social process of scientific investigation*, ed. K. D. Knorr, R. Krohn, and R. Whitley, 77–106. Dordrecht: Reidel.

———. 1985. Theory testing in science—the case of solar neutrinos: Do crucial experiments test theories or theorists? *Philosophy of the Social Sciences* 15:167–87.

———. 1986. *Confronting nature: The sociology of solar-neutrino detection*. Dordrecht: Reidel.

Plamondon, J. A. 1988. Telephone interview with K. Hufbauer (May 6). NHO.

Plamondon, J. A., and J. M. Kendall, Sr. 1965. Cavity-type absolute total-radiation radiometer. *JPL Space Programs Summary* 37–35, 4:66–68.

Plaskett, H. H. 1933. Solar research. *MNRAS* 93:277–83.

Plendl, [J. N.] 1931. Über den Einfluss der elfjährigen Sonnentätigkeitsperiode auf die Ausbreitung der Wellen in der drahtlosen Telegraphie. *Jahrbuch der Deutschen Versuchsanstalt für Luftfahrt*, 665–71.

———. 1985. Neue bisher nicht vermutete Gesetze in Physik und Chemie. Manuscript autobiography. Copy on deposit in NHO.

———. 1986. Letter to K. Hufbauer, NHO.

Plotkin, H. 1978. Edward C. Pickering, the Henry Draper Memorial, and the beginnings of astrophysics in America. *Annals of Science* 35:365–77.

Plummer, H. C. 1939. Presidential address on presenting the Gold Medal of the society to M. Bernard Lyot. *MNRAS* 99:538–40.

Porter, R. W. 1971. International scientific community: International Council of Scientific Unions and COSPAR. In *International cooperation in outer space: A symposium*, ed. E. Galloway, 527–57. Washington D.C.: Government Printing Office.

Powell, C. S. 1988. J. Homer Lane and the internal structure of the sun. *JHA* 19:183–99.

Ranyard, A. C. 1879. Observations made during total solar eclipses. *Memoirs of the Royal Astronomical Society* 41:1–792.

Reber, G. 1944. Cosmic static. *Astronomical Journal* 100:279–87.

———. 1983. Radio astronomy between Jansky and Reber. In *Serendipitous discoveries in radio astronomy: Proceedings of a workshop held at the National Radio Astronomy Observatory, Green Bank, West Virginia on May 4, 5, 6, 1983*, ed. K. Kellermann and B. Sheets, 71–78. Green Bank, W. Va.: National Radio Astronomy Observatory.

Referee's report for the *Journal of Geophysical Research* (August 5). 1965. JMW.

Reijnen, G. M. C. 1974. Concise report of the combined JOSO-CESRA meeting, 7 March 1974, Berne, Switzerland. *JOSO: AR 1973*, 38–40.

Reines, F. 1967. The search for the solar neutrino. *Proceedings of the Royal Society* A301:159–70.

Rickett, B. J., and W. A. Sime. 1979. Solar cycle changes in the high latitude solar wind. In *Study of the solar cycle from space: Proceedings of a symposium held at Wellesley, Massachusetts, June 14–15, 1979*, 233–43. NASA Conference Publication 2098 233–43.

Roberts, M. S., and R. J. Havlen, 1982. National Radio Astronomy Observatory. *Bull. AAS* 14:370–410.

Roberts, W. O. 1943. Preliminary studies of the solar corona and prominences with the Harvard coronagraph. Ph.D. diss., Harvard University.

———. 1945. Artificial eclipses of the sun. *Sky and Telescope* 4:3–6.

———. 1950. High Altitude Observatory of Harvard University and University of Colorado: Report of superintendent to board of trustees [for] 1 July 1948 through 31 December 1949. WOR.

———. 1952–57. Correspondence. WOR.

———. 1956. Quarterly research report, High Altitude Observatory, University of Colorado, no. 1, p. 18. WOR.

———. 1957. The director's quarterly report, High Altitude Observatory, University of Colorado (1 April to 30 September). WOR.

———. 1983. Interview with D. H. DeVorkin (July 26–28), Space astronomy oral history project, National Air and Space Museum, Smithsonian Institution, Washington, D.C.

Rochberg-Halton, F. 1984. New evidence for the history of astrology. *Journal of Near Eastern Studies* 43:115–40.

Roland, A. 1985. Science and war. *Osiris* 1:247–72.

Rood, R. T. 1972. A mixed-up sun and solar neutrinos. *Nature: Physical Science* 240:178–80.

Rösch, J., and J. Dragesco. 1980. The French quest for high resolution. *Sky and Telescope* 59:6–13.

Rosen, E. 1971. *Three Copernican treatises: The Commentariolus of Copernicus, the Letter against Werner, the Narratio prima of Rheticus.* 3d ed. New York: Octagon Books.

Rosenberg, R. L. 1970. Unified theory of the interplanetary magnetic field. *Solar Physics* 15:72–78.

Rosenberg, R. L., and P. J. Coleman, Jr. 1969. Heliographic latitude dependence of the dominant polarity of the interplanetary magnetic field. *Journal of Geophysical Research* 74:5611–22.

Rossi, B. B. 1960. Outer space. *Harvard Alumni Bulletin* 62:597–99.

———. 1984. Discovery of the solar wind. In *Plasma astrophysics: International School and Workshop on Plasma Astrophysics, held at Varenna, Italy, 28 August–7 September 1984,* ed. T. D. Guyenne and J. J. Hunt, 27–32. ESA: SP-207. Paris: European Space Agency.

Roth, G. D. 1976. *Joseph von Fraunhofer: Handwerker—Forscher—Akademiemitglied, 1787–1826.* Stuttgart: Wissenschaftliche Verlagsgesellschaft.

Rowland, H. A. 1882. Letters to C. A. Young (February 26, March 19 and 26). SHMA Young, reel 2.

Royal Astronomical Society Council. 1852. Report . . . to the thirty-second annual general meeting of the society. *MNRAS* 12:73–123.

———. 1861. Report . . . to the forty-first annual general meeting of the society. *MNRAS* 21:89–123.

Russell, H. N. 1919. On the sources of stellar energy. *Publications of the Astronomical Society of the Pacific* 31:205–11.

———. 1929a. Letter to D. H. Menzel (March 16). SHMA Russell, reel 10.

———. 1929b. Letter to A. Unsöld (March 23). SHMA Russell, reel 10.

———. 1929c. On the composition of the sun's atmosphere. *Astrophysical Journal* 70:11–82.

———. 1941. A puzzle solved? A new and promising interpretation of the old problem of the solar corona. *Scientific American* 165:70–71.

Russell, H. N., R. S. Dugan, and J. Q. Stewart. 1926. *Astronomy: A revision of Young's manual of astronomy.* Boston: Ginn.

Russell, J. L. 1989. Catholic astronomers and the Copernican system after the condemnation of Galileo. *Annals of Science* 46:365–86.

Rust, D. M. 1984. The Solar Maximum observatory. *Johns Hopkins University Applied Physics Laboratory Technical Digest* 5:188–96.

Sabine, E. 1852. On periodical laws discoverable in the mean effects of the larger magnetic disturbances—No. 2. *Phil. Trans.* 142:103–24.

Sachs, A. 1952. Babylonian horoscopes. *Journal of Cuneiform Studies* 6:49–75.

St. John, C. E. 1928. Commision de Physique Solaire. *Trans IAU* 3:50–76, 231–36.

Saito, T. 1975. Two-hemisphere model of the three-dimensional magnetic structure of the interplanetary space. *Science Reports of the Tohoku University,* ser. 5, 23:37–54.

Sampson, R. A. 1911. The sun. In *Encyclopaedia Britannica,* 11th ed., 26:85–91.

Sang, H. -P. 1987. Leben und Bedeutung Joseph von Fraunhofers. *Mitteilungen der Astronomischen Gesellschaft* 70:35–61.

Schaffer, S. 1978. The phoenix of nature: Fire and evolutionary cosmology in Wright and Kant. *JHA* 9:180–200.

———. 1980a. The great laboratories of the universe: William Herschel on matter theory and planetary life. *JHA* 11:81–111.

———. 1980b. Herschel in Bedlam: Natural history and stellar astronomy. *British Journal for the History of Science* 13:211–39.

———. 1981. Uranus and the establishment of Herschel's astronomy. *JHA* 12:11–26.

Schatten, K. H. 1968. Large-scale configuration of the coronal and interplanetary magnetic field. Ph.D. diss, University of California, Berkeley.

———. 1986. Telephone interview with K. Hufbauer (November 12). NHO.

———. 1988. A model for solar constant secular changes. *Geophysical Research Letters* 15:121–24.

Schatten, K. H., and H. G. Mayr, 1985. On the maintenance of sunspots: An ion hurricane mechanism. *Astrophysical Journal* 299:1051–62.

Schatten, K. H., H. G. Mayr, K. Omidvar, and E. Maier. 1986. A hillock and cloud model for faculae. *Astrophysical Journal* 311:460–73.

Schatten, K. H., N. Miller, L. Oster, and S. Sofia. 1981. Solar irradiance modulation by active regions from 1969 through 1980. *Bulletin of the American Astronomical Society* 13:877.

Schatten, K. H., N. Miller, S. Sofia, and L. Oster. 1982. Solar irradiance modulation by active regions from 1969 through 1980. *Geophysical Research Letters* 9:49–51.

Schatten, K. H., J. M. Wilcox, and N. F. Ness. 1969. A model of interplanetary and coronal magnetic fields. *Solar Physics* 6:442–55.

Schettino, E. 1989. A new instrument for infrared radiation measurements: The thermopile of Macedonio Melloni. *Annals of Science* 46:511–17.

Schröder, W. 1984. *Das Phänomen des Polarlichts: Geschichtsschreibung, Forschungs-ergebnisse und Probleme.* Darmstadt: Wissenschaftliche Buchgesellschaft.

Schroeder-Gudehus, B. 1978. *Les scientifiques et la paix: La communauté scientifique internationale au cours des années 20.* Montréal: Presses de l'Université de Montréal.

Schröter, E. H. 1974. Trends and problems in optical solar research. *JOSO: AR 1973,* 19–33.

———. 1984. Solar observational facilities on the Canary Islands—operational, under construction, planned. *JOSO: AR 1984,* 29–35.

Schulz, M. 1973. Interplanetary sector structure and the heliomagnetic equator. *Astrophysics and Space Science* 24:371–83.

Schwabe, S. H. 1852. Letter to J. R. Wolf (August 30). Quoted in Wolf 1876, 134.

Schwarzschild, M. 1969. Introduction. *Bull. AAS* 1:1.

Secchi, A. 1865. Les découvertes spectroscopiques dans leur rapports avec la recherche de la nature des corps célestes. *Archives des Sciences Physiques et Naturelles* 23:145–66.

———. 1870. *Le soleil: Exposé des principales découvertes modernes sur la structure de cet astre, son influence dans l'univers et ses relations avec les autres corps célestes.* Paris: Gauthier-Villars.

A second life for Mount Wilson Observatory? 1984. *Sky and Telescope* 68:203–4.

Severny, A. B. 1960. Untitled comments. In *Proceedings of the fourth symposium on cosmical gas dynamics: Aerodynamic phenomena in stellar atmospheres,* ed. R. N. Thomas et al., 278–79. Bologna: Zanichelli.

Shapiro, I. I. 1968. Spin and orbital motions of the planets. In *Radar astronomy,* ed. J. V. Evans and T. Hagfors, 143–85. New York: McGraw-Hill.

Shapley, H. 1945. *Astronomical observatory.* Cambridge: Harvard University. Reprinted from *Report of the President of Harvard College and Reports of Departments, 1944–45.*

———. 1946. *Astronomical observatory.* Cambridge: Harvard University. Reprinted from *Report of the President of Harvard College and Reports of Departments, 1945–46.*

———. 1947. *Astronomical observatory.* Cambridge: Harvard University. Reprinted from *Report of the President of Harvard College and Reports of Departments, 1946–47.*

———. 1948. *Astronomical observatory.* Cambridge: Harvard University. Reprinted from *Report of the President of Harvard College and Reports of Departments, 1947–48.*

———. 1950. *Astronomical observatory.* Cambridge: Harvard University. Reprinted from *Report of the President of Harvard College and Reports of Departments, 1949–50.*

Shea, W. R. 1972. Sunspots and inconstant heavens. In *Galileo's intellectual revolution: Middle period, 1610–1632,* 2d ed., 49–74. New York: Science History Publications.

Shirley, J. W. 1983. *Thomas Harriot: A biography.* Oxford: Clarendon Press.

Shklovskii, I. S., V. I. Moroz, and V. G. Kurt. 1961. The nature of the earth's third radiation belt. *Soviet Astronomy AJ* 4:871–73.

Siegel, D. M. 1976. Balfour Stewart and Gustav Kirchhoff: Two independent approaches to Kirchhoff's radiation law. *Isis* 67:565–600.

Simpson, J. A. 1985. Cosmic ray astrophysics at Chicago (1947–1960) (some personal reminiscences). In *Early history of cosmic ray studies: Personal reminiscences with old photographs,* ed. Y. Sekido and H. Elliot, 385–409. Dordrecht: Reidel.

———. 1989. Evolution of our knowledge of the heliosphere. *Advances in Space Research* 9, 4:5–20.

Simpson, J. A., H. W. Babcock, and H. D. Babcock. 1955. Association of a "unipolar" magnetic region on the sun with changes of primary cosmic-ray intensity. *Physical Review* 98:1402–6.

Singer, S. 1965. The Vela satellite program for detection of high-altitude nuclear detonations. *Proceedings of the IEEE* 53:1935–48.

Smerd, S. F., ed. 1969. Solar radio emission during the International Geophysical Year. *Annals IGY* 34.

Smith, C., and M. N. Wise. 1989. *Energy and empire: A biographical study of Lord Kelvin.* Cambridge: Cambridge University Press.

Smith, D. H. 1989. An observatory at 90° south. *Sky and Telescope* 77:598–602.

Smith, E. J., B. T. Tsurutani, and R. L. Rosenberg. 1978. Observations of the interplanetary sector structure up to heliographic latitudes of 16°. *Journal of Geophysical Research* 83A:717–24.

Smith, H. J., S. D. Jordan, S. Bauer, R. Bena, J. D. Bohlin, J. C. Brandt, E. Coleman, J. W. Evans, R. Fleischer, R. La Count, H. Lane, J. L. Linsky, R. Manka, R. W. Noyes, G. Oertel, E. Tandberg-Hanssen, A. F. Timothy, R. Tousey, and D. J. Williams. 1975. *Report of the Solar Astronomy Task Force to Ad Hoc Interagency Coordinating Committee on Astronomy.* Washington D.C.: Federal Council for Science and Technology.

Smith, R. J. 1979. Shuttle problems compromise space program. *Science* 206:910–14.

Smith, R. W. 1981. The heavens recorded: Warren de la Rue and the 1860 eclipse. Paper given at the Sixteenth International Congress of the History of Science, Bucharest, Rumania.

———. 1982. *The expanding universe: Astronomy's "Great Debate," 1900–1931.* Cambridge: Cambridge University Press.

———. 1989. *The space telescope: A study of NASA, science, technology, and politics.* Cambridge: Cambridge University Press.

Snyder, C. W. 1986. Letter to K. Hufbauer (November 24). NHO.

———. 1987. Letter to K. Hufbauer (August 30). NHO.

———. 1988. Letter to K. Hufbauer (February 28). NHO.

Snyder, C. W., and M. M. Neugebauer. 1964. Interplanetary solar-wind measurements by Mariner II. In *Space research IV: Proceedings of the Fourth International Space Science Symposium, Warsaw, June 4–10, 1963,* ed. P. Muller, 89–113. Amsterdam: North-Holland.

———. 1966. The relation of Mariner-2 plasma data to solar phenomena. In Mackin and Neugebauer 1966, 25–32.

Snyder, C. W., M. M. Neugebauer, and U. R. Rao. 1963. The solar wind velocity and its correlation with cosmic-ray variations and with solar and geomagnetic activity. *Journal of Geophysical Research* 68:6361–70.

Sofia, S. 1981a. Indirect methods for measuring variations of the solar constant. In Sofia 1981b, 73–79.

———, ed. 1981b. *Variations of the solar constant: Proceedings of a workshop held at Goddard Space Flight Center, Greenbelt, Maryland, November 5–7, 1980.* NASA Conference Publication 2191. Washington D.C.: NASA.

Sofia, S., L. Oster, and K. H. Schatten. 1982. Solar irradiance modulation by active regions during 1980. *Solar Physics* 80:87–98.

Solar and terrestrial relationships. 1926. *Observatory* 49:228–29.

Solar Physics Division: 1973. 1973. *Bull. AAS* 5:484.

Solar telescope canceled. 1986. *Sky and Telescope* 73:126–27.

Sonett, C. P., P. J. Coleman, Jr., and J. M. Wilcox, eds. 1972. *Solar wind: The proceedings of a conference sponsored by the National Aeronautics and Space Administration and*

held March 21–26, 1971, at the Asilomar Conference Grounds, Pacific Grove, California.
NASA SP-308. Washington D.C.: NASA.

Space Science Board. 1958a. Minutes of the first meeting . . . June 27, 1958. NHO.

———. 1958b. Preliminary table of experiments proposed for 1959–1960 by Simpson–Van Allen Committee [no. 8] of Space Science Board (July 10). Archives, National Academy of Sciences, Washington D.C.

———. 1958c. Minutes of the second meeting . . . July 19. NHO.

———. 1958d. Minutes of the second meeting [of committee no. 6] December 11. Archives, National Academy of Sciences, Washington, D.C.

Spencer-Jones, H. 1959. The inception and development of the International Geophysical Year. *Annals IGY* 1:383–413.

Spruit, H. C. 1977. Heat flow near obstacles in the solar convection zone. *Solar Physics* 55:3–34.

———. 1982. Effect of spots on a star's radius and luminosity. *Astronomy and Astrophysics* 108:348–55.

Stares, P. B. 1985. *The militarization of space: U.S. policy, 1945–1984.* Ithaca, N.Y.: Cornell University Press.

Statement of intentions (June 1970). 1970. *JOSO: AR 1970,* 4–7.

Stenflo, J. O. 1984a. LEST Foundation: Report of the council. *JOSO: AR 1983,* 7–9.

———. 1984b. Curriculum vitae and autobiographical statement (December). NHO.

———. 1984c. LEST Foundation: Report of the council. *JOSO: AR 1984,* 7–12.

———. 1986. Letter to K. Hufbauer (August 19). NHO.

———. 1988. Letters to K. Hufbauer (September 16, October 17). NHO.

Stenflo, J. O., O. Engvold, and K. -I. Hillerud. 1990. Decision on siting LEST on La Palma: Memo to the LEST community (January 17).

Sterne, T. E. 1942. Review of C. G. Abbot et al.'s *Annals of the astrophysical observatory of the Smithsonian Institution* 6 (1942). *Astrophysical Journal* 96:484–86.

Stratton, F. J. M. 1946. The International Astronomical Union. *Observatory* 66:246–48.

Strömgren, B. 1932. The opacity of stellar matter and the hydrogen content of the stars. *Zeitschrift für Astrophysik* 4:118–53.

Struve, O. 1946. The Copenhagen conference of the International Astronomical Union. *Popular Astronomy* 54:327–39.

Sturrock, P. A. 1974. Division on Solar Physics: 1974. *Bull. AAS* 6:494–95.

Sturrock, P. A., R. W. Noyes, L. J. Lanzerotti, and T. M. Donahue. 1976. Appendix E: Team report on solar physics. In Report on NASA's Office of Space Science Supporting Research and Technology and Data Research (May), ed. NASA Physical Sciences Committee, 35–43. NHQ

Sullivan, W. T., III. 1990. *A history of early radio astronomy.* Cambridge: Cambridge University Press. Forthcoming.

Summary of Mount Wilson magnetic observations of sun-spots for May and June, 1922. 1922. *Publications of the Astronomical Society of the Pacific* 34:224–25.

The sun. 1944. *Monthly Astronomical Newsletter* 23:1–3.

Svalgaard, L., and J. M. Wilcox. 1976. Structure of the extended solar magnetic field and the sunspot cycle variation in cosmic ray intensity. *Nature* 262:766–68.

―――. 1978. A view of solar magnetic fields, the solar corona, and the solar wind in three dimensions. *ARAA* 16:429–43.

Svalgaard, L., J. M. Wilcox, and T. L. Duvall, Jr. 1974. A model combining the polar and the sector structured solar magnetic fields. *Solar Physics* 37:157–72.

Swerdlow, N. M. 1968. Ptolemy's theory of the distances and sizes of the planets: A study of the scientific foundations of medieval cosmology. Ph.D. diss., Yale University.

Swerdlow, N. M., and O. Neugebauer. 1984. *Mathematical astronomy in Copernicus's De revolutionibus*. 2 vols. New York: Springer.

Swings, P. 1938. Joint meeting of commissions 11, 12, 23, 28, 29 and 36 for a discussion on emission lines. *Trans. IAU* 6:426.

―――. 1939. Une grande énigme de la spectroscopie astronomique actuelle: Le spectre de raies d'émission de la couronne solaire. *Scientia* 65:69–78.

―――. 1943. Edlén's identification of the coronal lines with forbidden lines of Fe X, XI, XIII, XIV, XV; Ni XII, XIII, XV, XVI; Ca XII, XIII, XV; A X, XIV. *Astrophysical Journal* 98:116–28.

―――. 1961. Report for commission 44. *Trans. IAU* 521–53.

Tatarewicz, J. N. 1990. Space technology and planetary astronomy. Bloomington: Indiana University Press.

Tayler, R. J. 1989. The sun as a star. *QJRAS* 30:125–61.

Thekaekara, M. P. 1973. Postscript: A tribute to Andrew J. Drummond. In Drummond and Thekaekara 1973, 161–62.

Tousey, R. 1953. Solar work at high altitudes from rockets. In Kuiper 1953, 658–76.

―――. 1955. Emission lines in the extreme ultra-violet spectrum of the sun. *Trans. IAU* 179–80.

―――. 1958. The extreme ultra-violet spectrum of the sun. *Trans. IAU* 708–10.

―――. 1986. Solar spectroscopy from Rowland to SOT. *Vistas in Astronomy* 29:175–99.

Tucker, W., and K. Tucker. 1986. *The cosmic inquirers: Modern telescopes and their makers*. Cambridge: Harvard University Press.

Turner, A. J. 1977. *Science and music in eighteenth century Bath: An exhibition in the Holburne of Menstrie Museum, Bath*. Bath: University of Bath.

―――. 1988. Portraits of William Herschel. *Vistas in Astronomy* 32.67–94.

Ulrich, R. K. 1975. Solar neutrinos and variations in the solar luminosity. *Science* 190:619–24.

Ulrich, R. K., and R. T. Rood. 1973. Mixing in stellar models. *Nature: Physical Science* 241:111–12.

U.S. Comptroller General. 1964. *Certain weaknesses in the management of orbiting solar observatory projects: National Aeronautics and Space Administration, for the period January 1959 through December 1962*. Washington D.C.: Government Printing Office.

U.S. Congress. Congressional Research Service. 1977. *World-wide space activities: National programs other than the United States and Soviet Union*. Washington D.C.: Government Printing Office.

―――. 1982. *Soviet space programs: 1976–80*. Part 1. Washington D.C.: U.S. Government Printing Office.

Van Allen, J. A. 1983. *Origins of magnetospheric physics*. Washington D.C.: Smithsonian Institution Press.

Van de Hulst, H. C. 1953. The chromosphere and the corona. In Kuiper 1953, 207–321.

Van Helden, A. 1977. The invention of the telescope. *Transactions of the American Philosophical Society* 67, no. 4.

———. 1985. *Measuring the universe: Cosmic dimensions from Aristarchus to Halley*. Chicago: University of Chicago Press.

Vlastos, G. 1975. *Plato's universe*. Seattle: University of Washington Press.

Waldmeier, M. 1939. Untersuchungen an der grünen Koronalinie 5303 A. *Zeitschrift für Astrophysik* 19:21–44.

Waldrop, M. M. 1981. Space science in the year of the shuttle. *Science* 212:316–18.

———. 1985. Sacramento Peak Observatory to close? *Science* 228:36.

———. 1986. The Challenger disaster: Assessing the implications. *Science* 231:661–63.

———. 1987. A crisis in space research. *Science* 235:426–29.

Wall, B. E. 1975. Anatomy of a precursor: The historiography of Aristarchos of Samos. *Studies in History and Philosophy of Science* 6:201–28.

Warner, D. J. 1971. Lewis M. Rutherfurd: Pioneer astronomical photographer and spectroscopist. *Technology and Culture* 12:190–216.

———. 1986. Rowland's gratings: Contemporary technology. *Vistas in Astronomy* 29:125–30.

Waterston, J. J. 1853. On dynamical sequences in the kosmos. *Athenaeum* 1351:1099–1100.

Weizsäcker, C. F. von. 1938. Über Elementumwandlungen im Innern der Sterne. 2. *Physikalische Zeitschrift* 39:633–46.

Westfall, R. S. 1980. *Never at rest: A biography of Isaac Newton*. Cambridge: Cambridge University Press.

———. 1985. Science and patronage: Galileo and the telescope. *Isis* 76:11–30.

White, O. R., ed. 1977. *The solar output and its variation*. Boulder: Colorado Associated University Press.

White, O. R., and J. T. Jefferies. 1973. Reports on solar instruments. *Trans. IAU* 15A:67, 75–76.

Wilcox, J. M. 1960. Professional history and goals (ca. August 31). JMW.

———. 1963a. Letter to C. Watson-Munro (March 15). JMW.

———. 1963b. Letter to H. Alfvén (December 6). JMW.

———. 1964a. Letter to N. F. Ness (March 20). JMW.

———. 1964b. Letter to J. Piel (October 2). JMW.

———. 1964c. Letter to N. F. Ness (November 2). JMW.

———. 1965a. Letter to N. F. Ness (February 17). JMW.

———. 1965b. Letter to N. F. Ness (August 27). JMW.

———. 1968. The interplanetary magnetic field. Solar origin and terrestrial effects. *Space Science Reviews* 8:258–328.

———. 1969. Referee's report for the *Journal of Geophysical Research* (April 10). JMW.

———. 1970. Referee's report for the *Astrophysical Journal* (April 27). JMW.

Wilcox, J. M., and N. F. Ness. 1965. Quasi-stationary corotating structure in the interplanetary medium. *Journal of Geophysical Research* 70:5793–5805.

Wild, J. P. 1980. The sun of Stefan Smerd. In *Radio physics of the sun, IAU symposium no. 86, held in College Park, MD, August 7–10, 1979,* ed. M. R. Kundu and T. E. Gergely, 5–21. Dordrecht: Reidel.

———. 1987. The beginnings of radio astronomy in Australia. *Proceedings of the Australian Astronomical Society* 7:95–102.

Williams, M. E. W. 1987. Astronomy in London: 1860–1900. *QJRAS* 28:10–26.

Willson, R. C. 1967. Radiometer comparison tests. Jet Propulsion Laboratory Technical Memorandum 33-371.

———. 1971. Active cavity radiometric scale, international pyrheliometric scale, and solar constant. *Journal of Geophysical Research* 76:4325–40.

———. 1972. Experimental comparisons of the international pyrheliometric scale with the absolute radiation scale. *Nature* 239:208–9.

———. 1973. Active cavity radiometer. *Applied Optics* 12:810–17.

———. 1974. Absolute radiometry and the solar constant. In Hickey and Klein 1974, 464–81.

———. 1975a. JPL absolute radiometry and solar constant measurements. In Zirin 1975, 317–32.

———. 1975b. Instrumentation for measurements of solar irradiance and atmospheric optical properties. *Proceedings of the Society of Photo-optical Instrumentation Engineers* 68(August): 31–40.

———. 1976. An experiment to monitor the total and spectral solar irradiance above the earth's atmosphere and to study the solar physics and climatological implications of the measurements: Active cavity radiometer irradiance monitor, a proposal for the Solar Maximum Mission. Jet Propulsion Laboratory Document 750-80 (May 24).

———. 1978. Accurate solar "constant" determinations by cavity pyrheliometers. *Journal of Geophysical Research* 83C:4003–7.

———. 1979. Active cavity radiometer type IV. *Applied Optics* 18:179–88

———. 1980. Solar irradiance observations from the SMM/ACRIM experiment (May 20). Courtesy of Richard C. Willson, Jet Propulsion Laboratory, Pasadena, Calif.

———. 1981. Solar total irradiance observations by active cavity radiometer. *Solar Physics* 74:217–29.

———. 1982a. Variations of total solar irradiance. *Trans. IAU* 18A:93–96.

———. 1982b. Solar irradiance variations and solar activity. *Journal of Geophysical Research* 87A:4319–26.

———. 1984. Measurements of solar total irradiance and its variability. *Space Science Reviews* 38:203–42.

———. 1985. Interview with K. Hufbauer (December 13). NHO.

———. 1988a. Telephone interview with K. Hufbauer (January 5). NHO.

———. 1988b. Letters to K. Hufbauer (January 25, June 9). NHO.

Willson, R. C., C. H. Duncan, and J. Geist. 1980. Direct measurement of solar luminosity variation. *Science* 207:177–79.

Willson, R. C., A. Gulkis, M. Janssen, H. S. Hudson, and G. A. Chapman. 1981. Observation of solar irradiance variability. *Science* 211:700–02.

Willson, R. C., and J. R. Hickey. 1977. 1976 rocket measurements of the solar constant and their implications for variation in the solar output in cycle 20. In White 1977, 111–16.

Willson, R. C., and H. S. Hudson. 1981a. Solar Maximum Mission experiment: Initial observations by the active cavity radiometer. *Advances in Space Research* 1, 13:285–88.

———. 1981b. Variations of solar irradiance. *Astrophysical Journal* 244:L185–89.

———. 1988. Solar luminosity variations in solar cycle 21. *Nature* 332:810–12.

Willson, R. C., H. S. Hudson, C. Fröhlich, and R. W. Brusa. 1985. Observation of a long-term downward trend in total solar irradiance. Paper given at the American Geophysical Union meeting, San Francisco (December 13).

———. 1986. Long-term downward trend in total solar irradiance. *Science* 234:1114–17.

Willson, R. C., H. S. Hudson, and M. F. Woodard. 1984. The inconstant solar constant. *Sky and Telescope* 64:501–3.

Wilson, A. 1774. Observations of the solar spots. *Phil. Trans.* 64:1–30.

Wolf, J. R. 1876. Erinnerungen an Heinrich Samuel Schwabe. *Vierteljahrschrift der Naturforschenden Gesellschaft zu Zürich* 21:129–45.

Woodard, M. F. 1984a. Upper limit on solar interior rotation. *Nature* 309:530–32.

———. 1984b. Short-period oscillations in the total solar irradiance. Ph.D. diss., University of California, San Diego.

———. 1987. Frequencies of low-degree solar acoustic oscillations and the phase of the solar cycle. *Solar Physics* 114:21–28.

Woodard, M. F., and H. S. Hudson. 1983a. Solar oscillations observed in the total irradiance. *Solar Physics* 82:67–73.

———. 1983b. Frequencies, amplitudes, and linewidths of solar oscillations from total irradiance observations. *Nature* 305:589–93.

Woodard, M. F., and R. W. Noyes. 1985. Change of solar oscillation eigenfrequencies with the solar cycle. *Nature* 318:449–50.

Woodgate, B. E. 1984. Initial results from the repaired Solar Maximum Mission and future prospects. *Advances in Space Research* 4:393–402.

Woolf, H. 1959. *The transits of Venus: A study of eighteenth-century science.* Princeton: Princeton University Press.

Wright, H. 1966. *Explorer of the universe: A biography of George Ellery Hale.* New York: Dutton.

Wright, H., J. N. Warnow, and C. Weiner, eds. 1972. *The legacy of George Ellery Hale: Evolution of astronomy and scientific institutions in pictures and documents.* Cambridge: MIT Press.

Wukelic, G. E., ed. 1968. *Handbook of Soviet space-science research.* London: Gordon and Breach.

Wyller, A. A. 1989. Yngve Öhman (1903–1988). *Solar Physics* 119:1–3.

Young, C. A. 1884. *The sun.* 2d ed. New York: Appleton.

———. 1891. Address at the dedication of the Kenwood Observatory. *Sidereal Messenger* 10:312–21.

———. 1895. *The sun.* 4th ed. New York: Appleton.

Zirin, H. 1970. The Big Bear Solar Observatory. *Sky and Telescope* 39:215–19.

———, ed. 1975. *Proceedings of the workshop: The solar constant and the earth's atmosphere.* BBSO 0149. Pasadena: Big Bear Solar Observatory.

———. 1986. Letter to K. Hufbauer (April 7). NHO.

———. 1988a. *Astrophysics of the sun.* Cambridge: Cambridge University Press.

———. 1988b. Letter to K. Hufbauer (February 1). NHO.

Zirin, H., R. L. Moore, and J. Walter. 1976. Proceedings of the workshop, The Solar Constant and the Earth's Atmosphere, held at Big Bear Solar Observatory . . . 19–21 May 1975. *Solar Physics* 46:377–409.

Zirin, H., and J. M. Mosher. 1988. The Caltech Solar Site Survey, 1965–1967. *Solar Physics* 115:183–202.

Zirker, J. B., ed. 1977. *Coronal holes and high speed wind streams: A monograph from Skylab solar workshop.* Boulder: Colorado Associated University Press.

Zöllner, J. C. F. 1865. *Photometrische Untersuchungen mit besonderer Rücksicht auf die physische Beschaffenheit der Himmelskörper.* Leipzig: Engelmann.

Zwaan, C. 1974. Why optical solar research. *JOSO: AR 1973,* 34–37.

INDEX

Designed by David denBoer
Composed by The Composing Room of Michigan, Inc.,
in Meridien text and Galliard display
Printed on 50-lb., MV Eggshell Cream Offset
by The Maple Press Company